Linking Australia's Landscapes

Lessons and Opportunities from Large-scale Conservation Networks

Editors:
James Fitzsimons
Ian Pulsford
Geoff Wescott

CSIRO PUBLISHING

National Library of Australia Cataloguing-in-Publication entry

Linking Australia's landscapes : lessons and opportunities
from large-scale conservation networks / edited by James Fitzsimons, Ian Pulsford and Geoff Wescott.
9780643107045 (pbk.)
9780643107052 (epdf)
9780643107069 (epub)
Includes bibliographical references and index.
Ecosystem management - Australia.
Ecological zones - Australia.
Endangered ecosystems - Australia.
Conservation of natural resources - Australia.
Wilderness areas - Management - Australia.
Fitzsimons, James.
Pulsford, Ian.
Wescott, G. C. (Geoff C.)
363.69994

Published by
CSIRO PUBLISHING
36 Gardiner Road, Clayton VIC 3168
Private Bag 10, Clayton South VIC 3169
Australia

Telephone: [+613] 9545 8555
Local call: 1300 788 000 (Australia only)
Fax: +61 3 9662 7555
Email: csiropublishing@csiro.au
Web site: www.publishing.csiro.au

Front cover: Habitat links extend from Woomargama National Park into surrounding private farm lands; part of the Slopes to Summit section of the Great Eastern Ranges Initiative, New South Wales. (Photo: Ian Pulsford)

Set in Minion Pro 10.5/12

Edited by Anne Findlay
Cover design by Alicia Freile, Tango Media
Text design by James Kelly
Typeset by diacriTech, Chennai, India
Index by Russell Brooks
Printed by Ingram Lightning Source

Feb26_RP_ILS

Contents

List of contributors

Matt Appleby
Bush Heritage Australia, PO Box 329, Flinders Lane, Melbourne VIC 8009, Australia, mappleby@bushheritage.org.au.

Lynn Baker
Biodiversity Assessment and Conservation Section, Conservation and Regulation Division, Office of Environment and Heritage, Department of Premier and Cabinet, PO Box A290, Sydney South NSW 1232, Australia, lynn.baker@environment.nsw.gov.au.

Georgia Beyer
Nature Conservation Trust of New South Wales, PO Box 1121, Lismore NSW 2480, Australia, georgiab@nct.org.au.

Keith Bradby
Gondwana Link Ltd, PO Box 5276, Albany WA 6332, Australia, bradby@gondwana link.org.

Andrew Bridges
Department of Natural Resources, Environment, the Arts and Sport, PO Box 1120, Alice Springs NT 0871, Australia, Andrew.Bridges@nt.gov.au.

Andrew Cameron
Tasmanian Land Conservancy, PO Box 2112, Lower Sandy Bay TAS 7005, Australia, acameron@tasland.org.au.

Ben Carr
Habitat 141°, C/- Greening Australia Victoria, PO Box 999, Horsham VIC 3402, Australia. Current address: PO Box 67, Kew East VIC 3102, Australia, bencarr@fibrefusion.com.au.

Stuart Cowell
Bush Heritage Australia, PO Box 329, Flinders Lane, Melbourne VIC 8009, Australia; and Tasmanian Land Conservancy, PO Box 2112, Lower Sandy Bay TAS 7005, Australia, cowellsg@gmail.com.

Rosemary Crane
Mosquito Productions, Wamboin NSW 2620, Australia, rcrane@tpg.com.au.

Jim Crosthwaite
Biodiversity and Ecosystem Services, Department of Sustainability and Environment, 8 Nicholson Street, East Melbourne VIC 3002, Australia; Current address: 6 Naroon Road, Alphington VIC 3078, Australia, jimxwaite@pacific.net.au.

Christine Cumming
Holbrook Landcare Network, GPO Box 121, Holbrook NSW 2644, Australia, chriscumming@holbrooklandcare.org.au.

Erik D. Doerr
CSIRO Ecosystem Sciences, GPO Box 1700, Canberra ACT 2601, Australia; and Division of Evolution, Ecology, and Genetics, Research School of Biology, Australian National University, Acton ACT 0200, Australia, erik.doerr@csiro.au.

Veronica A.J. Doerr
CSIRO Ecosystem Sciences, GPO Box 1700, Canberra ACT 2601, Australia; and Division of Evolution, Ecology, and Genetics, Research School of Biology, Australian National University, Acton ACT 0200, Australia, veronica.doerr@csiro.au.

Rob Dunn
Great Eastern Ranges Initiative, C/- Greening Australia, 142 Addison Road, Marrickville NSW 2204, Australia, rdunn@greeningaustralia.org.au.

Sarah Eccles
Bush Heritage Australia, PO Box 329, Flinders Lane, Melbourne VIC 8009, Australia, seccles@bushheritage.org.au.

Nick Edgar
NZ Landcare Trust, PO Box 4305, Hamilton, New Zealand 3247, nick.edgar@landcare.org.nz.

Robyn Edwards
Trust for Nature (Victoria), 156 McLeod Street, Bairnsdale VIC 3885, Australia, robyne@tfn.org.au.

Simon Ferrier
CSIRO Ecosystem Sciences, GPO Box 1700, Canberra ACT 2601, Australia, simon.ferrier@csiro.au.

James Fitzsimons
The Nature Conservancy, Suite 2-01, 60 Leicester Street, Carlton VIC 3053, Australia; and School of Life and Environmental Sciences, Deakin University, 221 Burwood Highway, Burwood VIC 3125, Australia, jfitzsimons@tnc.org.

Trish Fox
Trust for Nature (Victoria), 156 McLeod Street, Bairnsdale VIC 3885, Australia, patricia_mcnaughton@yahoo.com.au.

David Freudenberger
Greening Australia, 333 Bennetts Road, Norman Park QLD 4170, Australia; Current address: Fenner School of Environment and Society, Australian National University, Canberra ACT 0200, Australia, david.freudenberger@anu.edu.au.

Jody Gates
Department of Environment, Water and Natural Resources, GPO Box 1047, Adelaide SA 5001, Australia, Jody.gates@sa.gov.au.

Jane Greacen
113 Carpenter Street, Lakes Entrance VIC 3909, Australia, janegreacen@bigpond.com.

Lainie Grigg
Greening Australia, 333 Bennetts Road, Norman Park QLD 4170, Australia; Current address: Ausecology Pty Ltd, 24 Bridgenorth Street, Carindale QLD 4152, Australia, lgrigg@ausecology.com.

Gary Howling
Office of Environment and Heritage (NSW), PO Box A290, Sydney South NSW 1232, Australia, gary.howling@environment.nsw.gov.au.

Stella Kondylas
Department of Environment, Water and Natural Resources, GPO Box 1047, Adelaide SA 5001, Australia, stella.kondylas@sa.gov.au.

Judy Lambert
Community Solutions, 179 Sydney Road, Fairlight NSW 2094, Australia, judy@communitysolutions.com.au.

Harvey Locke
Yellowstone to Yukon Conservation Initiative, PO Box 4887, Banff, Alberta T1L 1G1, Canada, harvey@wild.org.

Duncan Mackenzie
BirdLife Australia Gluepot Reserve, Ecotourism Australia and South Australian Tourism Industry Council, 4 Edinburgh Avenue, Stonyfell SA 5066, Australia, dmackenzie@iname.com.

Brendan Mackey
Griffith Climate Change Response Program, Griffith University, Southport QLD 4222, Australia, b.mackey@griffith.edu.au.

Ian Mansergh
Research Centre for Applied Alpine Ecology, La Trobe University, Bundoora VIC 3084, Australia, ian.mansergh@gmail.com.

Toni McLeish
Biodiversity Conservation Section, NSW Office of Environment and Heritage, PO Box 733, Queanbeyan NSW 2620, Australia, toni.mcleish@environment.nsw.gov.au.

Peter Morison
Wedderburn Conservation Management Network, C/- Department of Sustainability and Environment, PO Box 3100, Bendigo VIC 3554, Australia; Current address: 155 Retreat Road, Bendigo VIC 3550, Australia, jandpz@bigpond.com.

Wendy Murphy
Wedderburn Conservation Management Network, C/- Department of Sustainability and Environment, PO Box 3100, Bendigo VIC 3554, Australia, wendy.murphy@dse.vic.gov.au.

Sam Niedra
Nature Conservation Trust of New South Wales, PO Box 789, Albury NSW 2640, Australia, sam@nct.org.au.

Lorraine Oliver
Biodiversity Conservation Section, NSW Office of Environment and Heritage, PO Box 733, Queanbeyan NSW 2620, Australia, lorraine.oliver@environment.nsw.gov.au.

Hannah Parris
CSIRO Ecosystem Sciences, GPO Box 1700, Canberra ACT 2601, Australia, hannah.parris@csiro.au.

Hugh P. Possingham
ARC Centre of Excellence for Environmental Decisions, University of Queensland, St Lucia QLD 4072, Australia, h.possingham@uq.edu.au.

Ian Pulsford
IUCN World Commission on Protected Areas, 204 Duffy Street, Ainslie ACT 2602, Australia, ianpulsford@homemail.com.au.

Ralf Regeer
Greening Australia, 333 Bennetts Road, Norman Park QLD 4170, Australia; Current address: Ausecology Pty Ltd, 24 Bridgenorth Street, Carindale QLD 4152, Australia, rregeer@ausecology.com.

Rainer Rehwinkel
Biodiversity Conservation Section, NSW Office of Environment and Heritage, PO Box 733, Queanbeyan NSW 2620, Australia, rainer.rehwinkel@environment.nsw.gov.au.

Angela Sanders
Bush Heritage Australia, PO Box 7103, Lower King WA 6330, Australia, asanders@bushheritage.org.au.

Peter G. Spooner
Institute for Land, Water & Society, and School of Environmental Sciences, Charles Sturt University, Albury NSW 2640, Australia, pspooner@csu.edu.au.

Daniel Sprod
Tasmanian Land Conservancy, PO Box 2112, Lower Sandy Bay TAS 7005, Australia, dsprod@tasland.org.au.

Julia Stanley
PO Box 202, Wodonga VIC 3689, Australia, jstanley01@gmail.com.

David Walker
National Landcare Network, PO Box 777, Gunnedah NSW 2380, Australia, dwalker@nln.org.au.

Philippa Walsh
Bush Heritage Australia, PO Box 329, Flinders Lane, Melbourne VIC 8009, Australia, pwalsh@bushheritage.org.au.

Geoff Wescott
School of Life and Environmental Sciences, Deakin University, 221 Burwood Highway, Burwood VIC 3125, Australia, wescott@deakin.edu.au.

Stuart M. Whitten
CSIRO Ecosystem Sciences, GPO Box 1700, Canberra ACT 2601, Australia, stuart .whitten@csiro.au.

Graeme L. Worboys
IUCN World Commission on Protected Areas; and Fenner School of Environment and Society, Australian National University; 3 Rischbieth Crescent, Gilmore, ACT 0200, Australia, g.worboys@bigpond.com.

Carina Wyborn
Fenner School of Environment and Society, Australian National University, Canberra ACT 0200, Australia, Carina.wyborn@anu.edu.au.

Charlie Zammit
ARC Centre for Excellence for Environmental Decisions, University of Queensland, St Lucia QLD 4072, Australia; Current address: 6 Grand Valley Court, Mullumbimby Creek NSW 2482, Australia, zammit.byron@gmail.com.

Acknowledgements

The second *Linking Landscapes* Summit in Melbourne, 2010, inspired two of the editors to contact the third and stimulated the ideas that became this book. For this we would like to thank the principle organisers of that summit, Penny Figgis, Graeme Worboys and Ian Walker. We would particularly like to thank the chapter authors for their contributions, not only in preparing the chapters, but also for their commitment to pushing the practical frontiers of the new and rapidly growing and evolving realm of connectivity conservation in Australia. John Manger, Tracey Millen and Deepa Travers at CSIRO Publishing provided guidance throughout the publishing process, while Anne Findlay undertook a final edit of the entire manuscript. Glenn Johnstone kindly prepared the location maps that head the chapters in Sections 2 and 3 as well as Figure 1.1. We would also like to acknowledge The Nature Conservancy (JF) and Deakin University (GW) for supporting us in this endeavour.

Foreword

Australia, like my country, has successfully focused on building a civilisation based on developing, exploiting, and connecting a few widely separated regions of fertile farmland, a vast coast, and a wild northern interior. This was hard work that required courage, innovation and perseverance. Great cities grew but, with the exception of national parks, nature was not considered much in its own right for the services it provided to us nor did we respect its importance to indigenous cultures. We saw this natural bounty as wealth only after we had converted it to our own use.

Today a new challenge lies before us. Humans have come to profoundly affect every aspect of the world in which we live. We are changing the climate through sending too much CO_2 into the atmosphere and the chemistry of the Earth by altering the nitrogen balance. Through land, coastline and seabed conversion we are unravelling the varied and beautiful tapestry of biodiversity. This is a big problem for humanity because these things provide the food, fresh water and oxygen on which we depend for life. The good news is that we are not helpless to confront these unintended consequences of our efforts. Through large landscape conservation we can take a big step to address both climate change and biodiversity loss. Large landscape conservation can also embrace the traditional knowledge of indigenous peoples.

It is difficult for a country with a small population to make an impact on the future direction of humanity. With the many initiatives already in operation and with the Australian Government's *National Wildlife Corridors Plan*, Australia has a chance to break out and lead the world in a positive new direction. Many of us have devoted the last 20 years to developing the idea of large landscape conservation and, in some regions like Yellowstone to Yukon, we are leading the way to implementing aspects of it. But, as this volume and its eminently qualified authors show, Australia as a country and a continent is now on track to do more large landscape conservation than anywhere else on Earth.

The pioneering work of the 21st century is to set ourselves on a course where progress is defined as keeping all of life and the Earth's natural processes intact and thriving along with us. This should be the context for meeting our social and economic development goals rather than a competing interest. To get there will require courage, innovation and perseverance from us all. Thanks for your leadership Australia. I hope we in North America and the rest of the world will soon take up the challenge of large landscape conservation as systematically as you have.

Harvey Locke
Co-founder and Strategic Advisor, Yellowstone to Yukon Conservation Initiative
Banff, Canada

SECTION 1.
SCENE SETTING

1

Linking Australia's landscapes: an introduction

James Fitzsimons, Ian Pulsford and Geoff Wescott

Australia has undergone a recent surge in interest in creating networks and initiatives that aim to 'link up' habitats and landscapes that have been fragmented by clearing or varying land uses or ownership. This increased interest has been stimulated by the involvement of a range of players – government, non-government organisations, landowners and the broader community.

In Australia, these initiatives include a range of activities that are promoted under a variety of names including 'biosphere reserves', 'wildlife corridors', 'conservation management networks', and 'biolinks', among others. These various initiatives are not just about achieving 'connectivity conservation' but are often couched in terms of 'landscape-scale' projects. There has been relatively little analysis of these diverse and *practical* initiatives in a single synthesis. Nor have the perspectives of those who have established or are coordinating and managing these projects received as much attention as they deserve. The result has been a gap in dialogue and understanding between the theoretical discussions of developing 'a network of corridors' across a continent (or state jurisdiction) and the practical reality of establishing and operating such initiatives in the landscape.

We aim in part to address these gaps through this book by bringing together the lessons from a diverse range of established connectivity initiatives and policy frameworks, together with a series of broader themed chapters encompassing aspects such as social, ecological and governance considerations. Initiatives featured include those from all Australian states and territories, and many initiatives which cross state and territory boundaries. The chapter authors are from universities, non-government organisations, state and national government agencies, research institutions and consultancies. Our intention, and hope, is that the descriptions in this book will not only inform policy makers, land managers, facilitators, and scientists, but it will also stimulate even greater efforts 'on the ground' by being of interest to the general public as well.

This book set out with the aim of canvassing a broad range of these multi-tenure conservation initiatives, described by the people directly involved in each project, not only to document the successes and challenges of these initiatives at specific geographies but also to begin to analyse if there were critical common lessons that have been learnt already in this new and evolving field. In particular we were keen to see if there were emerging models of approach that could potentially be adapted by new entrants into connectivity conservation initiatives, rather than have such groups repeat the mistakes of others.

In choosing a cross-section of examples of various approaches, we have focused on those with several years of experience behind them. We recognise there are other initiatives out there and many more emerging or proposed as this book goes to press (see Whitten *et al.* 2011 and NWCPAG 2012 for example). There have been of course several significant books already written on connectivity conservation both internationally (e.g. Soulé and Terborgh 1999; Crooks and Sanjayan 2006; Worboys *et al.* 2010; Hilty *et al.* 2012) and for Australia (e.g. Saunders *et al.* 1996; Bennett 2003; Lindenmayer and Fisher 2006). This book aims to be different in focus concentrating on the practical on-ground attempts at landscape-scale conservation.

We were concerned that there was some danger there would be significant repetition in the case studies but this has proved not to be the case. While there were several common themes, outlined in the synthesis at the end of this book, there were also many differences in approach.

During the course of preparing this book, policy on the creation of a *National Wildlife Corridors Plan* was released by the Australian Government (DSEWPC 2012a). The plan was broad in nature and the draft of the plan (NWCPAG 2012) met with a mixed response from public submissions (DSEWPC 2012b). Many were keen to see the concept on the national agenda and for potential federal financial support, but others were worried about the potential of the lessening of the role of the National Reserve System, proposed new legislation and the proposed process for nomination of National Wildlife Corridors. The authors of various chapters in this book reflect some of this diversity of views.

Before outlining the chapters that follow we would like to make it clear that the focus of this book is not a review of the ecological theory behind connectivity conservation, or related terms such as corridors (although some chapters do address this). It is a book about real, on-ground examples of trying to achieve multi-tenure conservation at scales from landscape to sub-continental by many different, and committed, departments, organisations, individuals and communities.

The book is divided into four main sections with a fifth acting as summary.

Section 1 provides an overview and international perspective on the state of connectivity conservation efforts as an introduction and background to Section 2, where the history, successes, lessons and application of those lessons for 14 case studies from around Australia are described by the practitioners of connectivity conservation. We have taken a 'geographic' approach to ordering the case study chapters, moving from west to east (Figure 1.1).

Section 3 outlines six policies and frameworks from different parts of Australia and New Zealand. These chapters consider approaches beyond the individual initiative; in effect efforts to create or coordinate 'networks of networks' or 'systems of corridors'.

Section 4 explores broader themes associated with multi-tenure connectivity approaches, namely connectivity conservation principles, social dynamics, governance arrangements, socio-economic considerations and the importance of interdisciplinary research.

Section 5 synthesises the key themes raised by the authors throughout the book and suggests areas for further research.

Figure 1.1 Case study conservation networks featured in Section Two of this book. Differences in shading differentiates overlapping networks. Numbers refer to chapter numbers: 3 – Gondwana Link, 4 – Fitzgerald Biosphere Reserve, 5 – Territory Eco-link, 6 – South Australian NatureLinks, 7 – Riverland (Bookmark) Biosphere Reserve, 8 – Habitat 141°, 9 – Tasmanian Midlandscapes, 10 – Wedderburn Conservation Management Network, 11 – Gippsland Plains Conservation Management Network, 12 – Grassy Box Woodlands Conservation Management Network, 13 – Great Eastern Ranges Initiative, 14 – Slopes to Summit, 15 – Border Ranges Alliance, 16 – Bunya Biolink

References

Bennett A (2003). *Linkages in the Landscape. The Role of Corridors and Connectivity in Wildlife Conservation.* IUCN, Gland and Cambridge.

Crooks KR and Sanjayan M (Eds) (2006). *Connectivity Conservation.* Cambridge University Press, Cambridge.

DSEWPC (2012a). *National Wildlife Corridors Plan: A Framework for Landscape-scale Conservation.* Department of Sustainability, Environment, Water, Population and Communities, Canberra.

DSEWPC (2012b). Draft National Wildlife Corridors Plan – public comment. Department of Sustainability, Environment, Water, Population and Communities. Canberra. Available: http://www.environment.gov.au/biodiversity/wildlife-corridors/consultation/index.html [Accessed 9 September 2012].

Hilty JA, Chester C, and Cross M (2012). *Climate and Conservation: Landscape and Seascape Science Planning and Action.* Island Press, Washington, DC.

Lindenmayer DB and Fisher J (2006). *Habitat Fragmentation and Landscape Change: An Ecological and Conservation Synthesis.* CSIRO Publishing, Collingwood.

NWCPAG [National Wildlife Corridors Plan Advisory Group] (2012). *Draft National Wildlife Corridors Plan.* National Wildlife Corridors Plan Advisory Group, Canberra.

Saunders DA, Craig JL, and Mattiske EM (Eds) (1996). *Nature Conservation 4: The Role of Networks.* Surrey Beatty and Sons, Chipping Norton.

Soulé ME and Terborgh J (Eds) (1999). *Continental Conservation: Scientific Foundations of Regional Reserve Networks.* Island Press, Washington, DC.

Whitten SM, Freudenberger D, Wyborn C, Doerr V, and Doerr E (2011). 'A compendium of existing and planned Australian wildlife corridor projects and initiatives, and case study analysis of operational experience'. Report to the Department of Sustainability, Environment, Water, Population and Communities. CSIRO, Canberra.

Worboys GL, Francis W, and Lockwood M (Eds) (2010). *Connectivity Conservation Management: A Global Guide.* Earthscan, London.

2

Connectivity conservation initiatives: a national and international perspective

Graeme L. Worboys and Brendan Mackey

Connectivity conservation

There has been a social and political revolution in the care and management of Australia's landscapes in the 21st century. As the 19th and 20th century cultural imperative for clearing and converting the Australian bush has diminished, a new urgency for the conservation of bush and wildlife has emerged across Australia's landscapes. This is manifesting among a growing and diversifying community of conservation practitioners as a commitment to protecting and restoring Australia's native biodiversity and achieving more ecologically sustainable land management. This is especially so through the establishment of many strategically and grassroots-organised large conservation corridors. Corridor initiatives that have commenced include Gondwana Link in Western Australia, the Trans-Australia Eco-Link of South Australia and the Northern Territory, South Australia's NatureLinks, Habitat 141° in Victoria–South Australia–New South Wales, the Great Eastern Ranges Corridor in Australia's eastern states and the Midlandscapes initiative of Tasmania (see Chapters 3, 5, 6, 8, 13 and 9). This pioneering connectivity conservation work has witnessed the involvement of individual corridor champions, landowners, community groups, business groups, non-government organisations (NGOs) and some governments in hitherto unseen partnerships and collaboration. Connectivity conservation and the establishment of large corridors have captured the imagination and commitment of many Australians. They have become important to many people in providing a focus for community engagement around a shared conservation vision for the land they own, lease or are involved in managing.

Corridors help conserve natural areas across private, leasehold, public and Indigenous land tenures, and across land uses, including land worked for pastoralism, forestry, cropping, mining, water supply and defence. They are not protected areas in their own right but ideally embed and (where appropriate) connect protected areas. They are a voluntary use of land that for its success requires the involvement of many sectors of society and multiple individual contributions. Private landowners, lessees, government agencies and Traditional Owners, all have vital roles to play in contributing to conservation goals including maintaining and restoring structural and functional connectivity. Corridors can help maintain and enhance the natural integrity of protected areas by enabling key processes to continue, including the natural adaptation responses of species (Mackey *et al.* 2008). Corridors help retain options and opportunities for species to disperse and migrate across landscapes, for example in response to climate change. Corridors can also minimise the potential for parks to become islands and ecological 'sinks' in a sea of developed lands. The Yellowstone to Yukon (Y2Y) corridor of the USA and Canada (for example) through the actions of dozens of conservation groups and land trusts is working to maintain natural interconnections between Yellowstone National Park and the Rocky Mountains national parks of Canada. The aim of this work is to help conserve habitats and movement corridors of the Grizzly Bear (*Ursus arctos horribilus*) and other species across private lands to help prevent the remnant Yellowstone bears being genetically isolated (Locke 2010).

Two great threats and a desire for better Australian responses to addressing climate change and conserving the environment have helped mobilise individuals, organisations, governments and communities to establish corridors. The first threat is land clearing and habitat degradation. Between 2000 and 2010 Australia's bushland was still being cleared by an estimated 1 million hectares per annum (State of the Environment 2011 Committee 2011). While recent (pre-2012) state legislation emerged to help slow land clearing, we remain faced with a monumental challenge of managing an array of threatening processes associated with habitat degradation. Calls in 2012 for a weakening of this protective legislation are of great concern. Climate change is the second great threat. The pervasiveness of actual and forecast impacts across the landscape is greatly disturbing (Steffen *et al.* 2009) and large corridors are one recognised response.

Corridors are a 21st century grassroots and (sometimes) government-inspired conservation use of land in the manner of their establishment, design, implementation and management. It has meant a greater sharing of conservation management accountabilities by governments and the further encouragement and empowering of communities, individuals, non-government and private organisations to take conservation action. Large connectivity conservation corridors represent a fundamental change in how we relate to our land. It is a paradigm shift that establishes *strategic conservation* of biodiversity at large scales in the landscape as a response to climate change, to the fragmentation and destruction of interconnected vegetation and habitats that span the continent, and as a response to the need to conserve whole-of-continent ecological processes. The establishment and management of corridors reflect a shared vision of the community and foster a shared responsibility. It is about partnerships and collaboration for long-term conservation outcomes.

In 2012, an important step forward occurred for the conservation of Australia's biodiversity in the face of current and future threats. The Australian Government formally announced that it had a national facilitation role in establishing and managing corridors. This was in addition to its stated commitments to an expanded National Reserve System and strategic investments in biodiversity conservation and corridor management enabled by its carbon dioxide pollution reduction legislation, the *Clean Energy Act 2011*, and the associated biodiversity conservation investment fund. The Australian Government's corridor facilitation role was identified through a *National Wildlife Corridors Plan* (DSEWPC 2012) which was informed by recommendations from an expert Advisory Group (NWCPAG 2012). To our knowledge, this Plan was the first whole-of-continent connectivity conservation corridor governmental policy response anywhere in the world and it confirmed an empowerment of communities and individuals in the process of establishing and managing corridors. It recognised whole-of-continent ecological processes and the potential role of large national wildlife corridors in sustaining Australia's flyways and ecological responses to its 'boom and bust' cycles of biological productivity that are in response to rainfall variability that so influences wildlife population distribution and abundance (DSEWPC 2012; see also Chapter 17).

Active conservation management of corridors is essential if they are to fulfil their mandate. Retention and restoration of interconnected and permeable landscapes are needed to provide suitable habitats and opportunities for biological and environmental flows. The removal or reduction of weeds and pest animals, ecological fire management, and addressing environmental hydrological flows improve natural resilience, reduce pressure on native species and help prepare for extreme wildfire events. With large corridors, such management actions are implemented across many degrees of latitude and longitude and thereby maintain connectivity for species responses at a continental scale (Steffen *et al.* 2009).

Areas that are subject to connectivity conservation management are described by many different terms in the literature including: landscape linkages; biolinks; corridors; conservation management networks; habitat corridors; connectivity corridors; ecological corridors; wildlife corridors; linear corridors; riparian corridors; ecological networks; greenways; stepping stones; and flyways (Worboys *et al.* 2010). However, some of these terms are more specific in their meaning. For example, the term *ecological networks* is commonly used in Europe to specifically describe systems of nature reserves and their interconnections that commonly include core areas, buffer zones and ecological corridors (Worboys *et al.* 2010). Connectivity conservation may be more than this. A *connectivity conservation area* is perhaps a more useful generic description that captures a comprehensive understanding of connectivity conservation. However, this three-word descriptor can be unwieldy and is not always used. More commonly, connectivity conservation areas are described as *corridors* and their active management is recognised as *connectivity conservation management* (Worboys *et al.* 2010).

Given their scale and cultural context, even in a relatively sparsely populated continent such as Australia, many people live and work in corridors, and these corridors support a diversity of human communities. These landscapes are often of great cultural and spiritual importance, particularly to Indigenous communities and corridors may, for example, be traditional pathways described by Indigenous songlines of country (DSEWPC 2012). For rural communities, corridors can hold resources important for farms such as shelter belts and riparian vegetation. Corridors include habitat areas that are retained on private lands that are otherwise the subject of very large, commercial and intense (industrial-scale) land uses such as logging, mining, cropping and grazing. Connectivity conservation management prescriptions for natural remnants in these intensively used landscapes can help achieve conservation outcomes and, for example, can help retain corridor permeability for species. Corridors can help maintain scenic amenity important for the health of local communities, they provide other ecosystem services such as potable water and clean air and provide welcome natural relief from urban environments. They can also play a critical role in water catchment protection and erosion control. Corridors are an intuitively appealing concept for people. In the eyes of many, it is just 'common sense' to care for the corridor and conserve rather than eliminate and degrade any further the natural features and assets of surrounding lands. In parallel with this community response, there has been increasing scientific documentation about the conservation benefits of corridors.

The science of connectivity

Connectivity conservation is defined using biodiversity conservation criteria, but it also recognises the human dimension in terms of social, cultural, economic and institutional factors. Included in the account of what constitutes connectivity conservation therefore are the ways in which people are connected to the natural and semi-natural corridor lands (Worboys *et al.* 2010).

The concept of connectivity is a geographically framed concept concerned with the spatial dimension, including the relationships between scale, pattern and process, and the biodiversity values of the broader landscape mix of natural, semi-natural and altered landscapes (the matrix), that have always informed conservation biology (Mackey *et al.* 2010). This matrix mix and opportunities for connectivity are illustrated in Figure 2.1. The relationship between connectivity conservation in the landscape and biodiversity conservation is further described below.

Whittaker *et al.* (2005) suggested that the theoretical underpinnings of conservation biology draw on four sources. First is the development and evaluation of biological theory spanning population, biological and genetic processes, and concerns with minimum viable populations, genetic erosion from small populations, competitive influence of invasive species, and behavioural ecology – i.e. processes in which biogeography is not usually prominent. Second is the theory concerning processes at the bioregional to landscape scale, including MacArthur and Wilson's (1967) equilibrium theory of island biogeography, the derivative 'Single Large or Several Small' reserves (SLOSS) debate (Diamond 1975), metapopulation theory (Hanski and Gaggiotti 2004), and nestedness (reviewed by Whittaker 1998) – i.e. themes that bridge ecology and biogeography.

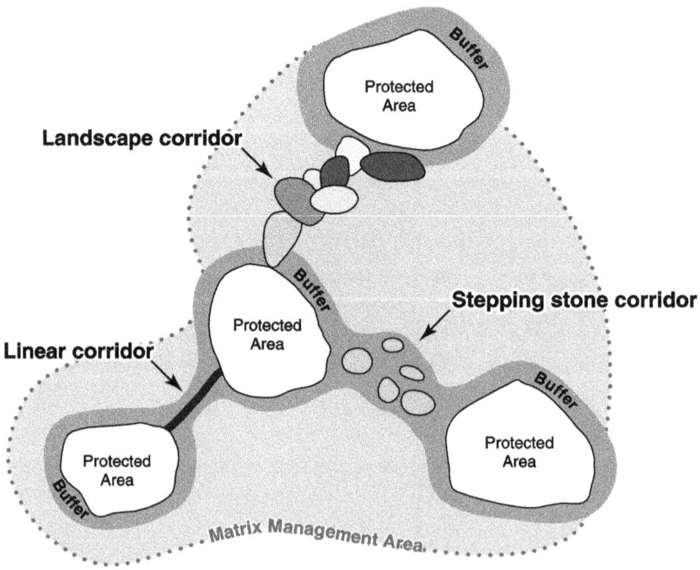

Figure 2.1 Some conceptual elements that comprise connectivity conservation spatial planning: core protected areas, the landscape-wide matrix management area, native vegetation that serves as 'stepping stones' and linear corridors of native habitat (Adapted from Bennett (2004) and Mackey *et al.* 2010).

Third are applications on a yet coarser scale, in part concerned with mapping and modelling biogeographic patterns and in part invoking historic biogeographic theories concerned with the distribution and explanation of geographic patterns in phylogeny, distributions and diversity at global, regional and continental scales. These reach back to the origins of modern biogeography (e.g. the work of Wallace 1859) and encompass the exploration of specific processes such as the peninsula effect (Means and Simberloff 1987). Fourth is macroecology (Brown 1995), which involves analysis of emergent statistical properties of ecological datasets and can address both ecological traits (growth rates, propagule size, breeding system, body size) and biogeographic traits (range size, region of origin).

Building upon the above conservation biology theory, a revised conservation science consensus is beginning to emerge in response to the limitations of conservation efforts to date, and the enormity of the challenge, given the scale of the current biodiversity extinction crisis and the potential synergistic impacts of climate change (e.g. Welch 2005). The retention and protection of unfragmented, large natural landscapes is an important objective for biodiversity conservation; this includes additional protected areas that are interconnected (not islands) with corridors and active management of corridors and protected areas to deal with threats. The term *connectivity conservation* has been widely used to describe this emerging scientific consensus among conservation researchers and practitioners (Crooks and Sanjayan 2006).

The IUCN World Commission on Protected Areas (WCPA) considered these theoretical and practical matters in a workshop at Papallacta, Ecuador, involving connectivity corridor specialists and managers and prepared a 'connectivity conservation declaration' (IUCN WCPA 2006) that stated:

> *'The maintenance and restoration of ecosystem integrity requires landscape-scale conservation. This can be achieved through systems of core protected areas that are functionally linked and buffered in ways that maintain ecosystem processes and allow species to survive and move, thus ensuring that populations are viable and that ecosystems and people are able to adapt to land transformation and climate change. We call this proactive, holistic, and long-term approach connectivity conservation.'*

Corridor structure and function

Natural connectivity for species in the landscape has a structural component, which relates to the spatial arrangement of habitat or other elements in the landscape and it has a functional (or behavioural) component, which relates to the behavioural responses of individuals, species or ecological processes to the physical structure of the landscape. These components and other aspects of connectivity conservation have been described in further detail in important texts by Bennett (2003), Anderson and Jenkins (2006), Crooks and Sanjayan (2006), Hilty *et al.* (2006), Lindenmayer and Fischer (2006) and Steffen *et al.* (2009). The descriptions of structure and function provided recognise four types of connectivity in the landscape that may be managed for conservation (Worboys *et al.* 2010). These are:

Landscape (vegetation) connectivity considers the extent to which the natural vegetation cover has been cleared, fragmented or degraded by human land use (see discussion in Lindenmayer and Fischer 2006). The ecological implications are considered in a general sense rather than in terms of the habitat needs of a particular species, consistent with

Southwood's (1977) habitat template theory. This theory depends on defining template axes appropriate to the organism(s) of interest, predicting the traits of species associated with different parts of the template and testing these predictions in a range of habitats whose positions in the template have been determined.

Habitat connectivity is the connectedness between patches of habitat that are suitable for a particular species based on the particular resources required by those species for food, shelter, reproduction and movement (Lindenmayer and Fischer 2006). There are landscape features that contribute to habitat connectivity and these include wildlife corridors, stepping-stones and a matrix of non-native vegetation that has similar attributes to patches of native vegetation (Bennett 2003 in Lindenmayer and Fischer 2006).

Ecological process connectivity is the connectedness of ecological processes at multiple scales, including processes related to highly dispersive species, highly interactive species, disturbance regimes, and hydro-ecological flows (Lindenmayer and Fischer 2006; Soulé *et al.* 2006; Mackey *et al.* 2007) (see also Chapter 23).

Evolutionary process connectivity refers to spatially based natural processes that pertain to both macro-evolution (leading to speciation) and micro-evolution including co-evolutionary interactions and local adaptions by a population to environmental conditions. The spatial dimension of evolutionary processes relate to the exchange of genetic material between populations, the extent to which populations are open or closed to inflows and outflows, the degree to which climate change will result in forced movements and the impact of other threatening processes. For many large animals and dispersive species, evolutionary processes involve the movement of these species over long distances (Soulé *et al.* 2006).

Benefits of connectivity conservation

Habitat destruction and degradation greatly impact species and ecosystems and lead to species decline and extinctions. Connectivity conservation is one critical response that aims to prevent such destruction, conserve biodiversity and to restore impacted areas within corridors. Ideally corridors need to be strategically selected for maximum biodiversity conservation benefit; they are very large (such as tens of kilometres wide and hundreds of kilometres long); and, if they are smaller (such as just hundreds of metres long and tens of metres wide), they should be large enough to allow movement, persistence and adaptation of the species or ecosystems they seek to benefit. All corridors need to be *actively* managed. This means, for example, responses to threats such as weeds, pest animals, soil erosion and changed fire regimes. A 2010 meta-analysis of 35 studies of corridor effectiveness found that 'corridors increase species movements in fragmented landscapes and that efforts spent on maintaining and creating corridors are worthwhile' (Gilbert-Norton *et al.* 2010, p. 660). Arguments by some scientists in the past have raised questions about the positive benefits of connectivity conservation, including movement of pest species and pathogens (e.g. Dunlop and Brown 2008). These arguments often relate to small, ill-planned and unmanaged corridors. Reviews of these critiques identify that when such concerns are met then worthwhile benefits usually accrue, especially where corridors are large and actively managed (Hilty *et al.* 2006; Chester and Hilty 2010; Doerr *et al.* 2011).

International perspectives of corridor initiatives

Internationally, high profile corridors such as Yellowstone to Yukon and the Terai Arc Landscape have been established to help conserve biodiversity. Conservation visionaries and leaders in many countries, working in meaningful partnerships with governments and local communities, combined with just plain hard work, have helped to achieve these corridors. The active management in these expanses of protected areas and interconnected landscapes is helping conserve many species otherwise threatened with extirpation and extinction (Gurung 2004; Locke 2010).

The benefits and necessity of these global connectivity conservation efforts were formally recognised in the *2011–2020 Strategic Plan for Biodiversity* prepared under the terms of the Convention on Biological Diversity (SCBD 2010a; b) in 2010 in Nagoya, Japan. The plan was an urgent response to help stem the loss of biodiversity on Earth and was agreed to by 188 countries. Target 11 of the plan stated (emphasis added):

> *'By 2020, at least 17 per cent of terrestrial and inland water, and 10 per cent of coastal and marine areas, especially areas of particular importance for biodiversity and ecosystem services, are conserved through effectively and equitably managed, ecologically representative* ***and well connected*** *systems of protected areas and other effective area-based conservation measures,* ***and integrated into the wider landscape and seascapes.****'*

Connectivity conservation is seen as an integral (supporting) part of an aspirational target for nations to expand their systems of protected areas to 17% of their terrestrial area and to interconnect them with larger landscapes. Both initiatives were seen as a critical response to climate change (Dudley *et al.* 2010).

The geographer Udvardy recognised eight Biogeographic Realms for Earth (Udvardy 1975; Figure 2.2), and connectivity conservation is an important conservation response in all these areas. Nations in some Realms are more advanced in corridor establishment and management and some of these are described here.

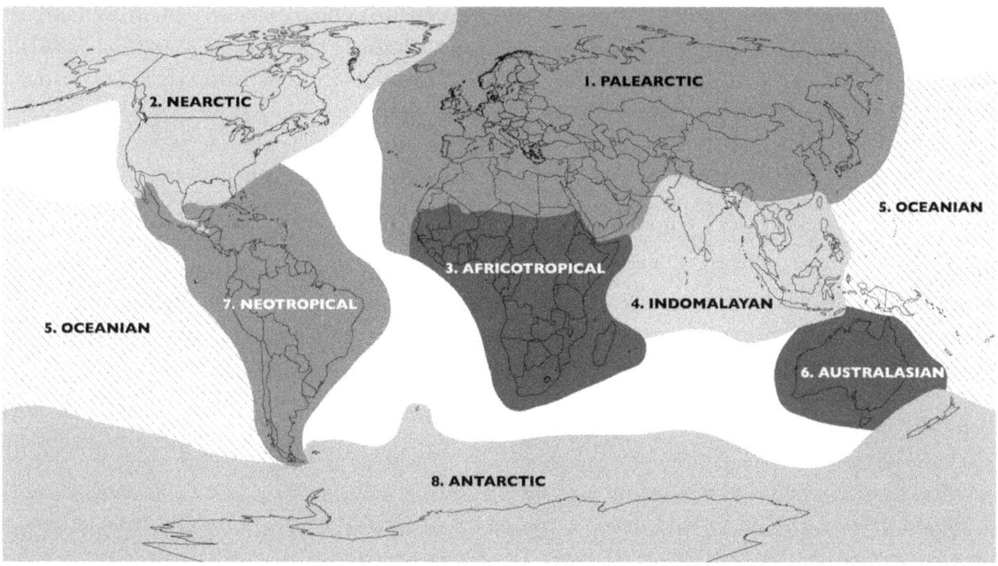

Figure 2.2 Udvardy's eight Biogeographic Realms of Earth (After Udvardy 1975).

Africotropical Realm

Important large connectivity corridors have been established in Africa. Three of these are located in the Cape Floristic Region of South Africa (Sandwith *et al.* 2010); the Greater Maloti-Drakensberg transboundary mountains area on the border between South Africa and Lesotho (Zunckel 2010) and the Greater Virunga landscape of the Albertine Rift Valley of central Africa that is a transboundary area of three nations, the Democratic Republic of Congo, Rwanda and Uganda (Plumptre *et al.* 2010). The great wildlife migration areas of Africa are also of special interest and a challenge for connectivity conservation management in the future given: (i) dynamically shifting environments with climate change; (ii) international boundaries; (iii) competition for land with expanding human populations; and (iv) illegal activities such as wildlife poaching for meat and ivory and human conflict in migration zones.

Experience from the South African corridors identified that farmers have been willing participants in connectivity conservation. A structured connectivity conservation management program by the South African Government, stewardship payments, stable and experienced corridor management staff and varying degrees of voluntary participation by farmers and stakeholders all assisted in successfully establishing the corridor work. The Virunga landscape connectivity conservation of the Albertine Rift Valley demonstrates the great benefits to wildlife conservation of cooperation across political boundaries at times of conflict as well as the perseverance and courage of professional staff (Plumptre *et al.* 2010).

Indomalayan Realm

Initiatives in Bhutan and Nepal led the way globally in connectivity conservation management practice. Three outstanding conservation leaders, Dr Chandra Gurung (WWF), Mingma Sherpa (WWF) and Dr Thirtha Maskey (Director of Nepal's Department of Parks and Wildlife), all helped to pioneer this work at the end of the 20th century. This has led to the establishment of a network of corridors in Bhutan and the Terai Arc corridor of Nepal. The Bhutan corridors (Sherpa *et al.* 2010) illustrate what is possible for a single nation to achieve in retaining its native forests and an interconnected protected area system. Based on a science assessment, the corridors focused on helping to conserve the Snow Leopard (*Panthera uncia*), the Tiger (*Panthera tigris*) and ecological processes along the elevation gradients such as bird migrations. The corridors were classified into 'intact corridors', 'bottlenecks' and 'critical corridors' based on their width and intactness as a basis for their management. Key lessons included the need for political support, the use of a science-based approach for effective corridor design, and that a combination of top-down and bottom-up approach can provide the greatest protection for connectivity (Sherpa *et al.* 2010).

Nepal's Terai Arc corridor (Gurung 2004) is managed by WWF in cooperation with the Nepalese Government. The corridor is well known for its science-based Tiger and Indian Rhinoceros (*Rhinoceros unicornis*) conservation work and direct work with local communities that help maintain economic support through conservation management-based employment and sustainable use through restoration and forest protection.

Nearctic Realm

The Yellowstone to Yukon (Y2Y) Conservation Initiative is a longstanding connectivity corridor that has provided inspiration and guidance for many people around the world. Established by community members, scientists and NGOs, it straddles the Rocky Mountains on the western side of the northern USA and southern Canada (Locke 2010). It has achieved a range of innovative approaches to management. The Y2Y corridor design and management is based on scientific knowledge concerning the conservation requirements of key species such as the Grizzly Bear and Wolf (*Canis lupus*) and their respective home ranges. The initiative is citizen driven, supported by philanthropic investment and it is coordinated and managed through a Board of Management. It has a small secretariat which helps to coordinate and prioritise the work of multiple volunteer organisations along the length of Y2Y (Locke 2010). Its ability to achieve large-scale wildlife conservation is supported through wildlife laws both in the US and Canada.

There are many aspirational connectivity conservation actions in the Nearctic. In 2012, the US National Parks Service was exploring and facilitating five interagency landscape-scale conservation projects. To the east, there are proposals for a Greater Northern Appalachian bioregion corridor (Hamilton and Trombulak 2010) and another in the Southern Appalachians (Irwin 2010), while to the west, connectivity along the greater Rocky Mountains chain is of importance to many experts, practitioners, communities and in 2012, the US Government.

Neotropical Realm

The Paseo Pantera or 'Path of the Panther Corridor,' pioneered by conservationist Mario Boza, is now known as the Mesoamerican Biological Corridor and it extends across seven Central American countries (Muller and Barborak 2010). The Corridor includes multiple protected areas and their interconnections and is administered through an institutional structure that includes the participating nations. Costa Rica is a very active participant. The San Juan–La Selva biological corridor within Costa Rica for example interconnects the tropical forests of the Braulio Carrillo National Park with the Nicaraguan border, and helps to conserve the Great Green Macaw (*Ara ambiguus*) flyway including its habitat trees. Farmers and landowners are encouraged to retain and restore the forests on their lands, and incentive payments are made for such actions. This is funded from a petrol tax and a levy on potable water and has successfully helped to restore forests to 51% of Costa Rica's lands in 2012 compared with ~20% in the 1980s (O. Chassot, pers. comm. 2012).

Palaearctic Realm

In Europe and Asia there is a great deal of connectivity conservation work happening and many proposals under consideration. Some are small corridors while others, such as the Altai-Sayan corridor, are very large. Cooperation and partnerships dominate these projects' governance arrangements. The Pan-European Ecological Network, for example, is supported by over 50 countries and has helped to halt the loss of biodiversity in Europe (SCBD 2010b). Other European networks and aspirational projects (Ehringhaus 2010a) include the Baltic network; corridor work in the Cantabric Mountains and Pyrenees (Mallarach *et al.* 2010); the Apennines of Italy (Romano 2010); the Carpathians, the Mount Blanc Massif (Ehringhaus 2010b) and the Alpine Ecological Network of protected areas (Kohler and Plassmann 2010). In Central Asia, the transboundary Mega Altai-Sayan corridor is being championed in the Altai Mountains along the borders of Kazakhstan, Russia, Mongolia and China by the Russian Academy of Science and NGOs including WWF (Badenkov 2010; Badenkov *et al.* 2012).

Lessons for connectivity conservation in Australia

The release of the Australian *National Wildlife Corridors Plan* (DSEWPC 2012) in November 2012 has created history both nationally and internationally. It has identified for the first time a whole-of-continent strategy for connectivity conservation as an essential contribution to the conservation of biodiversity (see Chapter 17). The original draft of this plan was prepared by a specially appointed Australian Government Advisory Committee. Significantly, the plan recognised the importance of the existing corridor initiatives in Australia described in greater detail elsewhere in this book (e.g. Chapters 3, 5, 6, 8, 9, 13). We discuss below some key concepts and broad guidelines for corridor planning drawn from the plan and Australian and international on-ground management experience.

Scale and design

Geographic extent is an important consideration in the design of corridors. Very large corridors that are located in areas of strategic importance for biodiversity conservation are especially important, given the challenges to be faced in the coming decades. The pervasive effects of climate change are of grave concern and are of continental scale in influence. Very large corridors are a necessary response to this threat, given that the adaptive response of many species will be long distance dispersal/migration. Consequently, corridors may need to span many degrees of latitude to provide poleward migration opportunities; or, they may span many degrees of longitude (such as across northern Australia) where east–west migrations (due to changed rainfall patterns) may be important. Large corridors will also be needed to interconnect protected areas, refugia, and low-lying natural areas with areas of higher altitude, providing enhanced permeability and opportunities for the movement of species. The Y2Y (Locke 2010); Great Eastern Ranges (Pulsford *et al.* 2012; Chapter 13) and Mesoamerican corridors (Muller and Barborak 2010) are good examples of large-scale corridor-conservation planning responses spanning broad latitudinal and elevation gradients. However, smaller-scaled regional and local corridors are also important as they may provide special interconnections critical for the survival and wellbeing of particular species and present the building blocks of future larger-scaled corridor initiatives.

Vision and leadership

Large corridor initiatives inspire people, help create conservation communities and provide opportunities where 'individuals make a difference' and where 'grand visions have a real place'. A guiding vision, that can be combined with more detailed goals for a large connectivity corridor, usually also establishes aspirational targets (through a corridor plan) for retaining and managing a degree of naturalness and for conserving and improving connectivity across many working landscapes. The vision should also recognise the importance of multiple individual efforts all contributing to a larger conservation outcome. The Great Eastern Ranges (Pulsford *et al.* 2010; 2012) and Y2Y (Locke 2010) corridors have both established vision statements. Many people and organisations need to step up and become leaders to help facilitate and implement this vision. They may be individual landowners retaining and restoring parts of their property, NGOs and Natural Resource Management groups helping to facilitate connectivity conservation across parts of landscapes, and leaders within government finessing policy and programs to facilitate connectivity conservation outcomes. Leadership also comes from those people who have typically voluntarily helped provide governance and direction for corridor initiatives.

Governance

Each large corridor typically establishes a form of governance that is suited to its local situation. This may include a representative citizen Board of Management such as the Y2Y

Corridor; an NGO governance body such as Gondwana Link in Western Australia (a partnership of many groups); a NGO-government body like the Terai Arc Landscape that involves WWF and the Nepalese Government; and government-facilitated corridors such as for the South Australian NatureLinks (Worboys and Pulsford 2011). Typically, the governance arrangement includes collaborative partnerships between many organisations.

Science

Scientists and scientific knowledge play a critical role in the identification, establishment, effective management and evaluation of the benefit of corridors for biodiversity conservation. Important scientific inputs include: identifying the strategic national and regional role large corridors play for biodiversity conservation during the establishment of corridors; determining threats to biodiversity; identifying short, medium and long-term connectivity conservation goals; defining priority management responses; and evaluating the efficacy and performance of management actions. Tiger and Snow Leopard researchers for example, helped guide the development of Bhutan's corridors to facilitate the movement of these species, and conservation of their populations (Sherpa *et al.* 2010).

Active management

Without active management, corridors will not successfully conserve biodiversity. Experience has shown that a systematic approach to management is needed and the IUCN World Commission on Protected Areas' *IUCN Connectivity Conservation Management Framework* was developed for this purpose (Worboys *et al.* 2010). Active management is more than just restoration and tree planting. It may take the form of working to build capacity with local communities, weed control, pest animal control, fire management, pollution management and the regulation of illegal activities. Typically, scientific data and information are used to help identify strategic management priorities.

In 2006 and 2008, the IUCN World Commission on Protected Areas convened international meetings of connectivity conservation managers to help develop guidance for corridor management. These experts developed best practice management for connectivity conservation and this was published in the IUCN book titled *Connectivity Conservation Management: A Global Guide* (Worboys *et al.* 2010). Among other recommendations, it was concluded that corridor management needs the following 10 components.

1 A clear, concise vision for the corridor;
2 An agreed governance model that is suitable and effective for the corridor;
3 A small secretariat that reports to the governance group and that provides a facilitation and support role;
4 An agreed science-based 'corridor plan', that appraises the connectivity conservation values, the threats, the priority management needs and provides clear goals and targets;
5 Active management of key threats through a range of different mechanisms, partnerships and novel collaboration arrangements;
6 Ongoing science-based evaluation of connectivity conservation performance through a process of evaluation that measures progress against agreed and effective indicators;
7 The active involvement of researchers as part of a corridor management partnership to track conservation progress and the nature of threats;
8 The active involvement and ownership of the community (from schoolchildren to seniors) in the celebration and conservation management of the corridor;

9 Corridor management leadership, planning and implementation that is concurrent at whole-of-corridor, landscape scale and site scale as per the IUCN Connectivity Conservation Management Framework; and
10 The promotion, education and information provision about the corridor.

Working with landowners and communities

Working with communities and individuals is a critical part of corridor management. WWF for example has worked extensively with local communities in managing the Terai Arc corridor in Nepal (Gurung 2004), and the Y2Y team has worked extensively with the Canadian First Nations peoples as well as private landowners such as coal companies to secure connectivity for the corridor and the continued movement of species such as the Grizzly Bear at critical choke points (Locke 2010).

Conclusion

Connectivity conservation contributes directly to the conservation of biodiversity and is recognised nationally and internationally for its role in helping to minimise the loss of species. Connectivity conservation corridors are a successful and grassroots, community-based partnership response to the threat of habitat destruction and species loss that usually involves collaboration with government agencies. The establishment and management of corridors are guided by a shared vision and they typically have context-dependent governance arrangements that are guided by general governance principles and the experience of other initiatives and practitioners. Wherever possible, corridors should interconnect and embed protected areas within the focus landscape or region. Corridors rely on voluntary collaboration, partnerships and leadership by many people and organisations for their success as they cross land tenures and active conservation management is usually needed. In the face of growing threatening processes, including climate change, corridors have a significant role to play in helping to ensure the long-term conservation of biodiversity and related natural heritage values.

Biographies

Graeme L. Worboys is Vice Chair (Mountains and Connectivity Conservation) for the International Union for the Conservation of Nature's (IUCN) World Commission on Protected Areas (WCPA); he is an Adjunct Fellow at the Fenner School of Environment and Society, Australian National University; a Board Member of the Tidbinbilla Nature Reserve Strategic Board and Managing Director of Jagumba Consulting Pty Ltd. Graeme provides WCPA's international leadership for connectivity conservation and mountain protected area management. His national consultant work includes specialist connectivity conservation, protected area, and National Heritage management advice. Internationally, he undertakes World Heritage Site evaluation and management advice work for IUCN, UNESCO and others.

Brendan Mackey is a Professor at Griffith University and Director of the Griffith Climate Change Response Program. He is a member of the Australian Climate Commission science advisory group and serves on the Council of the International Union for Conservation of Nature (IUCN). Brendan's research is currently focused on climate change science, ecosystem-based mitigation and adaptation, and the integration of biodiversity into regional assessments.

References

Anderson AB and Jenkins CN (2006). *Applying Nature's Design: Corridors as a Strategy for Biodiversity Conservation.* Columbia University Press, New York.

Badenkov YP (2010). Altai Mountain Knot: between conservation and development. In: *Connectivity Conservation Management: A Global Guide.* (Eds GL Worboys, W Francis and M Lockwood) pp. 246–250. Earthscan, London.

Badenkov YP, Yashina T, and Worboys GL (2012). Altai-Sayan Ecoregion: connectivity conservation of biological and cultural diversity in the Great Continental Divide of North Eurasia. In: *Climate and Conservation: Landscape and Seascape Science Planning and Action.* (Eds JA Hilty, C Chester, and M Cross) pp. 187–201. Island Press, Washington, DC.

Bennett A (2003). *Linkages in the Landscape: The Role of Corridors and Connectivity in Wildlife Conservation.* IUCN, Gland and Cambridge.

Bennett G (2004). *Integrating Biodiversity Conservation and Sustainable Use. Lessons Learned from Ecological Networks.* IUCN, Gland.

Brown JH (1995). *Macroecology.* University of Chicago Press, Chicago.

Chester C and Hilty JA (2010). Connectivity science. In: *Connectivity Conservation Management: A Global Guide.* (Eds GL Worboys, W Francis, and M Lockwood) pp. 22–33. Earthscan, London.

Crooks KR and Sanjayan M (Eds) (2006). *Connectivity Conservation.* Cambridge University Press, Cambridge.

Diamond RF (1975). The island dilemma: lessons of modern biogeographic studies for the design of nature reserves. *Biological Conservation* **7**, 129–146.

Doerr VAJ, Barrett T, and Doerr ED (2011). Connectivity, dispersal behaviour and conservation under climate change: A response to Hodgeson *et al. Journal of Applied Ecology* **48**, 143–147.

DSEWPC (2012). *National Wildlife Corridors Plan: A Framework for Landscape-scale Conservation.* Department of Sustainability, Environment, Water, Population and Communities, Canberra.

Dudley N, Stolton S, Belokurov A, Krueger L, Lopoukhine N, Mackinnon K, Sandwith T, and Sekhran N (2010). *Natural Solutions – Protected Areas: Helping People Cope with Climate Change.* IUCN WCPA, TNC, UNDP, WCS, World Bank and WWF, Doveton Press, Bristol.

Dunlop M and Brown PR (2008). 'Implications of climate change for Australia's National Reserve System: a preliminary assessment'. Report to the Department of Climate Change, and the Department of Environment, Water, Heritage and the Arts. CSIRO, Canberra.

Ehringhaus B (2010a). Overview of connectivity projects in the European Alps and adjacent mountains. In: *Connectivity Conservation Management: A Global Guide.* (Eds GL Worboys, W Francis, and M Lockwood) pp. 250–258. Earthscan, London.

Ehringhaus B (2010b). Tri-National Mont Blanc Massif: a crucial link in European alpine connectivity. In: *Connectivity Conservation Management: A Global Guide.* (Eds GL Worboys, W Francis, and M Lockwood) pp. 258–264. Earthscan, London.

Gilbert-Norton L, Wilson R, Stevens JR, and Beard KH (2010). A meta-analytic review of corridor effectiveness. *Conservation Biology* **24**, 660–668.

Gurung C (2004). Terai Arc Landscape: a new paradigm in conservation and sustainable development. In: *Managing Mountain Protected Areas: Challenges and Responses for the 21st Century.* (Eds D Harmon and GL Worboys) pp. 80–86. Adromeda Editrice, Colledara.

Hamilton LS and Trombulak SC (2010). Greater Northern Appalachian bioregion. In: *Connectivity Conservation Management: A Global Guide.* (Eds GL Worboys, W Francis and M Lockwood) pp. 152–160. Earthscan, London.

Hanski I and Gaggiotti EO (2004). *Metapopulation Biology: Past, Present and Future.* Oxford University Press, New York.

Hilty JA, Lidicker WZ, Jr, and Merenlender AM (2006). *Corridor Ecology: The Science and Practice of Linking Landscapes for Biodiversity Conservation.* Island Press, Washington, DC.

Irwin H (2010). Conservation network in the Southern Appalachian Mountains. In: *Connectivity Conservation Management: A Global Guide.* (Eds GL Worboys, W Francis and M Lockwood) pp. 140–152. Earthscan, London.

IUCN WCPA (2006). *The Papallacta Declaration, Papallacta, Ecuador, 17 November 2006.* Available: http://connectivityconservation.org/downloads/PapallactaDeclaration.htm [Accessed 5 July 2012].

Kohler Y and Plassmann G (2010). Ecological network of protected areas in the European Alps. In: *Connectivity Conservation Management: A Global Guide.* (Eds GL Worboys, W Francis, and M Lockwood) pp. 264–269. Earthscan, London.

Lindenmayer DB and Fischer J (2006). *Habitat Fragmentation and Landscape Change: An Ecological and Conservation Synthesis.* CSIRO Publishing, Collingwood.

Locke H (2010). Yellowstone to Yukon Initiative. In: *Connectivity Conservation Management: A Global Guide.* (Eds GL Worboys, W Francis, and M Lockwood) pp. 161–181. Earthscan, London.

MacArthur RH and Wilson EO (1967). *The Theory of Island Biogeography.* Princeton University Press, Princeton.

Mackey BG, Soulé ME, Nix HA, Recher HF, Lesslie RG, Williams JE, Woinarski JCZ, Hobbs RJ, and Possingham HP (2007). Towards a scientific framework for the WildCountry Project. In: *Key Topics and Perspectives in Landscape Ecology.* (Eds J Wu and RJ Hobbs) pp. 192–213. Cambridge University Press, Cambridge.

Mackey B, Watson JEM, and Hope G (2008). Climate change, biodiversity conservation, and the role of protected areas: an Australian perspective. *Biodiversity* **9**, 11–18.

Mackey B, Watson J, and Worboys GL (2010). *Connectivity Conservation and the Great Eastern Ranges Corridor.* Department of Environment, Climate Change and Water, Sydney.

Mallarach JM, Rafa M, and Sargatal J (2010). Cantabrian Mountains-Pyrenees-Massif Central-Western Alps Great Mountain Corridor. In: *Connectivity Conservation Management: A Global Guide.* (Eds GL Worboys, W Francis and M Lockwood) pp. 269–279. Earthscan, London.

Means DB and Simberloff D (1987). The peninsula effect: habitat correlated species decline in Florida's herpetofauna. *Journal of Biogeography* **14**, 551–56.

Muller E and Barborak J (2010). Mesoamerican biological corridor. In: *Connectivity Conservation Management: A Global Guide.* (Eds GL Worboys, W Francis and M Lockwood) pp. 183–191. Earthscan, London.

NWCPAG (2012). *Draft National Wildlife Corridors Plan.* National Wildlife Corridors Plan Advisory Group, Canberra.

Plumptre A, Kujirakwinja D, Owiunji I, Buhanga E, Mbula D, and Rwetsiba A (2010). Conserving connectivity in the Greater Virunga landscape. In: *Connectivity Conservation Management: A Global Guide.* (Eds GL Worboys, W Francis and M Lockwood) pp. 69–76. Earthscan, London.

Pulsford I, Worboys GL, and Howling G (2010). Australian Alps to Atherton Connectivity Conservation Corridor. In: *Connectivity Conservation Management: A Global Guide.* (Eds GL Worboys, W Francis and M Lockwood) pp. 96–105. Earthscan, London.

Pulsford I, Worboys GL, Howling G, and Barratt T (2012). Great Eastern Ranges, Australia. In: *Climate and Conservation: Landscape and Seascape Science Planning and Action.* (Eds JA Hilty, C Chester and M Cross) pp. 202–216. Island Press, Washington, DC.

Romano B (2010). The Apennines (European Alps to the Mediterranean). In: *Connectivity Conservation Management: A Global Guide.* (Eds GL Worboys, W Francis and M Lockwood) pp. 279–282. Earthscan, London.

Sandwith T, Ranger S, and Venter J (2010). Joining the dots: stewardship for connectivity conservation in the Cederberg Mountains, Cape Floristic region, South Africa. In: *Connectivity Conservation Management: A Global Guide.* (Eds GL Worboys, W Francis and M Lockwood) pp. 55–69. Earthscan, London.

SCBD (2010a). *Strategic Plan for Biodiversity Conservation 2011–2020.* Secretariat for the Convention on Biological Diversity. Available: http://www.cbd.int/decision/cop/?id=12268

SCBD (2010b). *Case Studies Illustrating the Socio-Economic Benefits of Ecological Networks.* Secretariat for the Convention on Biological Diversity, Montreal.

Sherpa MN, Wangchuck S, and Wikramanayake E (2010). Creating biological corridors for conservation and development: a case study from Bhutan. In: *Connectivity Conservation Management: A Global Guide.* (Eds GL Worboys, W Francis and M Lockwood) pp. 117–124. Earthscan, London.

Soulé ME, Mackey BG, Recher H, Williams JE, Woinarski JCZ, Driscoll D, Dennison WC, and Jones ME (2006). The role of connectivity in Australian conservation. In: *Connectivity Conservation.* (Eds KR Crooks and M Sanjayan) pp. 645–675. Cambridge University Press, Cambridge.

Southwood TRE (1977). Habitat, the templet for ecological strategies? *Journal of Animal Ecology* **46**, 337–365.

State of the Environment 2011 Committee (2011). *Australia, State of the Environment (2011).* Commonwealth of Australia, Canberra.

Steffen W, Burbidge AA, Hughes L, Kitching R, Lindenmayer D, Musgrave W, Stafford-Smith M, and Werner P (2009). *Australia's Biodiversity and Climate Change.* CSIRO Publishing, Melbourne.

Udvardy MDF (1975). *A Classification of Biogeographical Provinces of the World.* IUCN Occasional Paper No. 18. IUCN, Morges.

Wallace AR (1859). Letter from Mr Wallace concerning the geographical distribution of birds. *Ibis* **1**, 449–454.

Welch D (2005). What should protected area managers do in the face of climate change? *George Wright Forum* **22**, 75–93.

Whittaker RJ (1998). *Island Biogeography: Ecology, Evolution and Conservation.* Oxford University Press, Oxford.

Whittaker RJ, Araújo MB, Jepson P, Ladle R, Watson JE, and Willis KJ (2005). Conservation biogeography: assessment and prospect. *Diversity & Distributions* **11**, 3–23.

Worboys GL and Pulsford I (2011). *Connectivity Conservation in Australian Landscapes.* State of the Environment Committee, Department of Sustainability, Environment, Water, Population and Communities, Canberra.

Worboys GL, Francis W, and Lockwood M (Eds) (2010). *Connectivity Conservation Management: A Global Guide.* Earthscan, London.

Zunckel K (2010). Maloti-Drakensberg Transfrontier Conservation Programme: A South African perspective. In: *Connectivity Conservation Management: A Global Guide.* (Eds GL Worboys, W Francis, and M Lockwood) pp. 77–85. Earthscan, London.

SECTION 2. CASE STUDIES

3

Gondwana Link: 1000 kilometres of hope

Keith Bradby

Introduction

South-western Australia is still a wild place, at least in parts. Despite 180 years of settler impacts, the region still holds some of the resilience and adaptability developed through an unbroken biological lineage that dates back over 250 million years. Much of this land has been continuously vegetated and stable since dinosaurs roamed across it, and before both mammals and flowering plants evolved. Our mountains are merely the eroded remnants of much grander vistas, our once mighty river systems now barely recognisable as long chains of vaguely connected salt lakes.

Our plants and wildlife? Of what we had 250 years ago, many of the mammals are gone, as is the case across much of Australia, but the other animals and plants are mainly still here, and becoming better known. Several mammals once presumed extinct have been 're-discovered' (Morcombe 1967; Baynes *et al.* 1987; Sinclair *et al.* 1996), and since 1979 there has been a doubling in the number of plant species scientifically recognised from the area (Hopper and Gioia 2004). This, combined with an ecologically tragic recent history, which included large-scale clearing from the 1950s to 1990s, has led to international recognition of the south-west as a biodiversity hotspot 'where exceptional concentrations of endemic species are undergoing exceptional loss of habitat' (Myers *et al.* 2000).

The situation is so severe that in most places you could spend a lifetime saving some endangered species and countering some ecological threats and assaults, without making much impact. Instead we have developed an ambitious 1000 km initiative, to build on the opportunity that exists across the southern edge of the south-west of the Australian continent (Figure 3.1). This is where most of the remaining large habitats are, with numerous concentrations of the south-west's botanical richness and endemism, including an intriguing and specialised world of invertebrates and healthy populations of several of the rarest mammals and birds. It's also, arguably, the part of the south-west most buffered from climate change.

History and structure

Gondwana Link builds on earlier conservation achievements. These include a strong regional landcare movement, the end of old-growth logging and the end of government support for large-scale land clearing. Community-led conservation advocacy, from the 1970s through to the 1990s, resulted in the protection of large areas from which we can now build Gondwana Link. Despite these wins, most indicators of ecological health were still showing a decline.

In the late 1990s, The Nature Conservancy started to become more interested in Australia, and I invited them to come for some long drives through the south-western biodiversity hotspot. An agreed priority soon emerged – establishment of a large, opportunity-based approach, encompassing much of the biodiversity richness within south-western Australia. Before we could implement this, a small local group called Gondwana Link Inc. formed, with the intent of buying back and replanting properties

Figure 3.1 Gondwana Link encompasses much of the vegetated areas remaining in south-western Australia. Over 900 km of the 1000 km link is existing habitat.

between the Stirling Range and Fitzgerald River National Parks. Unfortunately, available funding proved inadequate and Gondwana Link Inc. disbanded.

In 2001, key Conservancy staff and I decided it was time to try again, this time aiming for a more cohesive and ambitious regional approach. It was also felt that launching a large landscape approach in the south-west was an opportunity to help The Nature Conservancy achieve some broader goals, including a strengthening of the involvement of Australia's non-government organisations (NGOs) and philanthropic sectors in large-scale, strategically placed ecological protection and restoration.

As a consequence, several groups were invited into a collective effort. Despite some nervousness at the scale of the vision, the enthusiasm was there and the presence of others provided reassurance. But it was still just an idea and there was no-one well positioned to make it happen.

In mid 2002, The Nature Conservancy, already providing the necessary catalysts of experience, credibility and enthusiasm, offered start-up funds to help support me as coordinator for 2 years. These funds were matched by Greening Australia and The Wilderness Society, an arrangement formalised through simple letters of agreement. A formal memorandum of understanding across the six original groups[i] was drafted and discussed, but put aside by agreement as some of the groups barely knew each other, and were uncertain what arrangements would be needed as the work progressed. The essential task was to build the relationships by starting to achieve the Link, together.

At this stage there was only a broad understanding of our vision. Satellite photos and Texta pens were used to delineate key habitat gaps and priority areas for protection and restoration. We chose not to invest in overall planning or prioritising, but to start securing habitats in an obvious gap within the 1000 km proposed corridor – between the Fitzgerald River and Stirling Range National Parks.

Within months I was contacted by two old friends, Eddy and Donna Wajon, who were considering the purchase of a conservation property. Realising that they could contribute to the achievement of a larger conservation goal while benefiting from their land being part of a supportive effort, they purchased two properties. In early 2003, with funds from The Nature Conservancy and its own donors, Bush Heritage Australia purchased another large property, Chereninup, and commenced a large replanting program. The Conservancy then introduced a donor, who supplied significant funds over a 3-year period for a range of on-ground projects by the collective effort.

The momentum and tangibility generated by significant land purchase and restoration in the first year was fundamentally important in turning Gondwana Link from a well-supported fuzzy idea to an ongoing and high achieving program. Groups and individuals continue to be attracted to Gondwana Link by both the vision and the ecologically supportive network built around tangible achievement.

There are now groups working to achieve Gondwana Link in all major habitat gaps, plus a focused effort underway in the Great Western Woodlands, which is mostly public land. While some of it has been a bumpy road, over a 10-year period there has been a steady growth and broadening of involvement.

Focused effort is now progressing in eight main sections (Figure 3.2). Strategies address local ecological decline in each while the collective effort achieves the larger vision. Sections are defined by the groups wanting to work there, with the exception of the Ravensthorpe Connection, where a lead group or coalition of groups has not yet emerged, and Gondwana Link Ltd is leading the initial scoping and planning work because of the area's critical

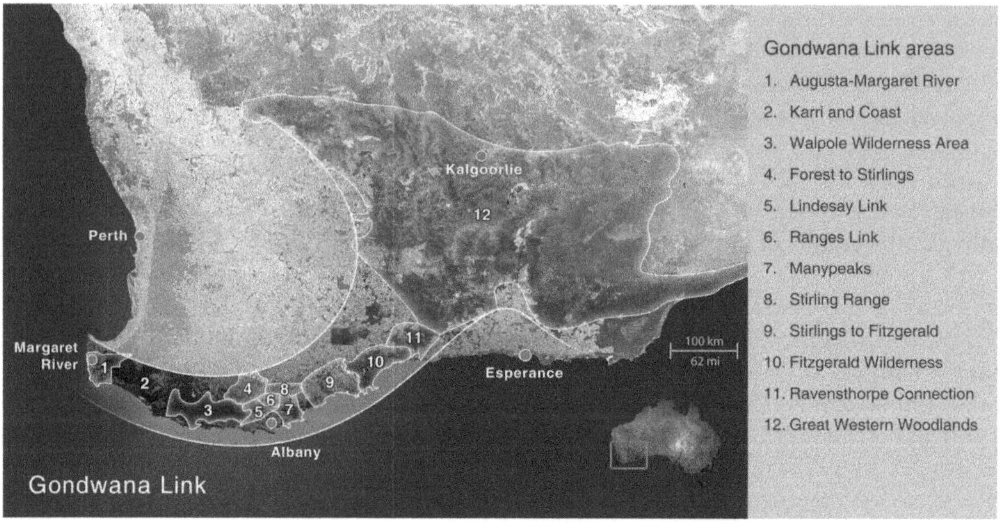

Figure 3.2 Ten years on, Gondwana Link programs are now underway in sections 1, 4, 5, 6, 7, 9, 11 and 12. Sections 3, 8 and 10 are part of the public conservation estate. (Source: Gondwana Link Ltd, June 2012).

ecological importance. Groups attract what funds they need, largely independently but with support from Gondwana Link Ltd as required.

Our collective vision has remained consistent: 'Reconnected country, from the wet forests of the far south-west to the woodland and mallee bordering the Nullarbor, in which ecosystem function and biodiversity are restored and maintained'. This was chosen carefully to reflect more than just the restoration of ecological linkages, as connectivity is only one of the many critical ecological functions in a healthy and resilient landscape.

Building the most effective support structure has not been quite as simple. We started with a 'Council' of involved groups who made decisions covering the range of operations in train, particularly those involving some funds jointly secured. It became apparent that groups were tripping over each other and, given the wide spectrum between advocacy groups and those operating corporate funded revegetation programs, were at risk of limiting each organisation's ability to get on and do what they did best. Legal issues concerning joint liability, when decisions are made jointly, were also raised. Consequently, joint decision-making meetings were abandoned and we moved to more collegial forums and networks.

Gondwana Link itself operated as a small (1.5 person) 'Coordination Unit' funded to help out wherever we could get positive and tangible results, which was often through supporting group efforts where the groups themselves had not yet built capacity. That phase lasted 3–4 years, with available effort largely absorbed in the Stirling to Fitzgerald section (Figure 3.3).

Then there was a slightly confused and overlapping transition period where larger groups were expected to pick up more of their operational roles, while core staff worked to build the larger Link. In August 2009, Gondwana Link Ltd was established to provide an institutional framework more independent of the foibles of particular groups. The structure was based on a reasonably standard company constitution, with the range of groups working to achieve Gondwana Link becoming the members. For the first 3 years funding

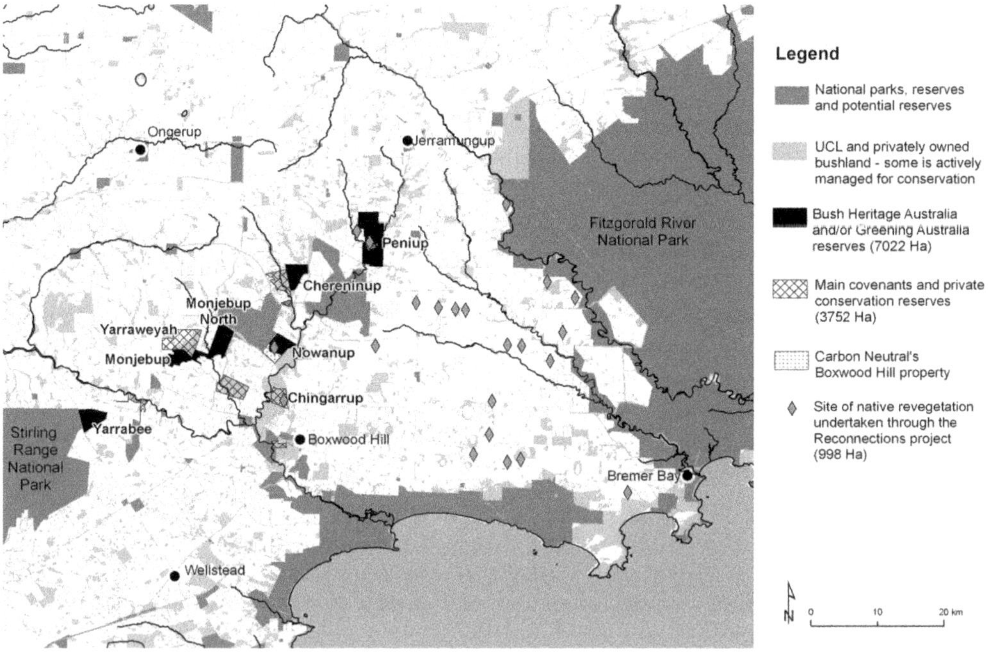

Figure 3.3 Conservation areas across the habitat gap between Stirling Range and Fitzgerald River National Parks at June 2012. Most non-government conservation areas have been established since Gondwana Link was formed.

was provided by three of the larger groups, and by some contract work undertaken that met broad Gondwana Link objectives.

Since 2007, there have been two core staff, with some additional contract positions on specific projects. In most sections of the Link groups make their cross-organisational arrangements without our involvement (Figure 3.4). Where groups have wanted joint decisions to go through a central 'hub', that hub can rapidly become a bottleneck.

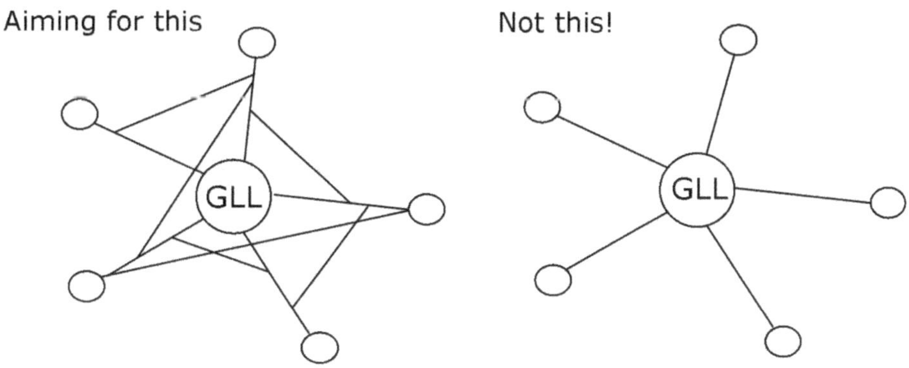

Figure 3.4 The preferred relationship between Gondwana Link Ltd (GLL) and the various groups involved in each particular section of the Link (Source: concept and drafting by Amanda Keesing).

Major successes

Ten years on, our vision has taken hold and is being steadily achieved. The involvement has broadened and now includes lead groups across most key sections of the Link, while our organisational structure has been adjusted as the program grew.

Large amounts of philanthropic and corporate funds have been raised and focused into high order ecologically strategic works in south-western Australia. Recent changes in Australian Government policy suggests that increased public funding will now be focused on large 'corridors' and will complement the work underway (DSEWPC 2012; Chapter 17).

Significant improvement in vegetation extent and condition attests to the benefits of thinking more ambitiously. In the Stirlings to Fitzgerald section, the purchase and ecological restoration of whole farms – once a radical step – is now on its way to becoming accepted practice. Through the work of groups like Bush Heritage Australia, Greening Australia and Carbon Neutral, over 11 000 ha of this critical habitat gap is now secured and under conservation management, with some 2400 ha of that land under restoration plantings. Many critical technologies have been developed, including planning for restoration (Jonson 2010), estimating the carbon benefits of biodiverse plantings (Jonson and Freudenberger 2011) and use of restoration standards to clarify the validity of confused claims about projects being 'biodiverse' (Gondwana Link 2010). The work has demonstrated the value of structural adjustment, albeit privately funded, in Australia's broadacre farming areas, with marginal properties stabilised and now being managed for the public good. In the face of ongoing decline in rural populations and social fabric, we have demonstrated the many social benefits that come from diversification, with a particular and timely success being the exciting and much needed programs now run in the area by the Noongar community (see www.nowanup.com.au).

On the eastern end of the Link the world's largest remaining temperate woodland is now on the conservation agenda. Now known as the Great Western Woodlands, this 16 000 000 ha region was, until 10 years ago, still largely unrecognised. There has been a clear sequence of success since 2003, when a donor to Gondwana Link first supported the involvement of The Wilderness Society (TWS) in the area. Members of TWS's Wild Country Science Council then embarked on a range of studies, and in 2008 a key document was launched highlighting the region's ecological values (Watson *et al.* 2008). Bipartisan political support was gained before the September 2008 Western Australian election and the new government then developed an overall strategy for the conservation of the area (Government of Western Australia 2010). Several major mining companies are now working with Gondwana Link and associated groups on conservation planning and education programs (e.g. Del Marco 2011; McDaniell 2012).

We are cautious of claiming too much success here though, as significant concerns exist at the impact of the Western Australian Government's accelerated program of fire management for the area, which is currently installing some 2300 km of bulldozed tracks, some over 5 metres wide and many of them through largely inaccessible areas (Herford *et al.* 2011).

Several locally based groups have started work on other sections of Gondwana Link, through developing Conservation Action Plans (CAPs), and gaining implementation funds. The first of these was Greenskills, who commenced working with us in 2004. They have steadily developed a program involving landholders and groups between the Stirling Range and the Walpole Wilderness Area. The Ranges Link program, between the Stirling Range and Porongurup National Parks, was established in 2009 by a collective of local groups. That same year the Margaret River-based Cape to Cape Catchments Group started on what is now a cohesive and ecologically focused plan spanning the gaps between the main forest areas and the west coast. In 2011, we started two other CAPs with locally based groups and one strategic plan for the critically important Ravensthorpe Connection, the tenuous connection between the Great Western Woodlands and the rest of the Link.

From the experience of developing locally based CAPs,[ii] and the ecological insights they provide, we are now well positioned, if not yet well funded, to develop the overall science-based plan that many people assume we should have started with.

By 2012, Gondwana Link was not the only such ambitious Australian program. The need to achieve ecological function at large scale is now an emerging wisdom, which looks set to strengthen. We have peers and colleagues to talk to and work with, and are often told that our early efforts helped encourage this growth in strategic thinking across Australia.

Major lessons

The fundamental lesson can be summarised as: 'The task of landscape restoration is simple – the hard bit is keeping it simple'. This is best explained by reference to three main areas where the learning curve has been steepest.

There is a critical need for an independent entity to lead (but not control) these programs

The team that eventually became Gondwana Link Ltd was initially hosted by one of the larger organisations involved. In that early period we also undertook considerable operational roles, helping them develop capacity and a large landscape perspective, while also trying to build an equitable and inclusive arrangement across all the other organisations. There was confusion as to whether we were part of the host organisation, and by inference favouring them, or independent of the host. This confusion led to numerous incorrect assumptions on the program's operation and development.

The journey to more effective structural arrangements was delayed by three factors. First, our focus on maintaining momentum and tangible achievement reduced the time spent on improving our own governance. Second, some organisations were concerned that a separate entity leading Gondwana Link would, or at least could, operate independently of their own decision making, and possibly become a competitor for funding. Third, there was concern that the cost of establishing and operating an independent entity would be out of proportion with on-ground funding. In the end, the establishment of Gondwana Link Ltd reduced running costs, increased the strategic focus on what was needed across the Link and supported significant growth in the number of groups involved and the funds they could obtain.

We learnt the importance of ensuring that the individual or even collective interests of the groups do not dominate leadership decision making. The whole really is more than the sum of its parts.

Key psychological drivers can put particular strain on collaborative ventures

Many people need to operate in their 'own territory' and with their 'own tribe', and to receive the recognition and rewards from their 'own efforts'. We've struggled with how to provide this energising freedom while also building cohesion and cooperation, and achieving equitable recognition for collective efforts. Similarly, the competitive drive which often leads to organisational success can be difficult to integrate into programs aimed at ecological success in a geographically specific area. Gender is another critical factor, with one working definition of collaboration being 'standing between alpha males while they conduct appendage measuring competitions'.

Our ecological ignorance remains profound

For much of Gondwana Link, species occurrence data is still poor, even for widespread and obvious species such as Carnaby's Black-Cockatoo (*Calyptorhynchus latirostris*). Major extensions of known species distributions have resulted from most surveys undertaken in the past decade. While all the new finds are exciting, it is a disturbing excitement as many decisions have to be made 'on the run', often without even baseline information. Guesswork increases exponentially when we come to fundamental questions of how ecosystems function, and what actions would 'improve' their resilience. Using Conservation Action Planning and other programs from the Open Standards approach[iii] we fill these gaps iteratively, while proceeding with what seems the 'best bet actions' at the time, but often ponder how a reasonably settled area in a booming first world economy can have received so little meaningful study.

Applying the lessons

Three core challenges face us over the next decade.

Moving to a more sustainable operating basis

We are now in the process of clarifying our functions and building the solid funding base needed to successfully operate a self-managing, multi-stakeholder structure.

There is no intention for the central hub – Gondwana Link Ltd – to control the range of groups, companies and individuals. Linear or pyramid-shaped structures that rely on power and control would stifle the dispersed energy and innovation needed. We need some 'rules' and 'standards' in place, based on agreed principles, but that is not the main focus of our efforts. Our aim is merely to provide the minimal degree of leadership and support necessary to stimulate and positively influence the exponential change still needed on the ground.

Further progress on the issue of inclusion and representation is critical. However, we have watched programs elsewhere in Australia stagnate through wrestling too hard with wider involvement in decision making. We are keen to be inclusive but remain cautious of the 'lowest common denominator consensus' decision making and stagnation that can result.

Supporting exponential growth

As much as visitors seem impressed by what has already been achieved on the ground, and the extent of Link-wide planning, energy and commitment, we have barely started achieving the immensity of ecological change needed to stabilise the downhill trends. But the groups have now accumulated a substantial foundation of knowledge, expertise and experience. People are not pottering at the edges – the actions researched, identified, agreed on and being undertaken are, by and large, the actions which will restore healthy and resilient ecosystems across the Link if undertaken effectively at a much larger scale.

Our best chance may be the carbon economy. Substantial ecologically based biosequestration through ecological restoration is already well underway in Gondwana Link (Jonson 2010; Jonson and Freudenberger 2011), some fundamental work has been done on the carbon values of our existing woodland and mallee (Berry *et al.* 2010) and future modelling of fire behaviour and impacts is planned. Additional work on carbon values of the forests is readily achievable, by building swiftly on work undertaken for the south-eastern forests (e.g. Mackey *et al.* 2008).

The bigger challenge is in developing business structures that deliver high quality ecological outcomes at scale in a competitive marketplace. In terms of carbon plantings, both Greening Australia and Carbon Neutral have made fine starts in this regard, but are facing uncertain national politics, consequent marketplace confusion, stiff competition from less ecologically focused efforts and an ecologically unsophisticated public policy framework.

Another significant opportunity to achieve significant progress is through support for the ongoing Native Title efforts underway across Gondwana Link. Despite court processes that have dragged on for over 14 years, it is likely that both Ngadju and Noongar claims will be resolved before the end of 2013, adding an important 'new' element to the management of public lands. We remain ready to support the development of land management programs, through the appropriate Ngadju and Noongar leadership.

Program support from the public sector

Our engagement with public sector programs remains a major challenge. We started as a proudly private sector effort, at a time when the concept of working at scale across a large landscape was not well accepted. Despite the high recognition we have, and some very useful public funds for projects, we have not yet built a mutually supportive relationship at program level with any state or national government initiatives.

While Gondwana Link Ltd needs to make changes in order to work closer with government, change may also be needed in some government attitudes. For example, the current Western Australian Government appears to only show support for landscape-scale programs where it has a strong controlling role and can directly receive Australian Government funding (Government of Western Australia 2012). Additionally, it is currently funding construction of a 500 km wildlife barrier fence running at right angles to the Gondwana Link pathway (WA Ministers for Agriculture and Environment 2010).

The support possible through the *National Wildlife Corridors Plan* for Australia (DSEWPC 2012) is a good opportunity for us to work closely with the Australian Government. However, fitting into a government-developed template is likely to bring substantial extra costs and may not be consistent with our entrepreneurial origins and main growth opportunities.

Conclusions

The deliberate strategy of starting the program and building momentum through tangible achievement, rather than through developing an overall prioritised plan, has established Gondwana Link. Cohesive and substantial ecological efforts are now progressing across 1000 km of biologically rich country. It has taken a lot of tenacity to get this far, and we remain a long way from the full glow of success. But we have built the foundation from which much needed exponential growth in ecological restoration of a critically important landscape can be secured.

Acknowledgements

The Gondwana Link program is only happening because of initial and ongoing support from The Nature Conservancy, the wisdom and generosity of the many individuals whose donations have provided the bulk of the funding to date, and the many wonderful people in the wide range of groups I have the pleasure of working with on this bold and necessary venture.

Biography

Keith Bradby is a longtime advocate for the ecological values of south-western Australia, starting as a community-based activist in the late 1970s. Initially working as a beekeeper and seed collector, he has since been part of some of Australia's earliest landcare groups, facilitated local enterprise development and managed a large catchment program. He worked inside government to tighten vegetation clearing controls in Western Australia during the 1990s, and for the last 10 years has had the privilege of working with the collaboration of private sector groups achieving Gondwana Link.

Endnotes

i. Bush Heritage Australia, Greening Australia (Western Australia), Malleefowl Preservation Group, The Wilderness Society, Friends of Fitzgerald River National Park, and Fitzgerald Biosphere Group (see Chapter 4).
ii. Most of these Conservation Action Plans are available at http://conpro.tnc.org/
iii. More information on Open Standards is available from the Conservation Measures Partnership: http://www.conservationmeasures.org/

References

Baynes A, Chapman A, and Lynam AJ (1987). The rediscovery, after 56 years, of the heath rat *Pseudomys shortridgei* (Thomas, 1907) (Rodentia: Muridae) in Western Australia. *Records of the Western Australian Museum* **13**, 319–322.

Berry S, Keith H, Mackey B, Brookhouse M, and Jonson J (2010). *Green Carbon: The Role of Natural Forests in Carbon Storage Part 2. Biomass Carbon Stocks in the Great Western Woodlands.* ANU E Press, Canberra.

Del Marco A (2011). 'Towards a land use plan for the Great Western Woodlands'. Report by Ironbark Environmental to the Woodlands Initiative.

DSEWPC (2012). *National Wildlife Corridors Plan: A Framework for Landscape-scale Conservation.* Department of Sustainability, Environment, Water, Population and Communities, Canberra.

Gondwana Link (2010). *Restoration Standards, Version 1.1.* Gondwana Link, Albany.

Government of Western Australia (2010). *A Biodiversity and Cultural Conservation Strategy for the Great Western Woodlands.* Department of Environment and Conservation, Perth.

Government of Western Australia (2012). Submission from the Western Australian Government on the Draft National Wildlife Corridors Plan. Department of Premier and Cabinet, Perth. Available: http://www.environment.gov.au/biodiversity/wildlife-corridors/consultation/sub missions/pubs/wag.pdf [Accessed 1 July 2012]

Herford I, Armstrong R, and Daniel G (2011). *Fire Management Plan for the Conservation of Biodiversity and Cultural Heritage Values in the Great Western Woodlands.* Department of Environment and Conservation, Perth.

Hopper SD and Gioia P (2004). The southwest Australian floristic region: evolution and conservation of a global biodiversity hotspot. *Annual Review of Ecology and Systematics* **35**, 623–650.

Jonson JH (2010). Ecological restoration of cleared agricultural land in Gondwana Link: lifting the bar at 'Peniup'. *Ecological Management & Restoration* **11**, 16–26.

Jonson JH and Freudenberger D (2011). Restore and sequester: estimating biomass in native Australian woodland ecosystems for their carbon-funded restoration. *Australian Journal of Botany* **59**, 640–653.

Mackey BG, Keith H, Berry SL, and Lindenmayer DB (2008). *Green Carbon: The Role of Natural Forests in Carbon Storage. Part 1. A Green Carbon Account of Australia's South-eastern Eucalypt Forests, and Policy Implications.* ANU E Press, Canberra.

McDaniell M (2012). *MK Project Explore: A Citizen Science Initiative Supported by BHP Billiton Nickel West.* Millennium Kids, Perth.

Morcombe MK (1967). The rediscovery after 83 years of the dibbler *Antechinus apicalis* (Marsupialia, Dasyuridae). *Western Australian Naturalist* **10**, 102–111.

Myers N, Mittermeier RA, Mittermeier CG, de Fonseca GAB, and Kent J (2000). Biodiversity hotspots for conservation priorities. *Nature* **403**, 853–858.

Sinclair EA, Danks A, and Wayne AF (1996). Rediscovery of Gilbert's potoroo, *Potorous tridactylus*, in Western Australia. *Australian Mammalogy* **19**, 69–72.

WA Ministers for Agriculture and Environment (2010). 'Government takes fight to wild dogs'. Government of Western Australia Ministerial Media Statements, Perth. Available: http://www.mediastatements.wa.gov.au/Pages/default.aspx?ItemId=133309&search=&admin=&minister=&portfolio=Agriculture+and+Food&region=&page=2 [Accessed 1 July 2012]

Watson A, Judd S, Watson J, Lam A, and Mackenzie D (2008). *The Extraordinary Nature of the Great Western Woodlands.* The Wilderness Society of WA Inc, Perth.

4

Fitzgerald Biosphere Reserve: a framework for achieving ecological and community sustainability … or is it?

Angela Sanders

Introduction

The south coast of Western Australia has two large-scale conservation initiatives; the Gondwana Link project (see Chapter 3) and the Fitzgerald Biosphere Reserve project. The two are not mutually exclusive – they share some overlap in geographical area, interest groups, local government areas, aspirations and challenges. In this chapter, I will give an outline of the Fitzgerald Biosphere Reserve, the nature of its landscape, social and cultural setting, its dynamic beginnings and how the locals are now rekindling their interest in the biosphere reserve concept.

In 1975–76 the Western Australian Government was asked to nominate areas for designation under UNESCO's Man and the Biosphere (MAB) program. The Fitzgerald River National Park (FRNP), together with 11 other areas, was nominated and was subsequently approved as an internationally recognised Biosphere Reserve in April 1978 (Davis and Drake 1983).

Although there have been several shifts in emphasis in the way biosphere reserves function since their inception in 1971, Fry (2010) describes their consistent interrelated functions as:

- conservation: landscapes, ecosystems, species and genetic variation;
- development: economic, human and culturally adapted; and
- logistic support: research, monitoring, environmental education and training.

Under the MAB program there is no statutory obligation to enforce compliance with the program's aims. Where declared over Crown land in Australia, individual biosphere reserves remain under the jurisdiction of the state where they are situated. The state determines the measures they deem necessary to improve the functioning of the individual sites (UNESCO 2008), although biosphere reserves are also currently listed under the Federal *Environment Protection and Biodiversity Conservation Act 1999.*

The FRNP is easy to pick out on a satellite image of the south coast – it is the large green rectangular patch of bush that hugs the coast roughly halfway between Albany and Esperance. The towns of Bremer Bay and Hopetoun are its coastal western and eastern gateways. Its inland boundaries lie sharp up against cleared farmland with a few ribbons of riparian vegetation linking it to inland bushland (Figure 4.1).

The land on which the FBR sits has been shaped by wind and water since the last major tectonic activity ~80 million years ago when Antarctica separated from the main landmass along near what is now the south coast. Its diverse landscape and resultant mosaic of soil types, many of which are nutrient deficient, coupled with millions of years of evolution have given rise to extraordinary plant diversity and high levels of endemism. The conservation values of the FRNP were well recognised by the 1970s and its plant diversity was the main reason for its designation as a biosphere reserve (Watson and Sanders 1997).

The official boundary of the FBR coincides with the national park boundary at the time of declaration in 1978, which incorporates only 242 739 ha. In order to make it more in line with the MAB program, the FBR now comprises a 'core' (the Fitzgerald River

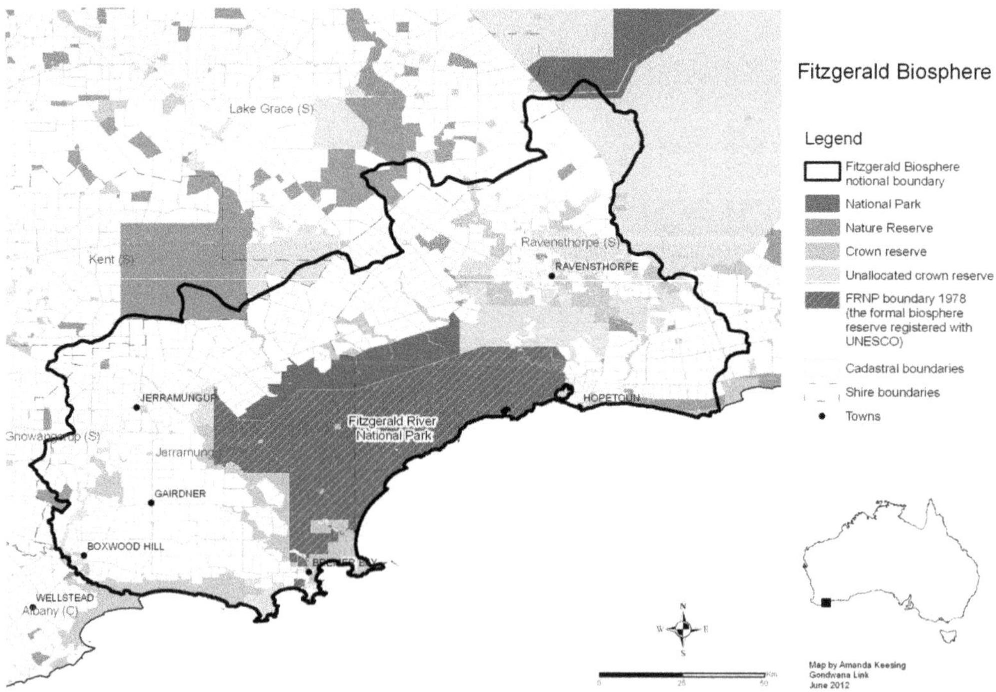

Figure 4.1 Fitzgerald Biosphere Reserve

National Park) and a 'zone of cooperation' covering some 1.3 million hectares.[i] This notional FBR now covers all the Shire of Jerramungup, approximately half of the Shire of Ravensthorpe and a small part of the Kent and Lake Grace Shires (Figure 4.1). The major land uses are broadacre farming, national park and nature reserves, unallocated Crown land and town sites.

The Gondwana Link project straddles the FBR and includes a swathe of land over 1000 km long stretching from the lower west coast around Margaret River and inland to Kalgoorlie and beyond (see Chapter 3). The vision for Gondwana Link is for:

> *'reconnected country across south-western Australia, from the Karri forest of the south-west corner to the woodlands and mallee bordering the Nullarbor Plain, in which ecosystem function and biodiversity are restored and maintained.'*

Gondwana Link contains several operational areas and three of these, Stirling-Fitzgerald, Fitzgerald Wilderness and the Ravensthorpe connection overlap the Fitzgerald Biosphere Reserve. Each of these operational areas require different strategies and in the Stirling-Fitzgerald for example, the strategy is to acquire and protect bushland and revegetate strategically located farmland. In addition to the overlap in FBR and Gondwana Link boundaries, there are local government, catchment, Department of Environment and Conservation and South Coast Natural Resource Management boundaries. This plethora of 'operational' boundaries can be confusing to those planning and implementing on-ground activities.

Aboriginal people have occupied the land that is now called the FBR for thousands of years and archaeological surveys have found ample evidence of their hunter-gatherer lifestyle (Dortch 2008; Hayward 2008; Applied Archaeology Australia 2010). White settlement began in a small way in the late 1800s and it was not until the 1950s that settlement and land clearing for agriculture began in earnest (Bradby 1989). In 2006, the population of Jerramungup Shire was 1128 and the main industry of employment was agriculture (48%). The Shire of Ravensthorpe had a population of 1950 and farming (20%) was the main industry of employment and construction (8%) was second (ABS 2006). While the profitability of the two main industries of agriculture and mining can be volatile, they are likely to continue to be the main economic base of the FBR. There is potential to increase the economic value of tourism, particularly that based around the natural and cultural values of the area (Fry 2010).

The Department of Environment and Conservation (DEC) is now the lead agency for the UNESCO Biosphere designation in Western Australia. The DEC also directly manages the 'core' area of the FBR (i.e. the Fitzgerald River National Park) and all the nature reserves within the FBR. The Department of Water is responsible for the protection of waterways and catchments within the FBR. The Department of Agriculture and Food Western Australia (DAFWA) aims for an integrated approach to soil, water and nutrient management on farm to minimise off-site impacts. The DAFWA has been involved in many sustainable agriculture initiatives in the FBR (Fry 2010).

After a 34 year history, the FBR is now considered by UNESCO as not functioning as a Biosphere Reserve under its criteria (UNESCO 2003, p. 19) and its status is under review. There is an effort by local groups to revitalise the FBR and address the UNESCO recommendations and a Biosphere Initiative Group has been formed (see below).

Evolution of the Fitzgerald Biosphere Reserve

The designation of the FBR galvanised the then Fitzgerald River National Park Association into seeking support to activate the FBR in line with the MAB program. A series of events during the 1980s occurred to facilitate this:

- in 1983, the Jerramungup Soil Conservation District Committee was formed as a response to the severe wind erosion problems being experienced in that Shire.
- in 1986, the Fitzgerald Biosphere Reserve Symposium resulted in the formation of the Fitzgerald Biosphere Project.
- in 1987, 'The Bush Comes to the City' conference was held at Murdoch University.
- in 1988, the Fitzgerald Biosphere Project became incorporated and launched its newsletter, *The Bush Telegraph.*
- in 1988, Dr Bernd von Droste (UNESCO's director of the Division of Ecological Sciences) visited the FBR.

During this time the Fitzgerald Biosphere Project successfully raised the profile of the FBR and gained support locally, nationally and internationally. Unfortunately this support came without dollars and the high level of activity by a few dedicated individuals took its toll and group burnout became apparent by the early 1990s.

The next step saw a commitment by the Department of Conservation and Land Management (CALM) through the Fitzgerald River National Park Management Plan to 'ensure ongoing CALM involvement in the Fitzgerald Biosphere Project committee' (CALM 1991).

At about the same time the natural resource management (NRM) model of regional planning was being developed with the aim of facilitating strategic investment by government. As part of this process the south coast of Western Australia was divided into several subregions and the Fitzgerald Biosphere subregion became the largest of them. Also during this time the Fitzgerald Biosphere Project sought funding from the Federal Government for a coordinator to manage the support network of the FBR; it was not forthcoming and much of the momentum achieved during the past 10 years was lost.

In March 1997, the FBR subregion *Land and Water Care Strategy* was prepared by the South Coast Regional Assessment Panel (RAP) and the South Coast Regional Initiative Planning Team (SCRIPT). The primary objective of the strategy was to assist the RAP in ensuring that National Landcare Program funds were directed to priority issues and areas (RAP/SCRIPT 1997).

At this time, the FBR subregion had an area of 1 354 630 ha and encompassed the Land Conservation District Committees (LCDCs) of Jerramungup and Ravensthorpe and their associated catchment groups. When the LCDCs were changed from statutory groups to independent incorporated groups an opportunity for the two LCDCs to amalgamate into a single regional group to coordinate NRM across the FBR was missed. The communities decided to form two separate groups; the Fitzgerald Biosphere Group which covered the Shire of Jerramungup, and the Ravensthorpe Agricultural Initiative Network (RAIN) covering the Shire of Ravensthorpe. At that time, the naming of the FBG by the Jerramungup community was not popular with those in Ravensthorpe.

The Fitzgerald Biosphere Marketing Association (FBMA) was formed in 2001 as a joint effort between the Fitzgerald Biosphere Group and RAIN to identify marketing opportunities that increased recognition of the biosphere concept and its sustainable development ethos. Despite the huge effort put into the FBMA project, there were no obvious outcomes and this has resulted in wariness by the farming community to become involved in similar large projects under the biosphere banner.

In 2003, the UNESCO Man and the Biosphere International Coordinating Council Bureau conducted a review of biosphere reserves and concluded that the FBR was not functioning as a biosphere reserve. There was little response from CALM or the community and it was not until 2010 that action was taken to address the recommendations given in the review. The Biosphere Initiative Group was formed comprising representatives from the Fitzgerald Biosphere Group, Friends of the Fitzgerald River National Park, Ravensthorpe Agricultural Initiative Network and key state and local government stakeholders. The group received funding in 2012 from the Australian National Commission for UNESCO to develop a Biosphere Action Plan to address its 2003 recommendations.

In 2008, UNESCO released the Madrid Action Plan which aimed to raise biosphere reserves to be the principal internationally designated areas dedicated to sustainable development in the 21st century (UNESCO 2008).

A 2010 report on the current situation of the FBR and recommendations for its future was produced for the Great Southern Development Commission, Gondwana Link Ltd and the DAFWA (Fry 2010). The recommendations included support for the revitalisation of the FBR and the redesignation of its boundaries; exploration of the inclusion of a marine component; incorporation of widespread consultation with Aboriginal people; and increased educational, research and monitoring opportunities and communication with other biosphere reserves.

Major successes

Although there have been many major successes in conservation management in the FBR over the past 30 years few of them have contributed to the aims of the MAB biosphere reserve program. This has not been through lack of trying, however, as illustrated by the many attempts by various groups since the 1980s to activate the program. In relation to agricultural sustainability there have been, coincidentally, some advances in techniques but with little reference to ecological sustainability. The major successes that have, in my view, contributed to the area's biosphere reserve status include those shown below.

1 Communication of the MAB Biosphere Reserve concept in the 1980s and 1990s helped to raise awareness of an interconnected landscape where everything affects everything else.
2 The FBR was recognised as a model biosphere reserve in the early 1990s owing to community-led development of zonation, which recognised the FRNP as the core, and a surrounding zone of cooperation (Parker 1993).
3 Some groups in the FBR have successfully used the biosphere reserve designation to obtain funding from a variety of sources and have linked research and demonstration on sustainable development with protection of the core (e.g. Fitzgerald River Catchment Project).

Major lessons

Despite an enormous effort by dedicated local people during the 1980s and 1990s, the Fitzgerald Biosphere Reserve now functions as an 'in name only' network. There has been little integration of knowledge between economic development and conservation efforts and even fewer examples of the area acting as a demonstration site for maintaining ecological and cultural diversity and securing ecosystem services for human wellbeing.

One major lesson has been that the original biosphere reserve nomination should have been thought through more thoroughly, and the federal and state governments should have provided financial and advisory support to assist the development of the FBR. After all, it was considered a model biosphere reserve in the 1990s not least owing to the high level of local energy that was expended trying to activate it (Parker 1993; Watson *et al.* 1995; Watson and Sanders 1997; West 2001). If this support had been forthcoming as requested by the local community in the 1980s we could, quite likely, be closer to a truly sustainable agricultural base that is more closely linked to the protection of biodiversity (and less bushland may have been cleared as a result). The lesson here is to thoroughly think through any large-scale conservation initiative (e.g. biosphere reserves) and ensure that planning for sourcing long-term funding is completed early in the project's development.

The name 'Biosphere Reserve' has been a particularly difficult label and right from the designation of the FBR, groups have struggled with it. The 'reserve' part of the title was dropped locally to try and address the perception that it meant restrictions on privately owned land. The delineation of the Fitzgerald Biosphere NRM subregion was a positive move, however the re-naming of the Jerramungup LCDC to the Fitzgerald Biosphere Group not only angered the Ravensthorpe community it further complicated attempts to unify activities under the FBR umbrella. So the lesson here is naming can be everything and getting it right is paramount.

But it is not just the label that has been difficult, it is also the biosphere concept that has been a major challenge to explain to just about everyone involved, including the original FBP members who sought clarification back in the 1980s. Generally people understand wildlife corridors, although creating them is dogged by complexity, the concept of linking the landscape is a relatively simple one that most people can understand. However, wildlife corridors are only part of what biosphere reserves are about. The challenge is to communicate the biosphere reserve concept in a locally relevant social, ecological and economic context that can be easily understood by those involved with its implementation.

The final major lesson learnt from examining the history of the FBR network is that there needs to be a person (or persons), who is fully funded, whose task is to specifically coordinate the network under the umbrella of the MAB Fitzgerald Biosphere designation. The attrition of agency staff and burnout of volunteers over the years has meant more pressure on fewer people and, if a renewal of the FBR is to succeed, the need for dedicated personnel is urgent.

Applying the lessons

It appears, for the past decade at least, that the biosphere reserve designation has been used to leverage funds for projects that have only paid lip-service to the aims of the MAB biosphere reserve program. Its designation has also been used to promote the region's high conservation value and to oppose inappropriate developments within its notional boundaries. This last point would be quite acceptable if, concurrently, groups were working towards a truly integrated and functioning biosphere reserve; however, there is little evidence to support this happening. This must change if the FBR has a future within the MAB network of biosphere reserves.

The Fitzgerald Biosphere Reserve designation is regularly included as just another dot point on NRM funding applications. Using the designation to leverage funds without any real or measurable contribution to the MAB biosphere reserve objectives has become the norm, but – as evidenced by the 2003 UNESCO review – is considered unacceptable.

For a chance at success we need to demystify the biosphere concept and meet the challenge of taking what seem to be several separate projects and place them in a broader context. The concept needs to be communicated in a locally relevant way and wholeheartedly integrated into all the activities we do as stakeholders in the FBR. This includes educational and interpretive activities, national park and nature reserve management, making a living and conservation activities.

The Madrid Action Plan may be useful here as it articulated well the use of the conceptual framework of ecosystem services that can be superimposed on the multiple functions of biosphere reserves.[ii] In the case of the FBR it would be a matter of identifying those parts of the network responsible for each service, i.e. supporting, provisioning, regulating and cultural opportunities for collaboration towards common goals.

Current predictions are that the economy of the FBR is likely to continue to be based largely on agriculture in Jerramungup and agriculture and mining in Ravensthorpe, at least in the short term. Also, there is likely to be an increase in farm size and cropping that will result in a continuation of depopulation and subsequent loss of community services (Fry 2010). So what comes next?

Given the uncertainty in the capacity of the region to lead a renewed level of activity in the FBR (Fry 2010) and the general uncertainty about future economic and climatic conditions, should the FBR network be revitalised? As part of answering this question there are a few things that should be considered, including those shown below.

1 Is the biosphere reserve framework the best one to use for our region?
2 If yes, then how are the difficulties associated with the name and communicating the concept going to be successfully addressed?
3 Who is going to drive the revitalised initiative?
4 Where are the dollars coming from?

In the process of answering these questions I think a way forward can be developed, and we should not be afraid to consider relinquishing the biosphere status and take heed from its successes and lessons learned and develop a more suitable framework.

Conclusions

The area we call the Fitzgerald Biosphere Reserve will continue functioning with or without a biosphere designation. At present all the predictions for the region point towards declining levels of population and ecosystem health. Its future could include more of the same or we can try something new to try and arrest the decline. I am not convinced, however, that in the case of the FBR the biosphere concept is the one we should continue to use to develop a diversity of sustainable enterprises and vibrant healthy communities and to improve and maintain ecological resilience. We need to seriously reconsider revitalising the FBR designation and take a good look at alternatives. Given the nature of human psychology, it seems that we may simply be more successful focusing on smaller chunks closer to home – and we need to be quick smart about it.

Biography

Angela Sanders' first introduction to the Fitzgerald Biosphere was at a seminar held at Murdoch University in Western Australia in 1988 titled 'The Bush Comes to the City'.

It was here that she was inspired by locals Kaye Vaux, Bob Twigg, Keith Bradby and others to become involved in this large-scale conservation initiative. She began work as an ecologist with Department of Conservation and Land Management in 1993 and conducted biological surveys in the zone of cooperation of the Fitzgerald Biosphere Reserve. She was involved in the early days of the Gondwana Link project and was involved in the Conservation Action Plan for the Stirling to Fitzgerald operational area of Gondwana Link. This plan remains the guiding document for management in the Gondwana Link Stirling to Fitzgerald link and Angela is currently employed as an ecologist by Bush Heritage Australia to work across the link.

Endnotes

i. Biosphere Reserves should have adequate zonation; a core area devoted to long term protection, a zone surrounding the core where only activities compatible with the conservation objectives occur, an outer transition area where sustainable resource management practices are promoted and developed (*The Seville Criteria*, UNESCO 1996).
ii. The Millennium Ecosystem Assessment (2005) has articulated and described ecosystem services in a manner that has gained widespread acceptance among public and private sector and civil society organisations. It recognises four distinct categories of services: supporting (nutrient cycling, soil formation and primary production); provisioning (food, fresh water, wood and fibre and fuel); regulating (climate, flood and disease regulation and water purification) and cultural (aesthetic, spiritual, educational and recreational).

References

ABS (2006). 2006 Census data. Australian Bureau of Statistics, Canberra. Available: www.abs.gov.au

Applied Archaeology Australia (2010). 'Cultural Heritage Assessment of Monjebup North, Pallinup Catchment, Great Southern Western Australia'. Report to Bush Heritage Australia.

Bradby K (1989). *A Park in Perspective.* A report on the past, present and future of Fitzgerald River National Park. Fitzgerald River National Park Association, Albany.

CALM (1991). *Fitzgerald River National Park Management Plan 1991–2001.* Department of Conservation and Land Management, Perth.

Davis BW and Drake GA (1983). *Australia's Biosphere Reserves: Conserving Biological Diversity.* Australian National Commission for UNESCO, Canberra.

Dortch J (2008). 'Archaeological surveys on Bush Heritage Australia reserve (Kent Location 1419) on Peniup Creek, Shire of Jerramungup, Western Australia'. Report to Bush Heritage Australia.

Fry J (2010). *The Fitzgerald River Biosphere Reserve: An Assessment of the Current Situation.* Centre of Excellence in Resource Management, Report 115. University of Western Australia, Perth.

Hayward K (2008). 'Bush Heritage Australia: cultural value assessment of Chereninup Project'. Report to Bush Heritage Australia.

Millennium Ecosystem Assessment (2005). *Ecosystems and Human Well-being: Synthesis.* Island Press, Washington, DC.

Parker P (1993). *Biosphere Reserves in Australia: A Strategy for the Future.* Australian Nature Conservation Agency, Canberra.

RAP/SCRIPT (1997). *South Coast Regional Land and Water Care Strategy: the Fitzgerald Biosphere Sub-region.* Soil and Land Conservation Council, Perth.

UNESCO (1996). *Biosphere Reserves: The Seville Strategy and the Statutory Framework of the World Network.* UNESCO, Paris.

UNESCO (2003). 'Man and the Biosphere (MAB) Programme Meeting of the Bureau of the International Coordinating Council UNESCO Headquarters, Paris, Room XVI (Bonvin Building) 8–11 July 2003'. Final Report.

UNESCO (2008). *Madrid Action Plan for Biosphere Reserve (2008–2013).* UNESCO, Paris.

Watson J and Sanders A, (1997). Fitzgerald River National Park biosphere reserve 1978–1997: the evolution of integrated protected area management. *Parks* **7**(1), 9–19.

Watson J, Lullfitz W, Sanders A, and McQuoid N (1995). Networks and the Fitzgerald River National Park Biosphere Reserve, Western Australia. In: *Nature Conservation 4: The Role of Networks* (Eds DA Saunders, JL Craig and EM Mattiske) pp. 482–487. Surrey Beatty, Chipping Norton, NSW.

West G (2001). Biosphere reserves for developing quality economies: the Fitzgerald River Biosphere Reserve, Australia. *Parks* **11**(1), 10–17.

5

Territory Eco-link: 21st century conservation

Andrew Bridges

Introduction

Territory Eco-link is one of Australia's most important conservation initiatives. It has an ambitious target of delivering more than 2000 km of linked conservation areas from the Arafura Sea to Alice Springs in the Northern Territory, and doing so at a landscape scale (Figure 5.1). The corridor, running from north to south, covers the majority of the Northern Territory's 67 sites of conservation significance and includes main population centres such as Darwin, Katherine and Alice Springs. In doing so, Territory Eco-link provides an opportunity for the majority of Northern Territory landowners to participate in the initiative.

Through a Memorandum of Understanding with the South Australian Government, South Australia is continuing the corridor to create the Trans-Australia Eco-Link (see also Chapter 6), the world's first transcontinental wildlife corridor extending from the Arafura Sea in the north to Port Augusta on the Spencer Gulf in the south, a distance of over 3000 km.

It is a new and more holistic approach to achieving sustainable biodiversity conservation outcomes in the Northern Territory and success relies on a 'whole of community' effort.

National parks and reserves, while being the foundation for biodiversity conservation, are typically conservation islands in the landscape. They are rarely large enough on their own to hold self-sustaining populations of all the plants and animals they contain. In many cases they do not provide enough room for species to move in response to climate change or other environmental changes or for populations to recolonise after local extinctions from fires and other disasters.

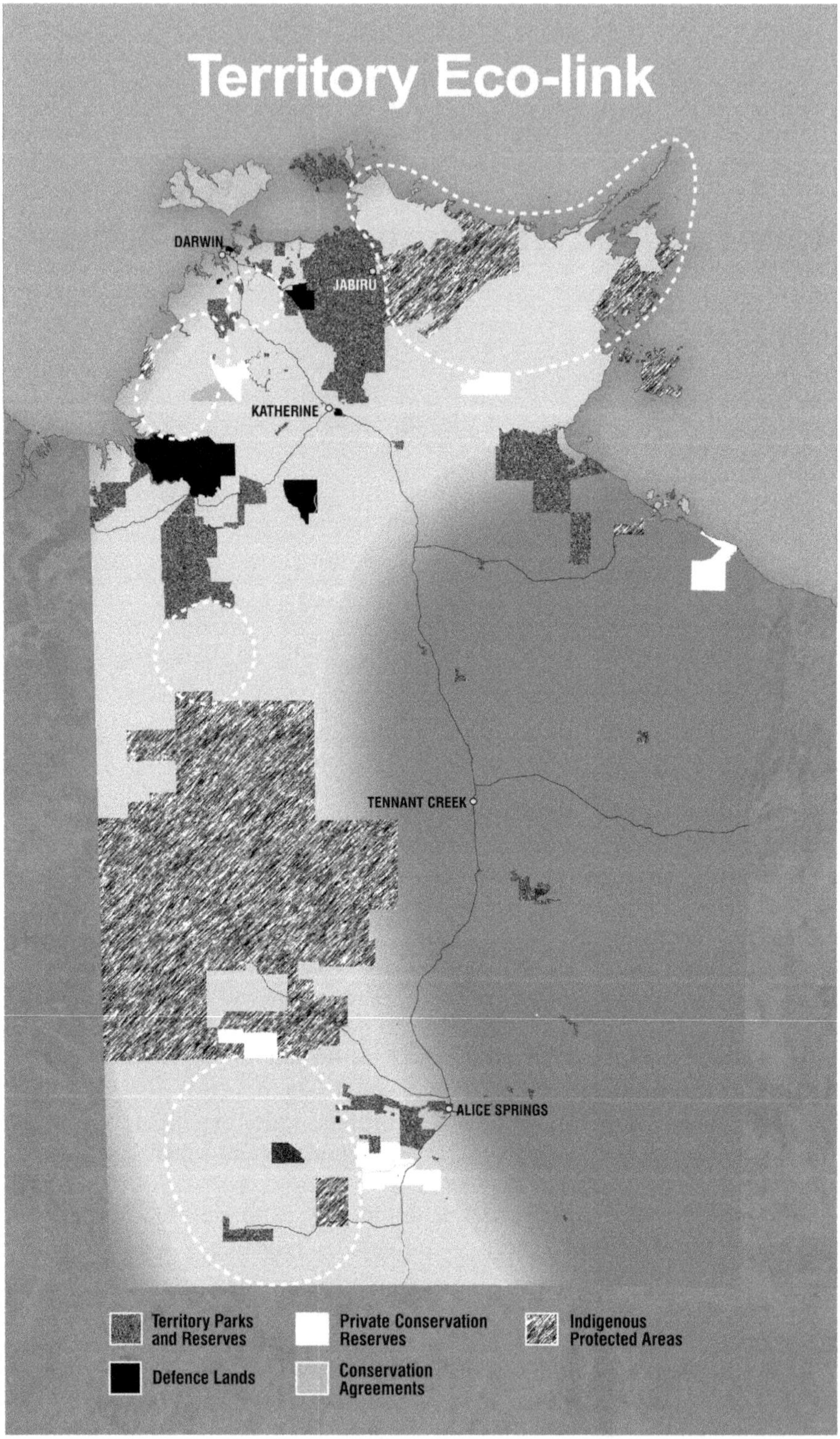

Figure 5.1 Territory Eco-link. The dashed circles indicate priority linkage areas.

Linking parks across the landscape with other lands that are managed for conservation is essential to protect the integrity and resilience of ecosystem function and reduce the potential for many more native plants and animals to become extinct. 'Landscape-scale operation' and 'connectedness' are rapidly becoming recognised both nationally and internationally as principles associated with best practice in conservation (Soulé *et al.* 2004; Mackey *et al.* 2008). Creating these links requires a coordinated and collaborative effort by many different landowners and flexibility in the approach taken.

While Territory Eco-link has a focus on increasing the size of the protected area estate, it has a greater focus on connecting these protected areas at a landscape scale. As such, Territory Eco-link has created a framework and mechanisms to encourage and support landowners who want to preserve natural values on their lands and to enable them to do so in a manner that meets their interests and aspirations, supports government and community aspirations for the sustainable protection of biodiversity and contributes to the creation of a transcontinental wildlife corridor.

Territory Eco-link can provide information and advice regarding the areas of most conservation value but it is the landowners themselves who determine those parts of their properties they wish to manage for conservation. Territory Eco-link then assists the landowner to choose an approach to conservation management that meets the landowner's aspirations. The options include non-binding Land for Wildlife-style commitments, through to 'in perpetuity' conservation covenants that are attached to title and bind future owners of the land.

The Territory Eco-link framework also provides opportunities for the broader community, largely in a volunteer capacity, to contribute to the protection of biodiversity by assisting with the management of threats and monitoring of biodiversity values. The Territory Eco-link framework has been developed with the understanding that a 'whole of community' effort is required to maximise the persistence of biodiversity into the future.

History and structure

In the Northern Territory, there has long been a realisation that the conventional system of protected areas would not be adequate for the persistence of the Territory's many unique species (Woinarski *et al.* 1992). Mobility and massive population fluctuations of wildlife in response to the influence of the monsoon in the north, and aridity with its unpredictable and extremely variable rainfall in the south, make it necessary to aim to protect as much of the still extensive natural vegetation outside of the national park and reserve estate as is possible. This need is becoming increasingly imperative with the additional influences of climate change.

With climate change being very much in the forefront of national interest in the late 2000s, Territory Eco-link became one of several election commitments designed in response to this threat in the lead-up to the 2008 Northern Territory election. As a result Territory Eco-link is a Northern Territory Government initiative and is funded by the Northern Territory Government. The initiative commenced in July 2009, with funding of $1.8 million spread over 3 years.

Territory Eco-link established an Expert Reference Group and Stakeholder Reference Group to ensure Territory Eco-link progressed in line with other community and conservation initiatives. The Expert Reference Group was established to administer the standards for inclusion within Territory Eco-link.

The Stakeholder Reference Group was established to ensure that the Territory Eco-link project develops in collaboration with other conservation, sustainable development and community-centred initiatives. Critical to the success of the initiative is the building and maintenance of strong working relationships with many stakeholder groups and organisations, and a framework that is sufficiently flexible to accommodate the needs of a variety of land uses while encouraging conservation outcomes.

Territory Eco-link has two staff dedicated to delivering its targets and outcomes, but is able to draw on the broad range of expertise within the Department of Natural Resources, Environment, the Arts and Sport to achieve these targets. Efforts to increase the size of the protected area estate relies heavily on sound working relationships with the Australian Government and particularly its National Reserve System section, along with a variety of conservation-focused non-government organisations.

Equally, working closely with Aboriginal land councils, supporting and encouraging Traditional Owners in their efforts to 'look after Country', and supporting their aspirations for the declaration of Indigenous Protected Areas have also been essential components of the Territory Eco-link initiative.

To successfully create linkages between protected areas, the support and cooperation of pastoral leaseholders is required, as much of these linkage areas are under pastoral lease. This requires a good understanding of the industry to identify opportunities that limit loss of production or better still provide an alternative income stream for any biodiversity conservation outcomes they are working to deliver.

Six key links were initially identified as priorities for achieving connectivity (Kakadu Link, Darwin Link, Litchfield Link, Gregory Link, Tanami Link, Central Desert Link) (Figure 5.1). Two have been all but completed and substantial progress has been made on a further two.

Major successes

Critical to the success of Territory Eco-link has been the building and maintenance of strong working relationships with many stakeholder groups and organisations, and gaining a sound understanding of the goals and aspirations of each. This then allows Territory Eco-link to seek out the opportunities that both meet stakeholder aspirations and contribute to Territory Eco-link targets. By selecting an appropriate 'opportunity' and then working closely with the stakeholder (or stakeholders) to help them realise their goal, Territory Eco-link achieves its targets. Not all opportunities are a protective conservation mechanism – for example the pilot of the Indigenous Ranger Program.

On occasions a small investment of funds is required to clarify or encourage interest in the opportunity to take up a conservation instrument. As an example Territory Eco-link has a sound working relationship with the Central Land Council and an understanding of Traditional Owners' aspirations.

Traditional Owners at the communities of Papunya and Daguragu expressed interest in starting Indigenous Ranger programs to manage the natural and cultural values on their lands. The Central Land Council (CLC) wanted a greater degree of confidence that the Traditional Owners' interest was sufficient and likely to persist into the future before committing to establishment of the program and seeking the necessary Australian Government funding. Territory Eco-link provided the CLC with sufficient funding to pilot the establishment of these ranger groups for three months. Both ranger groups were enthusiastic, demonstrated a keenness to continue into the future, and this has now led to the

CLC securing longer-term funding from the Australian Government. The groups are carrying out conservation management on their lands, lands that provide key linkages for Territory Eco-link.

At Papunya, the community is now expressing an interest in declaring a significant portion of their land as an Indigenous Protected Area. In the interim they are considering entering into a conservation agreement through Territory Eco-link to demonstrate their commitment to and to gain formal recognition for their conservation effort. The land surrounding Papunya borders West MacDonnell National Park and has internationally significant biodiversity values. The small investment in a pilot Indigenous Ranger Group is resulting in an additional 5 million hectares with significant biodiversity values being managed to protect these values. In doing so they are improving connectivity between protected areas and making a significant contribution to the transcontinental corridor.

The Henbury Conservation Project is another exciting and innovative initiative. R.M.Williams Agricultural Holdings purchased Henbury Station for inclusion in the National Reserve System with funding assistance from the Australian Government (see also Pearse 2012). Henbury Station borders Finke Gorge National Park and Owen Springs Reserve which in turn borders West MacDonnell National Park, creating linked protected areas covering more than 2 million hectares and including the headwaters and over 100 km of the upper reaches of the Finke River.

R.M.Williams Agricultural Holdings, a 'for profit' company is now working towards establishing a model for carbon farming in the arid rangelands that will deliver a company profit at the same time as it delivers long-term biodiversity conservation outcomes. If successful, it will provide a model that can be applied to portions of other pastoral properties in the arid rangelands and provide the opportunity for an alternative income stream for pastoral leaseholders. Finding a means to derive income from achieving biodiversity conservation outcomes, rather than biodiversity conservation being a cost liability, will make it financially possible for many more pastoral leaseholders to consider delivering biodiversity conservation outcomes as an integral part of their business.

Major lessons

The first 3 years of the Territory Eco-link initiative have provided much learning. The need to build trust and strong relationships with all stakeholders, landowners and partners is critical to success. Understanding their 'business', 'what's in it for them' and 'what success looks like' from their perspective has been critical to successfully achieving voluntary participation in Territory Eco-link.

Also important has been having a clear and powerful vision, and the 'stories' that make that vision compelling to each stakeholder group or individual. Equally necessary has been building community understanding of the value of biodiversity and the need for a 'whole of community' effort at a 'landscape scale' to maximise the chance that biodiversity will persist into the future. With much of the Northern Territory's vegetation intact, it is changed fire regimes and invasive species (such as camels, rabbits, cats and cane toads) that have had the greatest impact and will do so into the future with increases in average temperatures and changes in rainfall. As such, active management of these threats is essential.

However, the single most important lesson has been the need to develop mechanisms to assist the funding of ongoing conservation management of private and pastoral leasehold lands.

Applying lessons

It is quite evident that the greatest deterrent to a landholder, in considering whether to make a long-term commitment to managing a portion of land for biodiversity outcomes, is the cost of ongoing management.

On Aboriginal land, the Australian Government's Indigenous Protected Area and Working on Country programs are greatly enhancing conservation management. For example, some 15 million hectares of the Tanami Desert, an area of more than twice the size of Tasmania, are managed with funds from these two programs.

Figure 5.2 Traditional Owner Duncan Gallagher burning in the Tanami Desert on protected rock-wallaby habitat. (Photograph courtesy of the Central Land Council).

By replacing hot season wildfires in the Tanami with cool season patchier burns from the ground and using aerial incendiaries on a landscape scale, significant benefits to biodiversity are obtained. Research has shown that patchy early burning results in a doubling of the prevalence of Greater Bilby (*Macrotis lagotis*) activity. Traditional Owner guidance is a key component of planning and implementation of the Central Land Council's fire program on Aboriginal land (Figure 5.2). However, without Australian Government funding, conservation management in the Tanami Desert would decline significantly.

Equally, pastoral leaseholders may well have an interest in protecting some of the special areas with high biodiversity value on their leases, but without financial assistance their businesses may not be in the position to allow a long-term commitment to these costs.

Volunteer effort can help to remove some of this cost. The Land for Wildlife program has provided a useful pool of skilled volunteers, with many willing to travel to pastoral leaseholdings to assist with weed and fire control. However, the uncertainty of whether this assistance will continue into the long-term future is still a deterrent to leaseholders making a long-term commitment to biodiversity conservation.

Opportunities that may provide an alternative income stream such as abatement and sequestration of carbon, the Australian Government's Biodiversity Fund, environmental offsets or small grants programs, along with increased effort to link volunteer effort with pastoral leaseholders, need to be further developed to give greater confidence that the financial burden of conservation management can be reduced.

Philanthropic investment in conservation in the Northern Territory has largely focused on direct investment in property acquisitions for inclusion in the National Reserve System through NGOs such as the Australian Wildlife Conservancy with funding assistance from the Australian Government. However, more recently The Nature Conservancy has financially assisted start-up management for three large Indigenous Protected Areas (Warddeken, Djelk and Southern Tanami) and assisted the Indigenous Land Corporation acquire Fish River Station for conservation and hand back to Traditional Owners (Fitzsimons and Looker 2012). The opportunity exists to develop mechanisms that would allow philanthropic investment in recurrent conservation management on pastoral leasehold lands. Any increase in the likelihood of longer-term funding to assist them to achieve biodiversity conservation outcomes will increase the ability of pastoral leaseholders to commit to the delivery of these outcomes in the future.

Conclusions

Territory Eco-link is an important conservation initiative. It is a new and more holistic approach to achieving sustainable biodiversity conservation outcomes in the Northern Territory. Several factors have been essential to the successes achieved through the Territory Eco-link initiative so far, including establishing strong relationships with stakeholders, a clearer vision, building community understanding and the need for a 'whole of community' effort. However, by far the greatest limiting factor has been the lack of available recurrent funding for ongoing conservation management. Developing mechanisms to reduce the costs to landholders for ongoing biodiversity management will be important to the continued success of the Territory Eco-link initiative.

Acknowledgements

My sincere thanks go to Dr Diana Leeder for her valuable comments, Simon Love for preparing the map, the Central Land Council for the photograph of Duncan Gallagher burning in the Tanami Desert and to the editors for their guidance and extreme patience in waiting for these words.

Biography

Andrew Bridges is a park management practitioner with over 30 years' experience in parks, wildlife and cultural resource management in the Northern Territory. After starting as a ranger at Uluṟu-Kata Tjuṯa National Park he worked in several remote national parks before moving into a leadership role. Andrew has initiated change and innovation in conservation and protected area management, led the development of some of the Northern Territory's major national parks, and overseen the implementation of joint management arrangements with Aboriginal owners for several iconic central Australian national parks. Based in Alice Springs, Andrew currently leads the Territory Eco-link initiative as a Director within the Northern Territory Department of Natural Resources, Environment, the Arts and Sport.

References

Fitzsimons J and Looker M (2012). Innovative approaches to land acquisition and conservation management: The case of Fish River Station, Northern Territory. In: *Innovation for 21st Century Conservation* (Eds P Figgis, J Fitzsimons and J Irving) pp. 78–85. Australian Committee for IUCN, Sydney.

Mackey BG, Watson JEM, Hope G, and Gilmore S (2008). Climate change, biodiversity conservation, and the role of protected areas: An Australian perspective. *Biodiversity* **9**(3–4), 11–18.

Pearse R (2012). 'Henbury Station' – Financing conservation for carbon and biodiversity markets: An industry perspective. In: *Innovation for 21st Century Conservation.* (Eds P Figgis, J Fitzsimons and J Irving) pp. 172–179. Australian Committee for IUCN, Sydney.

Soulé ME, Mackey BG, Recher HF, Williams JE, Woinarski JCZ, Driscoll D, Dennison WC, and Jones ME (2004). The role of connectivity in Australian conservation. *Pacific Conservation Biology* **10**, 266–279.

Woinarski JCZ, Whitehead PJ, Bowman DMJS, and Russell-Smith J (1992). Conservation of mobile species in a variable environment: The problem of reserve design in the Northern Territory, Australia. *Global Ecology and Biogeography Letters* **2**, 1–10.

Editor's postscript

Shortly before going to press, the newly-elected Northern Territory Government decided to cease funding for Territory Eco-link. It stated this would result in 'savings of $380,000 in 2012–13 and $450,000 ongoing from 2013–14. The Government will continue to fund the Land for Wildlife component of the program … The changes reflect a sharper focus to "on park" responsibilities before considering "off park" initiatives' (Conlan M (2012). 'Parks and Wildlife to receive funding increase'. Minister for Parks and Wildlife Media Release, 4 December 2012. Available: http://newsroom.nt.gov.au/index.cfm?fuseaction=viewRelease&id=10126&d=5 [Accessed 7 December 2012]).

6

Ten years of NatureLinks in South Australia

Jody Gates and Stella Kondylas

Introduction

The NatureLinks initiative was developed by the South Australian Government as a strategy for tackling biodiversity decline at the landscape scale. This chapter explores the origins of NatureLinks from a 'top-down' strategy to establish five formally recognised biodiversity corridors for landscape-scale restoration and conservation, to a practical program with a strong emphasis on collaboration and connectivity between stakeholders.

Location

NatureLinks incorporates five 'NatureLinks' corridors across South Australia (Figure 6.1). These corridors – 'East meets West', 'Arid Lands', 'Flinders–Olary', 'Cape Borda to Barossa' and 'River Murray–South East' – vary in extent and occur across the range of ecosystems found in the state, including arid stony plains, rivers and wetlands, mallee and woodlands, grassy woodlands and forests. Each corridor is generally aligned with a particular bioregion or landscape with similar ecological and social systems.

In 2010, the South Australian and Northern Territory Governments collaborated to develop the Trans-Australia Eco-Link (TAEL), which aims to establish a corridor extending more than 3500 km from Arnhem Land in the Northern Territory to Port Augusta in South Australia (see also Chapter 5). In South Australia, TAEL incorporates much of the stony deserts bioregion where pastoralism is the major land use and mining and tourism are also important industries. The TAEL overlaps with the Arid Lands and Flinders–Olary NatureLinks.

Although the emphasis of the NatureLinks program is on the original five corridors and the Trans-Australia Eco-Link, similar principles are applied to the management of all landscapes in South Australia. NatureLinks has been embedded in state policy and planning frameworks, including Regional NRM Plans and planning strategies (e.g. *The 30-year Plan for Greater Adelaide*, Department of Planning and Local Government 2010). South Australia was the first government in Australia to incorporate the concept of large corridors into these policy and planning frameworks (Stokes and Leaman 2007).

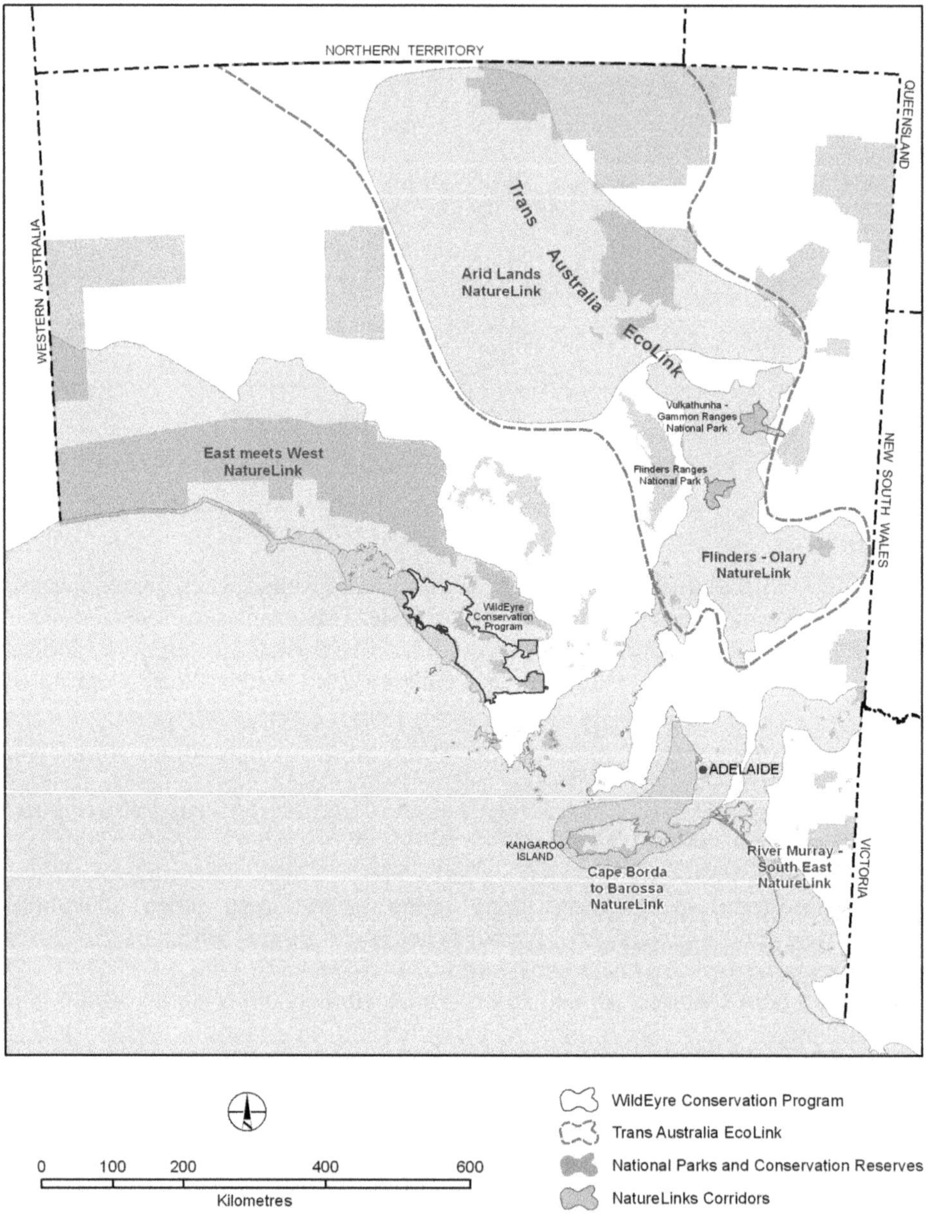

Figure 6.1 Location of NatureLinks in South Australia. Dark grey shading = NatureLinks; Light grey shading = protected areas.

Why was NatureLinks established?

NatureLinks was developed as a policy initiative in 2002 as part of its support for The Wilderness Society's WildCountry philosophy (Stokes and Leaman 2007). Modelled on the North American Wildlands Project (Noss 1992) the WildCountry aim is to establish an Australia-wide system of interconnected core protected areas, each surrounded and linked by other lands managed under conservation objectives (Recher 2003).

NatureLinks was established to provide a vision for ecological sustainability in South Australia with a focus on nature conservation, framed within the context of integrating conservation with regional development and natural resource management (Department for Environment and Heritage 2003; Stokes and Leaman 2007).

The original goal of the NatureLinks was 'To enable South Australian species and ecosystems to survive, evolve and adapt to environmental change', to be achieved via an objective of 'Connected habitat across South Australia, comprising a comprehensive system of core protected areas buffered and linked by lands managed for conservation objectives' (Department for Environment and Heritage 2003). Over the 10 years that NatureLinks has been operating the initiative has evolved but the underlying vision has remained the same.

The current composition of NatureLinks

NatureLinks is a landscape-scale initiative that encourages collaboration across tenures. As a result it incorporates a wide range of projects operating on both private and public land. Some of the best projects have been those created when public and private partners have come together around common goals for a particular landscape and implemented coordinated efforts to achieve these goals. For example, the WildEyre Project, in the East meets West NatureLink (WildEyre Working Group 2009) formed when staff from The Wilderness Society, Greening Australia, Nature Conservation Society of South Australia, the Eyre Peninsula Natural Resources Management Board and the then named Department for Environment and Heritage (DEH; predecessor to the Department of Environment and Natural Resources, DENR; and now Department of Environment, Water and Natural Resources, DEWNR) came together to undertake conservation action planning for a 1.2 million hectare area on Eyre Peninsula. This area encompasses public conservation reserves, private conservation reserves (Heritage Agreements) and mixed farming properties, and includes diverse mallee, red gum, sheoak, coastal, granite outcrop, saline wetland and many other important vegetation communities and wildlife habitats. The Conservation Action Plan covers 13 different environmental asset types and management actions include everything from pest control, revegetation and establishing vegetation buffers, to supporting volunteers and landholders. A similar network has recently been established in the southern Flinders Ranges ('Living Flinders'; Greening Australia 2010).

Long-term vision for NatureLinks

From inception it was acknowledged that NatureLinks would not provide a 'quick fix' for South Australia's degraded ecosystems or declining species (Department for Environment and Heritage 2003). While there is a need for short-term conservation priorities and targets, including ongoing land purchase for public conservation reserves, effective restoration is likely to take a generation. To this end, one of the original principles of NatureLinks was that restoration should be planned over long time scales.

The 50 year vision is that NatureLinks will have resulted in landholders implementing ecologically sustainable land management practices as part of their day to day businesses, and there is much greater community appreciation for natural systems and the services they provide. NatureLinks is not a prescriptive program, but a vision for an ecologically sustainable future. The idea is that in 100 years South Australia will have five landscape-scale conservation areas of connected habitat. These will consist of a comprehensive system of core protected areas buffered and linked by areas of land and sea managed for conservation. Within these areas South Australian species and ecosystems will be able to survive, evolve and adapt to environmental challenges such as climate change.

History and evolution of NatureLinks

NatureLinks was launched with three major projects: 'East Meets West' on Eyre Peninsula; 'Cape Borda to Barossa' covering the Mount Lofty Ranges to Kangaroo Island; and 'Bounceback', an expansion of an existing restoration project in the Flinders Ranges (now part of the Flinders–Olary NatureLink). Projects were also planned for the existing protected area network, threatened species recovery and marine protected areas.

In 2004, the establishment of five biodiversity corridors (i.e. NatureLinks) was set as a target in *South Australia's Strategic Plan* (Government of South Australia 2004). At the same time the *Natural Resources Management (NRM) Act 2004* passed through the South Australian Parliament, establishing eight NRM regions, and the requirement for a state NRM Plan (Government of South Australia 2006) and regional NRM plans (e.g. Adelaide and Mount Lofty Ranges Natural Resources Management Board 2008). NatureLinks, with its focus on landscape-scale planning, implementation and integration, community engagement and partnerships, aligned well with the principles underpinning NRM planning. The first regional NRM plans, completed in 2008–09, incorporated NatureLinks as a key policy initiative, but the high level aims had not yet been translated into specific priorities to guide on-ground activities in all corridors. At this point, people on the ground were seeking a narrative for each corridor to explain what the high level principles meant in practical terms for priority conservation needs on the ground. A Landscape Assessment Framework was developed by the then named DENR to provide this information (see 'Major successes' below).

When the target to establish five NatureLink corridors was first included in *South Australia's Strategic Plan* the boundaries of the corridors had not been finalised. Measures of progress and success are challenging for ecological projects, especially projects such as NatureLinks that have large spatial and temporal scales (Hughes *et al.* 2011). For pragmatic reasons the key measure of progress for the Strategic Plan target became the completion of a plan for each of the five corridors by 2010. The resulting plans were high level visionary documents, with limited prescriptive detail.

The locations of each corridor were finalised following a series of regional workshops with stakeholders (Figure 6.1). The corridors encompassed large areas of similar ecosystems that also aligned with social systems. For example, the Flinders–Olary NatureLink covers the Flinders and Olary Ranges, an iconic area of outback South Australia recognised for scenic and recreational values, where tourism and grazing enterprises coexist.

The original principles of NatureLinks (Department for Environment and Heritage 2003) were reviewed and updated in 2010 to simplify the science messages and place more emphasis on people and social systems as core elements (Table 6.1). This reflects the current focus on framing nature conservation within the context of production systems and people's livelihoods, and not purely on environmental, moral and spiritual grounds.

Table 6.1. The evolution of the underpinning principles for NatureLinks

In 2003, the 'Elements' covered the range of issues still considered critical today to ensure a successful long-term project. Initially the 'Principles' were based solely on the best science at the time. In 2010, these were renewed to include the social elements, and to expand and simplify the science messages.

Elements and principles 2003	Guiding principles 2010
Elements • Connectedness – connected habitat facilitating ecological flows across the land and sea. • No Species Loss – our native species and ecological communities surviving and continuing to evolve. • Integration and partnerships – natural resources management across the land meeting common biodiversity objectives. • People in nature – South Australians sharing the benefits of ecological sustainability. **Principles** • Biodiversity conservation activities should be planned at a landscape scale. • Habitat restoration should be undertaken at large spatial scales. • Species in fragmented landscapes should be managed as metapopulations (populations linked by exchange of individuals). • An 'ecological community' approach to biodiversity conservation should be encouraged. • Ecological restoration should be planned over long time scales. • Biodiversity conservation activities should be underpinned by sound ecological knowledge.	**NatureLinks** • Applies a landscape approach to plan conservation activities. • Delivers conservation activities at local and regional scales across public and private land. • Works cooperatively, flexibly and in partnership with people, communities and organisations. • Acknowledges that society and the economy will also influence and benefit from the conservation of nature and the services it provides. • Builds local ownership and support for conservation by acknowledging and helping with grassroots efforts. • Supports Aboriginal people to be involved in caring for country. • Makes decisions based on the best available scientific, traditional and local knowledge. • Strives to only undertake activities that result in lasting improvements for conservation. • Sets and measures landscape-specific goals and targets to achieve conservation outcomes. • Shares knowledge with others and evaluates success to guide future work. • Recognises that protecting and restoring landscapes is a long-term commitment.

Governance and coordination of NatureLinks

Governance of NatureLinks has been an evolving process. Although South Australia's Strategic Plan is community owned, the lead role for achieving the NatureLinks target, and accountability, rests with DEWNR. As a result NatureLinks has been perceived as a government initiative. Early in the process it was proposed that a range of committees be established to guide the implementation of NatureLinks. However, such arrangements were unlikely to adequately engage non-government organisations (NGOs) or the wider community.

Over time, governance arrangements have shifted from a centralised model of coordination by the then named DEH to a regionalised model that engages NGOs and the wider community. Within each corridor, governance arrangements remain relatively informal.

For example, in the 'East meets West' corridor the nested WildEyre project has its own governance group and a Memorandum of Understanding to guide the involvement of the partner organisations (WildEyre Working Group 2009). At the statewide level a NatureLinks NRM Alliance Working Group[i] was established in 2011 to facilitate shared ownership with key partners such as environmental NGOs. Importantly, the members of this group are already involved in projects in multiple corridors and can therefore bring a statewide perspective to implementation issues. This reflects one of the benefits of working in a state with a small population such as South Australia, where it is relatively easy to get all key players together. An important role provided by the NRM Alliance Working Group is to increase community support and ownership of NatureLinks beyond DEWNR.

To support NatureLinks at the whole-of-state level, overall coordination and facilitation is provided by a dedicated statewide manager in DEWNR. Five dedicated NatureLinks coordinator positions are also funded by DEWNR to provide a point of contact for each of the corridors. These positions are based within the region where each corridor occurs.

Most implementation is occurring through grant funds provided by the South Australian and Australian Governments, and to a lesser extent non-government grant bodies. Partnerships that increase non-government investment in the implementation of NatureLinks will become increasingly important in the future.

Major successes

The original published information on NatureLinks (Department for Environment and Heritage 2003) is still remarkably contemporary. The South Australian Government was a national leader in formally adopting the concept of large-scale corridors and underpinning landscape-scale restoration principles as policy. This focused attention on the importance of planning for multiple scales and in particular the landscape scale, which is also fundamental to NRM planning and delivery. Building on the foundation established by NatureLinks, the second iteration of South Australia's State NRM Plan is seeking to link the concepts of sustainable production, connectivity and ecosystem services. Connecting these concepts and translating them into programs for delivery is seen as critical to achieving conservation outcomes within the corridors.

To underpin NatureLinks a sound planning process was required to assess landscapes within corridors and identify landscape-scale (referred to as the 'coarse-filter'; Noss 1987; Groves 2003) restoration priorities. These landscape-specific priorities are essential to convert the high-level policy and principles into tangible actions for on-ground projects. Ecological scientists in DEWNR developed a Landscape Assessment Framework, which relies on several lines of evidence to identify birds (a useful indicator of environmental health and the only taxa for which adequate data and expertise exist) that are declining within landscapes. Birds can be grouped according to their broad habitat types, and declines can be related to systemic problems relating to ecosystem function within those habitats (such as disproportionate loss and decline in condition of remnants), which in turn informs priorities for restoration. There is real demand for this type of information among project implementation officers because it helps to provide a narrative for what restoration actions are required, where and why, thereby underpinning the ultimate purpose of the NatureLinks corridors.

NatureLinks has directly helped to initiate projects that are achieving good results (e.g. WildEyre; WildEyre Working Group 2009). It has also helped existing projects to expand by ensuring ongoing support and funding from within government (e.g. Bounceback; Department for Environment and Heritage 2006) and by promoting successful projects (e.g. threatened habitat restoration on eastern Kangaroo Island, in the Cape Borda to

Barossa NatureLink). These projects are not 'owned' by NatureLinks *per se*, but are recognised as contributing to the NatureLinks initiative. This is an important distinction to the people on the ground who have developed these projects, which embody NatureLinks principles: they are all based on collaboration and partnerships; all operate within appropriate landscape scales; they incorporate sound conservation science and assessments; and have all built local ownership and support by helping with grassroots efforts.

WildEyre has only been in place for a few years but has already attracted nearly $7 million in investment and is supporting stewardship programs and other works on some 30 properties. On eastern Kangaroo Island a large-scale restoration project was initiated through a threatened plant recovery project. A requirement to consider the landscape context for the threatened plants led to a focus on two key activities: ecological burning to stimulate the soil seed bank and regenerate small senescing remnants; and large-scale habitat reinstatement on private property. Since 2007 over 300 ha of high quality soils have been revegetated with over 400 000 seedlings from 125 species. An annual Kangaroo Island Planting Festival has been established, which has attracted thousands of volunteers from all walks of life. In 2011 alone, around 600 volunteers planted over 120 000 seedlings across 53 ha in under 3 days. The project is also based on a robust scientific and experimental framework.

The Bounceback project is primarily focused on fox, rabbit and goat control and began on a relatively small scale in two core conservation areas (Flinders and Vulkathunha–Gammon Ranges National Parks). It has since expanded to operate across a broad area including ~30 pastoral stations in the surrounding area (Department for Environment and Heritage 2006). The expansion was possible in part due to ongoing support from the South Australian Government through the NatureLinks initiative, which helped to maintain a mandate for the program and resulted in ongoing funding. Importantly, over 20 partner organisations have also contributed since the inception of Bounceback. A key indicator of the success of this program is the substantial increase in population size and extent of the nationally threatened Yellow-footed Rock-wallaby (*Petrogale xanthopus xanthopus*) (e.g. over a 10-fold increase in abundance in the Flinders Ranges National Park). This has occurred due to control of predators (foxes) and competitors (goats), which has essentially increased functional connectivity within the landscape for this species. Bounceback has also benefited several other nationally and locally important species.

Finally, NatureLinks is evolving to provide a framework for a partnership approach to nature conservation and to engage more of the community to get involved. The new communications campaign will reinforce this role, by including a website that provides a portal to help connect land managers, business, organisations and government.

Major lessons

Two of the major lessons from the NatureLinks initiative relate directly to the question 'What is NatureLinks?' First, clear and consistent messaging needs to be communicated to ensure that people have the same understanding of what the initiative is and what it is trying to achieve in practical terms. This was particularly critical for an initiative such as NatureLinks, which did not initially have a budget to directly fund projects, and which set an overarching vision and approach to nature conservation, but did not specifically prescribe the types of local projects required to achieve the aims. Many practitioners, both within and outside of government, sought more detail to understand what it meant for them on the ground. This leads directly to the second key lesson.

From the outset, robust restoration and conservation planning was required to translate landscape-scale thinking and related ecological principles into landscape-specific conservation priorities to guide on-ground projects (i.e. what species or communities might be at risk of local extinction and what specific actions were required to address this). The general principles of conservation planning are not considered adequate for guiding specific actions at any particular location (e.g. Lambeck 2002; Bennett and Mac Nally 2004; Cale 2010). However, while a successful planning approach is in place today, it would have been beneficial if more focus were placed on this in the early stages of the initiative. Having landscape-specific priorities for nature conservation, based on sound science, would have gone a long way to help translate the vision for NatureLinks into action on the ground, which is what many stakeholders were looking for.

A third lesson is centred on the fact that NatureLinks is still perceived as a state government initiative by some people. However, it is acknowledged that its success is largely dependent upon community ownership. While establishing targets in South Australia's Strategic Plan provided important leverage and momentum for NatureLinks at the government level, it also presented a tension between the intent to have greater community ownership and the need to formally report on progress to the government. There is an ongoing need to devolve the responsibility from government to seek more shared ownership and responsibility for delivery with the community. The evolution of the governance arrangements for NatureLinks reflects ongoing attempts to effectively involve environmental NGOs who are active in on-ground implementation projects, community members, and other key stakeholders, so they have a genuine say in the directions of the program. For this to ultimately be effective it is critical that there is a shared understanding of what NatureLinks is seeking to achieve. Essentially, the three lessons shared here are interrelated.

Applying the lessons

A new communications campaign was implemented in 2012 to further clarify and promote the role of NatureLinks and invite people to get involved. A key aim of the campaign is to provide clear messages about what NatureLinks seeks to achieve, why and how. The key message is now focused around 'bigger and better habitat' (i.e. 'bigger and better habitats help our environment become more resilient in the face of environmental challenges, from weeds and feral animals, to water security and climate change. Better habitats mean a better environment for us all, helping our communities to prosper'). In addition to the communications campaign, the creation of the NatureLinks NRM Alliance Working Group in 2011 was specifically aimed at increasing shared ownership with key partners and giving them a greater role in shaping how the program is delivered. The members of the working group have also played a key role in communicating the message.

The Landscape Assessment Framework continues to be a priority for DEWNR. The focus is on providing clear and accessible information to assist regional planners and project officers to determine where to invest limited resources within each of the NatureLinks corridors. Importantly, the Landscape Assessment Framework clarifies what to monitor and why, to determine success from an ecological perspective. This goes a long way to providing a narrative that explains what each corridor is seeking to achieve.

The lessons learned from NatureLinks are being applied to the Trans-Australia Eco-Link project. The social system in the rangelands of South Australia consists of very few people who all live remotely and rely on natural resources for their livelihoods. Maintaining the natural environment in a productive and sustainable state is a clear goal for natural resource management in this area, but also provides significant opportunities for integrating nature conservation. However, the mechanisms for supporting nature conservation in the region are limited and land purchase for conservation is not always supported by pastoralists. Therefore, when planning the TAEL initiative it was acknowledged that a range of stewardship mechanisms that enabled conservation to be integrated with pastoralism are required, with flexibility to incorporate desired conservation outcomes. The TAEL project is putting considerable effort into engaging with local stakeholders about what the project is trying to achieve, and to discover the interests, motivations, constraints and barriers for potential stewardship providers and investors. This is building awareness and ownership of the project. A landscape assessment was also initiated early in the project to identify conservation priorities to effectively target stewardship programs.

As the NatureLinks program matures, opportunities for NGOs to host NatureLinks coordinator positions are being investigated. For some time this has been recognised by stakeholders and DEWNR as an effective way of sharing ownership of the program, allowing DEWNR to continue to support coordinator positions through funding but to step back from a primary lead role to focus on other critical roles, such as being a knowledge-broker (i.e. provision of information from the Landscape Assessment Framework), statewide coordination, including communications and promotion, and developing new policy mechanisms and instruments for achieving landscape-scale conservation outcomes.

Conclusions

NatureLinks has provided a high level framework for biodiversity conservation and related NRM initiatives across the state. In particular, the original focus on landscape-scale conservation through partnerships and community engagement continues to be fundamental to achieving nature conservation outcomes. A key to the ongoing success is a clear acknowledgement that large-scale open-ended initiatives require an explanatory narrative that the community can understand and engage in an evolving process of governance.

Acknowledgements

NatureLinks has had many contributors and supporters over the last 10 years. The authors thank Greg Leaman, Brenton Grear, Glenn Gale, Andrew Willson, Lisien Loan and Clare Nicolson for valuable comments on this chapter.

Biographies

Jody Gates currently works on conservation policy and planning for the Department of Environment, Water and Natural Resources. He brings to this role over 15 years of experience in the delivery of conservation programs in South Australia, including working as a park ranger and threatened species ecologist, and a manager of regional conservation programs teams.

Stella Kondylas works as the NatureLinks Project Manager for the Department of Environment, Water and Natural Resources in South Australia. She brings to her role experience in communication, community engagement and strategic planning within both government and non-government organisations in Australia and the United Kingdom.

Endnote

i. NatureLinks NRM Alliance Working Group consists of Nature Foundation SA, Nature Conservation Society SA, Trees for Life, Greening Australia, Friends of Parks Inc., The Wilderness Society, Conservation Volunteers Australia, Conservation Council SA and Zoos SA.

References

Adelaide and Mount Lofty Ranges Natural Resources Management Board (2008). *Creating a Sustainable Future. An Integrated Natural Resources Management Plan for the Adelaide and Mount Lofty Ranges Region. Volume B – Ten Year Plan for the Region.* Government of South Australia, Adelaide.

Bennett AF, Mac Nally RC (2004). Identifying priority areas for conservation action in agricultural landscapes. *Pacific Conservation Biology* **10**, 106–123.

Cale B (2010). *Literature Review of Specific Topics relating to Science Directions 2010–2015.* Department of Environment and Natural Resources, South Australia, Adelaide.

Department for Environment and Heritage (2003). *NatureLinks: Implementing* WildCountry *Philosophy in South Australia.* Government of South Australia, Adelaide.

Department for Environment and Heritage (2006). *Bounceback Progress Report 2006.* Government of South Australia, Adelaide.

Department of Planning and Local Government (2010). *The 30-year Plan for Greater Adelaide. A Volume of the South Australian Planning Strategy.* Government of South Australia, Adelaide.

Government of South Australia (2004). *South Australia's Strategic Plan.* Government of South Australia, Adelaide.

Government of South Australia (2006). *State Natural Resources Management Plan.* Government of South Australia, Adelaide.

Greening Australia (2010). *Living Flinders: From Peaks to Plains.* Greening Australia, Adelaide.

Groves CR (2003). *Drafting a Conservation Blueprint. A Practitioner's Guide to Planning for Biodiversity.* Island Press, Washington DC.

Hughes FMR, Stroh PA, Adams WM, Kirby KJ, Mountford JO, Warrington S (2011). Monitoring and evaluating large-scale, 'open-ended' habitat creation projects: A journey rather than a destination. *Journal for Nature Conservation* **19**, 245–253.

Lambeck RJ (2002). Focal species and restoration ecology: response to Lindenmayer *et al. Conservation Biology* **16**, 549–551.

Noss RF (1987). From plant communities to landscapes in conservation inventories: a look at Nature Conservancy (USA). *Biological Conservation* **41**, 11–37.

Noss RF (1992). The Wildlands Project: land conservation strategy. *Wild Earth* **(Special Issue)**, 10–25.

Recher HF (2003). WildCountry. *Pacific Conservation Biology* **8**, 221–222.

Stokes A and Leaman G (2007). NatureLinks: Protected areas, wilderness and landscape connectivity in South Australia, Australia. In: *Science and Stewardship to Protect and Sustain Wilderness Value: Eighth World Wilderness Congress Symposium.* September 30-October 6 2005, Anchorage. (Eds A Watson, J Sproull and L Dean) pp. 212–217. Department of Agriculture, Forest Service, Rocky Mountains Research Station, Fort Collins, Colorado, USA.

WildEyre Working Group (2009). *WildEyre: Conservation through Collaboration.* WildEyre Working Group, Adelaide.

7

From Danggali to Riverland: experiences from the Bookmark Biosphere Reserve, South Australia

Duncan Mackenzie and James Fitzsimons

Introduction

The ~900 000 ha Riverland Biosphere Reserve (formerly Bookmark Biosphere Reserve) is located in the semi-arid Murray Mallee region of South Australia, adjoining the Victorian and New South Wales borders (Figure 7.1). It includes riverine and floodplain ecosystems associated with the River Murray, including River Red Gum (*Eucalyptus camaldulensis*) and Black Box (*Eucalyptus largiflorens*) woodlands, chenopod shrublands, and extensive areas of long-unburnt mallee in the semi-arid zone. Renmark and Berri are the major regional population centres.

The 253 000 ha Danggali Conservation Park was one of nine Australian protected areas to be designated as a Biosphere Reserve under the UNESCO Man and the Biosphere Program in 1977. At that time, the Danggali Conservation Park had itself only recently been declared (in 1975), with Commonwealth Government funds having been made available to facilitate the purchase of four pastoral properties (DENR 1995) (Figure 7.1). In 1993, the Biosphere Reserve changed its name to Bookmark to reflect an expansion of sites (see below). The primary reason for the declaration of the Bookmark Biosphere Reserve was to halt livestock grazing to allow the native vegetation to regenerate, control the feral animal population and to manage native wildlife, such as kangaroos and emus, in a sustainable manner (Bookmark Biosphere Trust 1995). The region contains important areas of long-unburnt mallee (Bennett *et al.* 2010), which are crucial for species such as the endangered Black-eared Miner (*Manorina melanotis*), as well as other ecosystems that are not well represented in the reserve system (Fitzsimons and Wescott 2008a). The reserve was, at the time, being used as a 'testing ground' for new reserves all over the country (Bookmark Biosphere Trust 1995).

The Riverland Biosphere Reserve contains some 49 individual sites, and two local government areas of Berri Barmera and Renmark Paringa (Figure 7.1; Table 7.1). Between 1993 and 2001 both the total area and number of individual components increased, but patterns of growth have been quite different (see Figure 7.2). The major increases in total area were

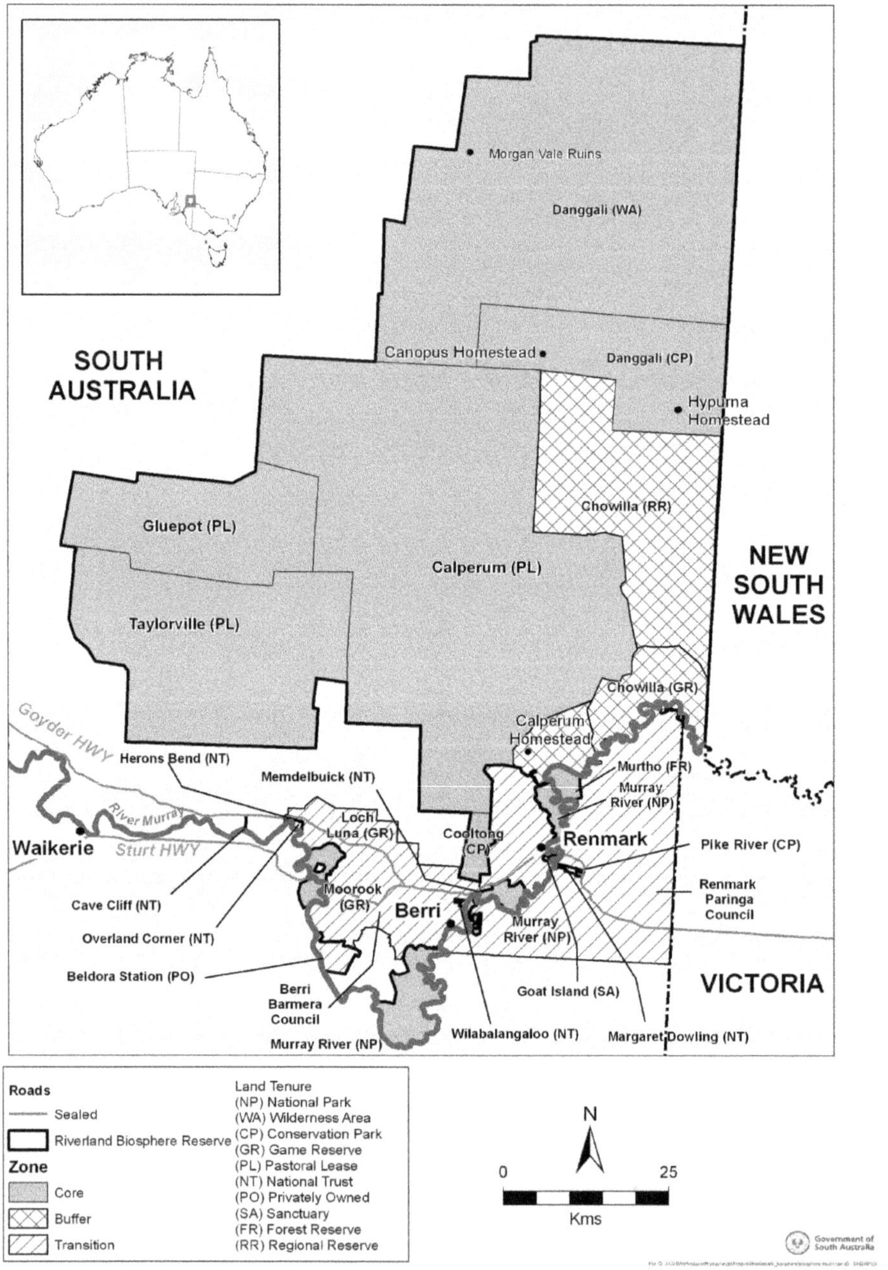

Figure 7.1 Riverland Biosphere Reserve management zones. Not all small properties shown.

the addition of three large pastoral leases, namely Calperum Station (along with various public protected areas) to the expanded biosphere reserve in 1993 (some 362 000 ha), Gluepot Station in 1997 (54 000 ha) and then Taylorville Station in 1999 (94 000 ha). In contrast, while increases in the number of properties were sporadic between 1993 and 1997, from 1997 to 1999 relatively high and steady increases have occurred (Figure 7.2). This latter growth is due mainly to the participation of private landowners and the inclusion of reserves owned by the National Trust of South Australia, which together contributed over 90% of the total number of components added since 1997. There were no additions to Bookmark/Riverland between 2001 and 2012. Individual land components within the biosphere reserve are physically well connected with 18 (36.7%) components physically adjoining other sites in the biosphere and a further 13 (26.5%) separated from others only by a linear feature, most often by the Murray River (Fitzsimons and Wescott 2008b).

Table 7.1. Number, area and type of components within the Riverland Biosphere Reserve (as at January 2002)

Component type	Number	Area (ha)
National Park	1	13 134
Conservation Park	6	256 044
Game Reserve	3	21 788
Regional Reserve	1	75 148
Forest Reserve	3	2631
Pastoral Leasehold	3	394 188
Local Government Reserve	2	167
National Trust Reserve	12	689
Private Land (Conservation)	8	5153
Private Land (Sustainability)	10	411

Note: (i) Some Private Land components have both conservation and sustainable use objectives: categorisation is based on the management objective for the majority of the property. (ii) The Berri Barmera Council (50 845 ha), and Renmark Paringa Council (92 100 ha) are also part of the transition zone of the biosphere reserve.

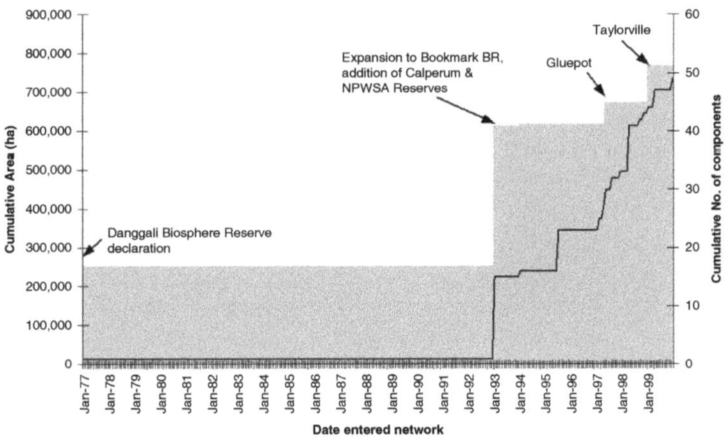

Figure 7.2 Chronological development of the Bookmark Biosphere Reserve since establishment and up to 2001. Shaded area represents cumulative area in Biosphere Reserve, while single line represents cumulative number of sites. Major additions in area are annotated. Note that area and number of sites do not include the two local government areas in the transition zone. (Source: Fitzsimons 2004)

The vision of the Riverland Biosphere Inc. (the current coordinating body for the biosphere) 'is to have a community that holds values and behaviours which ensure environmentally sustainable and balanced management of our natural resources, inspired by a love of the Riverland landscape and a desire to see the region thrive' (Riverland Biosphere Incorporated 2003a).

To achieve this vision, a Business Plan was developed. The major objectives of the Business Plan (Riverland Biosphere Incorporated 2003b) are shown below.

1 To define and expand the Biosphere concept, lifting its profile;
2 The sustainable development of the Riverland District;
3 To develop ecotourism/nature tourism as a revitalising component for the region; and
4 To provide education – particularly in subjects concerned with local environmental issues. This education is to be delivered through the Riverland schools and media.

History and structure

Biosphere Reserves are concerned primarily with integrating biodiversity conservation with ecologically sustainable development across a variety of land tenures and uses (UNESCO 1995; Brunckhorst *et al.* 1997). The theoretical biosphere reserve model revolves around a 'core' protected area managed primarily for nature conservation, a 'buffer' zone where activities that impact on the biodiversity of the core are minimised, and a 'transition' zone, where the sustainable use of natural resources is encouraged (see Parker 1993; Brunckhorst *et al.* 1997). The Man and the Biosphere Program, which coordinates the world network of biosphere reserves, was launched by UNESCO in 1971, and has since seen 580 biosphere reserves declared in 114 countries (UNESCO 2012).

As with other Australian biosphere reserves declared at the time, Danggali was restricted to the 'core' protected area until late 1993. The purchase of a 240 000 ha pastoral lease, Calperum Station, adjoining the southern boundary of Danggali, by the Commonwealth Government, with support from the Chicago Zoological Society, heralded the formation of the renamed Bookmark Biosphere Reserve. The purchase initially caused 'consternation amongst the local community', with fears it would become an underfunded and unmanaged national park (Punturiero 2002, p. 6). Other public reserves, both adjoining and nearby, were also included in the expanded biosphere in 1993.

The Bookmark Biosphere Trust (formerly the Murraylands Community Trust), established under the South Australian *National Parks and Wildlife Act 1972*, was formed to manage the biosphere reserve and included representatives from key stakeholder groups (Bookmark Biosphere Trust 1995; Milliken 1995).

The Bookmark Biosphere Trust was dissolved in June 2002, following 'disagreements (which) led to the South Australian Government being convinced that the Biosphere Trust model, which was a statutory body under the South Australian parliament, was not the best way to promote the continued growth of the Man and Biosphere program' (Cottam 2003, p. 9). Cottam (2003) alluded to apparent disharmony in the Biosphere Trust structure with the remark that the 'provision of management becomes a difficult task as individuals who disagree with the central management structure may decide to do their own thing' and a later reference to the need to rebuild the relationships between the land partners. Existing and future difficulties such as maintaining the autonomy of the individuals within a governance model that complies with the Man and Biosphere objectives were also identified by Cottam (2003) (see also Fitzsimons and Wescott 2008c; Pfueller 2008).

An interim planning committee evaluated several governance options and decided upon a 'community committee', not allied to any government or non-government identity but including representatives of the Commonwealth and state nature conservation agencies, Indigenous and some nine community representatives. The interim committee prepared a constitution (required for incorporation of the new body) and a Business Plan. On 27 June 2002, a public meeting was held in Berri, where a Bookmark Biosphere Community Committee was elected to undertake future management of the Biosphere. The Community Committee subsequently became a legal entity: Riverland Biosphere Inc. (although still known locally as the Community Committee).

Riverland Biosphere Inc. does not have any staff and has no ongoing funding. Despite regular meetings from 2002 to 2009, the Committee has not met for the past 3 years. Thus current governance is close to non-existent. The Australian Government's Director of National Parks is responsible for the management of Calperum and Taylorville Stations within the Riverland Biosphere Reserve and is the nominal focal point for biosphere reserves more generally in Australia (as they are an international designation). Individual Land Partners continue to adhere to the three basic functions of a biosphere that are complementary and mutually reinforcing. They are:

1 a conservation function (relating to the management of biodiversity and natural ecosystems);
2 a development function (relating to the demonstration and implementation of sustainable development); and
3 a logistics and support function (relating to community activation, education and support).

Some Land Partners, such as BirdLife Australia's Gluepot Reserve, have provided leadership in the development of feral and pest animal control systems, weed management, fire control systems, environmental research and monitoring and 13 annual 2-day environmental education courses.

Major successes

Bookmark Biosphere Reserve was in the late 1990s one of the few examples of a truly multi-tenure network of conservation lands in Australia. It was also one of the few biosphere reserves in Australia that attempted to implement the biosphere reserve model of core, buffer and transition zones. As such it was held up as a model biosphere reserve (e.g. Brunckhorst 2001). The collaborative, landscape-scale vision also proved attractive for philanthropic and non-government organisations to 'buy-in' (e.g. Chicago Zoological Society, Australian Landscape Trust, Birds Australia). Ultimately, this resulted in the significant increase in total area and legal protection of land managed for conservation in the Murray–Mallee. More broadly, this investment was an early forerunner to the significant increase of the role of these sectors in Australia's conservation scene, and the number of private conservation areas.

More recently, in an effort to better engage the Riverland community in the activities of the Biosphere, the new committee undertook several programs directly linked to conservation and the natural resources of the Biosphere. These included: (i) development of a Constitution and Business Plan for the Biosphere Committee; (ii) in collaboration with the Berri Barmera District Council, a 'Cool Communities Program' funded by the then named Environment Australia, that was run over a 2-year period; (iii) a very successful arts/science 'River Festival' that attracted significant funding from 'Country Arts SA' and featured close liaison between the Biosphere and several arts and conservation bodies in the Riverland; (iv) participation in an environmental education program that included youth forums, environmental debates and community projects; and (v) guidance provided by biosphere experts on conservation issues, in several forums across the Riverland district.

Three major projects developed by the Committee are outlined below.

1. A schools writing program

A schools writing program called 'Eye on the Biosphere – Our Vision for the Future' had as its objective a means for young people to help the Riverland community see the importance of sustaining the natural landscape through the next generation's eyes. The program was a partnership between the Riverland newspapers, Riverland secondary schools and the Biosphere Committee. Thirty newspaper columns were produced in each of the 2 years that the program operated. A comprehensive resource kit was provided to each participating school that contained a wide range of useful information including contact details of experts from relevant organisations and programs, industry bodies, people and Land Partners in the Biosphere and experts in biodiversity. Students were encouraged to contact these experts when writing their articles. The Biosphere provided a program coordinator who liaised with the newspapers and pre-edited all articles before submission to the newspaper. Each article was asked to conclude with a vision for the future, highlighting how the biosphere would be faring in 50 years' time.

The program was extremely successful and was operational for just over 2 years. During this period the published articles generated a renewed community interest in the biosphere and stressed the need for stronger protection of the unique regional biodiversity. The program also built a high level of self-confidence in many of the participating students as they became 'involved' in the biosphere through conducting interviews with key contacts.

2. Nature based tourism strategy

The development of the Riverland Nature Based Tourism Strategy – an Ecotourism Initiative by the Biosphere Committee, in partnership with the Riverland Tourism Association and several other stakeholders, specifically focused on the development of ecotourism as a key niche market for the Riverland recognising the need to develop sustainable nature-based tourism in the region. The implementation of the strategy was seen as a means of guiding the development of ecotourism to a mature and viable industry, with a focus on positioning the region as one of the most sought after ecotourism destinations in the world.

Although the Initiative was sound in its objectives and purpose, success depended on each of the stakeholders providing ongoing support, including resources. Unfortunately, with such a widely disparate group of partners, the forward momentum slowed and came to a halt after 2–3 years. Some of the major stakeholders (for a variety of reasons) are no longer in existence, and changes in stakeholder direction has meant that it is improbable that the Initiative will ever be revived in its present form. During development of the Initiative, several new projects came to fruition and the profile of the region as a great ecotourism destination was beginning to form. Again, a lack of resources with which to take it forward was one of the major stumbling blocks. An example of a highly successful ecotourism venture is Gluepot Reserve which has developed an international reputation as an ecotourism destination (particularly for birdwatchers and conservationists) and as a leader in the development of environmental management programs.

3. Ecotourism certification course

Allied to the Ecotourism Strategy and Biosphere Business Plan was an Ecotourism Certification course developed by the Murray Institute of TAFE in consultation with knowledgeable biosphere members. The course offered a series of linked certificates in ecotourism to educate a diverse range of students including those seeking a career in ecotourism, tourism industry workers, ecotourism operators and their employees, school leavers, people seeking to return to work and Riverland residents. This course was the first of its kind in Australia aimed at providing career pathways and business opportunities in ecotourism for Riverland residents, plus intrastate, interstate and international participants. The courses (covering Certificates II, III and IV) were linked to the Degree course in Ecotourism offered by Flinders University. Unfortunately TAFE was under-resourced for this fledgling course (to ensure that it was both coordinated and promoted to potential participants). In the first year a preliminary budget was organised (to foster community involvement) and a coordinator appointed. TAFE did a marvellous job during the 3 years that the course was operational, however, the resignation of the coordinator and course responsibility being shifted to a different TAFE department in a different region signalled its demise. Several students successfully completed through to Certificate IV, including a group of Indigenous students. The course was a great boon to the participating students, as it provided a potential means of employment and filled a definite gap in the regional tourism industry. Globally, Eco/nature-based tourism is by far the fastest growing sector of tourism and the Eco Certification course provided a vital element – trained people – to what is a maturing and viable industry.

Major lessons

The 'Community' Committee formed in 2002 was a great improvement over the previous governance model (a Trust) in that the community was truly involved in all aspects of shaping the future of the Biosphere and were an integral part of the decision-making process. The new Riverland Biosphere vision went beyond a land conservation and management focus, to one that was about individual and community behaviour within the Riverland. The Biosphere was the perfect place to demonstrate sustainability for the Riverland region, because the community members of the Biosphere are committed to the Biosphere vision.

Having a strong and dedicated community committee is probably the easiest part of the 'governance equation'. A committee becomes almost impotent unless adequate long-term financial and human resources are available. Usually these resources are provided by several stakeholders at the start-up of the venture but dissipate as the project settles down. The expectation is that in time, the committee will become self-sustaining.

In the case of the Riverland Biosphere, all major stakeholders were represented on the Committee and this was the case for the first few years. As time progressed, interest by both the South Australian and Commonwealth Governments waned and most of their input became 'in-kind' (postage, photocopying and so on). Finance for the ongoing administration of the Committee was minimal and the Biosphere did not have the funding to employ a full or part-time administrator.

In today's atmosphere of ongoing financial crisis, obtaining grants from entities such as state and Commonwealth governments, regional councils, regional development corporations and tourism organisations (all of them stakeholders in the Biosphere) is virtually impossible. Several of these are Land Partners in the Biosphere but a large number of stakeholders are non-government organisations and private citizens who will probably not be able to support financially the ongoing activities of the Biosphere. Certainly it is possible to obtain grants for a range of well developed and justifiable projects, however, grant funds available for administrative purposes (as opposed to on-ground works) are virtually non-existent. A further problem has arisen with the advent of NRM Boards, and in South Australia, the amalgamation of NRM with the now Department of Environment, Water and Natural Resources. Funding for bodies such as the Biosphere does not fit the 'new' criteria which has tended to narrow to very specific environmental areas of concern.

A community committee composed of volunteers and other stakeholders can certainly work, but to succeed it must have a chairperson who is dedicated to the vision, is knowledgeable, a good communicator and motivator, and is prepared to commit the time to help make the biosphere model work. But more importantly, the chair and committee need to have the full-time support of a paid administrator who can handle the myriad of day-to-day activities that a successful committee generates.

Applying the lessons

As outlined above, the Riverland Biosphere model of governance can work. However, there needs to be long-term commitment from the major stakeholders to ensure financial support for the administration. In today's environment, where budgets of state and Commonwealth environment departments are being significantly reduced and local and regional councils have their own woes in regard to budgets, mergers and so on, it is unlikely that funding for biosphere activities will be high on their lists of priorities.

In addition, major changes have been made to regional organisations such as tourism associations, development corporations and the like, which at one time supported biosphere activities, but now are not in a position to do so. As one of the lead stakeholders in the Riverland Biosphere, the Commonwealth Government has all but disappeared 'off the scene' and for them to re-commit to support the Biosphere would need a major shift in thinking.

At this stage it is unlikely that the Riverland Biosphere will be reinvigorated. No Committee meeting has been held for 3 years and several of the members are no longer in the Riverland district. Without adequate resources being available and guaranteed for the administration of the Biosphere, it would be a foolish move to contemplate 'raising the Phoenix', no matter how good the concept and vision. New approaches to sustainable funding and fundraising clearly need to be considered. This includes securing more lasting commitments from the Commonwealth, state and local governments, which are all part of or have land in the Biosphere, but also looking more broadly to the business and philanthropic sector. Better highlighting the relative value for money and leverage the Riverland Biosphere offers, such as the amount of volunteer time donated by partners, is also important. As an example, Gluepot Reserve volunteers donate an average of 27 750 hours per year.

Greater consideration of the role of Riverland Biosphere, and biosphere reserves more generally, needs to be given in the recent push to develop large-scale connectivity projects in Australia. For example, parts of Riverland Biosphere fall within Habitat 141° and the Murray River/South East NatureLink (Chapters 8 and 6, respectively, and Figure 1.1 in Chapter 1), both of which were recently highlighted in the *National Wildlife Corridors Plan* (DSEWPC 2012), yet the presence of the biosphere itself is not mentioned. As National Wildlife Corridors are likely to receive preference for grants under the Australian Government's Biodiversity Fund, the Riverland Biosphere Reserve could strategically align itself more closely with these other corridors to potentially capitalise on this new source of conservation funding.

Conclusions

Bookmark/Riverland Biosphere Reserve was one of Australia's first multi-tenure reserve networks and considered a model biosphere reserve in Australia. The formation of a

Community Committee provided greater local involvement in coordination of the biosphere and broadened its function to better align education and tourism with the Reserve's objectives. However, a lack of funding to support this coordinating body has seen the Committee become moribund. While the owners of properties within the biosphere reserve are still managing their properties for conservation and/or sustainable development purposes, the concept of a coordinated and integrated 'network' of lands is likely to be losing momentum.

Acknowledgements

This chapter brings together research by James Fitzsimons and perspectives of Duncan Mackenzie in his former role of chair of the Riverland Biosphere Committee and current role of chair of BirdLife Australia's Gluepot Reserve Committee, a key Riverland Biosphere Land Partner. Duncan Mackenzie acknowledges all volunteers and stakeholders who dedicated their efforts to making the Riverland Biosphere one of Australia's 'showcase' reserves. James Fitzsimons received financial assistance from a Deakin University Postgraduate Research Scholarship for part of this research.

Biographies

Duncan Mackenzie is currently Chairman of BirdLife Australia's Gluepot Reserve, Ecotourism Australia and the South Australian Tourism Industry Council. He is a past Chairman of the Riverland Biosphere (2002–2006) and Deputy Chairman (2006–2008) and past Deputy Chairman of the Riverland Tourism Association. He is Chief Scientist for the Scientific Expedition Group and in the 1960s spent several years managing Australia's biological programs in the Antarctic. Duncan's background is as a senior manager for the Ford Motor Co., Alcoa of Australia, Tubemakers of Australia, Managing Director of Data Conversion Corporation and Bell & Howell Australia/New Zealand and the Executive Director of Nature Foundation SA. He is the author of over 400 papers, articles and books.

James Fitzsimons is the Director of Conservation for The Nature Conservancy's Australia Program and an adjunct associate professor at the School of Life and Environmental Sciences, Deakin University. His particular research interests are in the fields of protected area policy, practical conservation planning and mechanisms to integrate conservation outcomes on public and private lands. He has worked in the fields of conservation policy and planning for government environment departments and agencies, and for non-government environment organisations.

References

Bennett A, Clarke M, Avitabile S, Brown L, Callister K, Haslem A, Kelly L, Kenny S, Nimmo D, Spence-Bailey L, Taylor R, Watson S and Holland G (2010). *Fire and Wildlife in the Mallee. Insights for Conservation and Management.* La Trobe University and Deakin University, Melbourne.

Bookmark Biosphere Trust (1995). *Introducing Bookmark: The Bookmark Biosphere Reserve Action Plan.* Bookmark Biosphere Trust, Berri.

Brunckhorst D (2001). Building capital through bioregional planning and biosphere reserves. *Ethics in Science and Environmental Politics* **2001**, 19–32.

Brunckhorst DJ, Bridgewater P, and Parker P (1997). The UNESCO Biosphere Reserve Program comes of age: learning by doing; landscape models for sustainable conservation and resource use. In: *Conservation Outside Nature Reserves.* (Eds P Hale and D Lamb) pp. 176–182. Centre for Conservation Biology, University of Queensland, Brisbane.

Cottam E (2003). Bookmark Biosphere Reserve – A governance model. *Australian Biosphere Reserve News* **July**, 9–10.

DENR (1995). *Danggali Conservation Park Management Plan.* National Parks and Wildlife Service, Department of Environment and Natural Resources, Murraylands Region.

DSEWPC (2012). *National Wildlife Corridors Plan. A Framework for Landscape-scale Conservation.* Department of Sustainability, Environment, Water, Population and Communities, Canberra.

Fitzsimons JA (2004). The contribution of multi-tenure reserve networks to biodiversity conservation. PhD thesis. School of Ecology and Environment, Deakin University, Melbourne.

Fitzsimons JA and Wescott G (2008a). Ecosystem conservation in multi-tenure reserve networks: the contribution of land outside of publicly protected areas. *Pacific Conservation Biology* **14**, 250–262.

Fitzsimons JA and Wescott G (2008b). The role of multi-tenure reserve networks in improving reserve design and connectivity. *Landscape and Urban Planning* **85**, 163–173.

Fitzsimons JA and Wescott G (2008c). Evolving governance arrangements in multi-tenure reserve networks. *Environmental Conservation* **35**, 5–7.

Milliken ST (1995). *Calperum and the Bookmark Biosphere Reserve: A Model for the Future.* Australian Nature Conservation Agency, Canberra.

Parker P (1993). *Biosphere Reserves in Australia: A Strategy for the Future.* Australian Nature Conservation Agency, Canberra.

Pfueller S (2008). Role of bioregionalism in Bookmark Biosphere Reserve, Australia. *Environmental Conservation* **35**, 173–186.

Punturiero M (2002). The business case for biosphere reserves. In: *Biospheres: Creating Sustainable Regional Communities.* Summary of Proceedings of a Seminar held 20 September 2002, Shearwater Cape Schanck Conference Centre, pp. 5–7. Earthwatch Institute, Melbourne.

Riverland Biosphere Incorporated (2003a). *Riverland Biosphere Incorporated Constitution.* Riverland Biosphere Committee, Berri.

Riverland Biosphere Incorporated (2003b). *Riverland Biosphere Incorporated Business Plan.* Riverland Biosphere Committee, Berri.

UNESCO (1995). *Biosphere Reserves: The Seville Strategy and the Statutory Framework for the World Network.* UNESCO, Paris.

UNESCO (2012). Directory of the World Network of Biosphere Reserves (WNBR). UNESCO, Paris. Available: http://www.unesco.org/new/en/natural-sciences/environment/ecological-sciences/biosphere-reserves/world-network-wnbr/wnbr/ [Accessed 21 February 2012]

8

Habitat 141°: linking outback to ocean

Ben Carr

Introduction

Habitat 141° is a large-scale, multi-partner collaboration that aims to promote conservation and restoration of natural habitats in south-eastern Australia. The establishment of Habitat 141° recognised the unique set of opportunities existing in the region and the strong desire of local people and communities to act for increased conservation across all tenures. Habitat 141° works through aligning and focusing the efforts and investments of contributing organisations, and recognises the advantages this has for achieving significant and enduring biodiversity conservation outcomes. The term 'Habitat 141°' refers to both the general region in which the initiative operates, the broad alliance of organisations, as well as the proposed legal entity that will plan, coordinate and align efforts.

Habitat 141° straddles longitude 141° east along the Victorian–South Australian border and extends 500 km from the Southern Ocean in the south to north of the Murray River including a small portion of far south-west New South Wales. The region is predominantly an agricultural landscape largely cleared of native vegetation for cropping and grazing except for some large areas of public land in north-west Victoria and eastern South Australia, largely dominated by mallee (Figure 8.1).

Habitat 141° was instigated in 2005 to: (i) improve the connectivity between large areas of public land through establishing corridors; (ii) increase the amount of natural habitat through extensive restoration; and (iii) improve the functionality of existing natural habitats through management.

The vision of Habitat 141° is 'to work with communities to conserve, restore and connect habitats for plants and wildlife on a landscape scale from the outback to the ocean' (Dunn and Jancauskas 2008). The alliance recognises that it will take at least 50 years to achieve this vision in the majority of nine landscape zones. The on-ground work of partner organisations

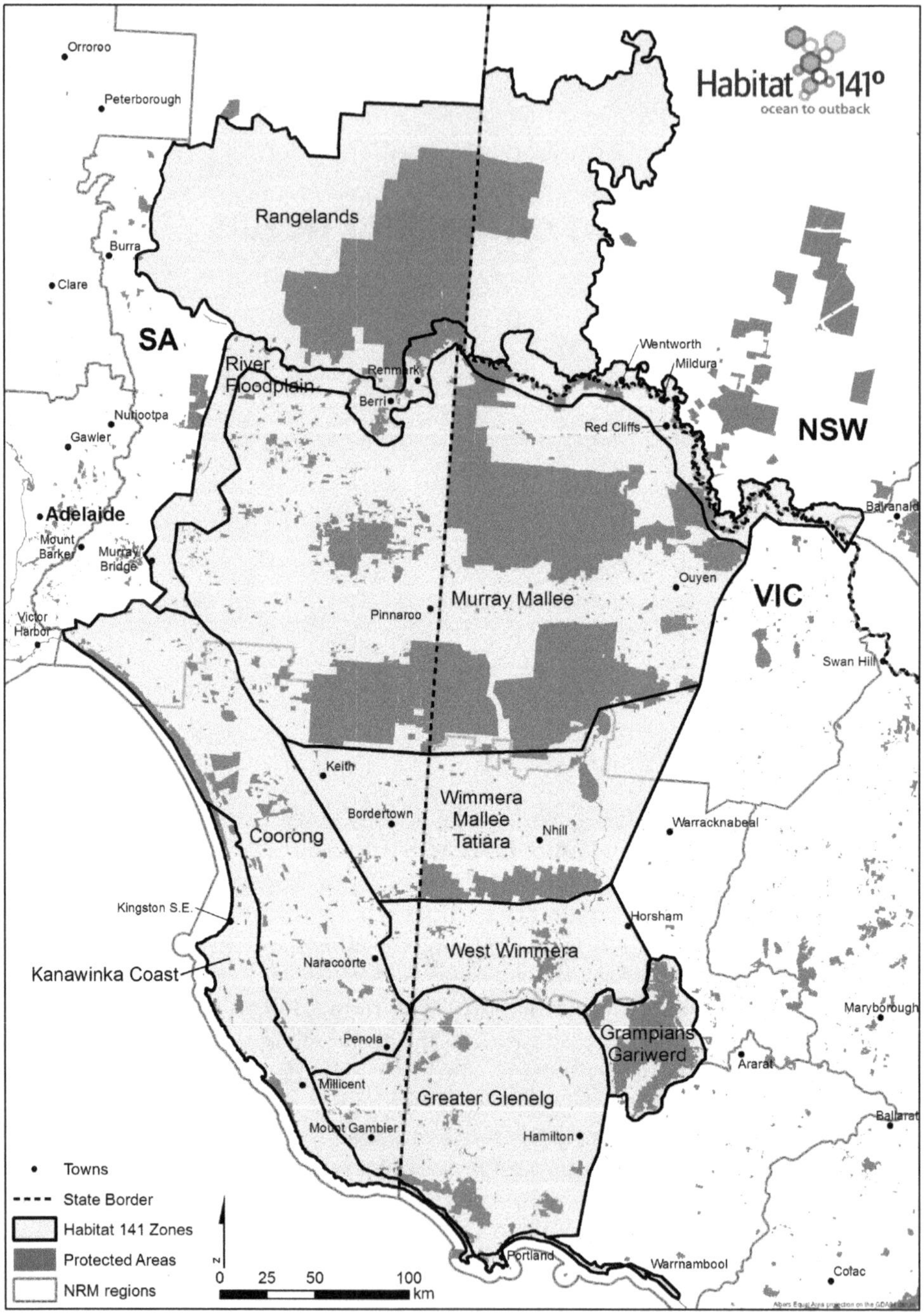

Figure 8.1 The location and extent of Habitat 141°.

will build on and maintain the outcomes of previous projects and programs. Habitat 141°'s objectives included slowing the decline of all threatened species and the recovery of some. Key threats will be managed at a landscape scale within and across landscape zones.

Habitat 141° involves many groups and organisations including landholding and land management organisations and advocacy, special interest, environmental and natural resource management (NRM) bodies. Landholders who have contributed to the Habitat 141° vision include: (i) private landholders who are undertaking conservation as well as grazing and cropping on their land, i.e. producers; (ii) private groups who own and manage land for conservation; (iii) private individuals who manage land for conservation; (iv) non-government organisations who manage land for conservation; (v) statutory authorities who manage land for conservation; and (vi) public agencies who manage land for conservation.

Habitat 141°'s conservation objectives have been gradually developed and refined from an initial focus on revegetation and habitat management. In addition to the broad objective of corridors there is a focus on recovery of threatened species and communities and an increased focus on functional connectivity. Objectives are determined through consultation and discussion between biodiversity conservation experts and groups and organisations that implement on-ground action at the landscape level. Current partners and other contributing groups are active across hundreds of individual sites on private freehold and leasehold land, public and private conservation reserves, as well as public land reserved for non-conservation purposes.

The boundaries of Habitat 141° have evolved as the initiative has grown and developed. The boundaries, originally based on biogeographic regions and subregions, have been modified to reflect catchment, land use and social/community boundaries where these influenced conservation planning. The nine zones are not intended to be rigid but are indicative delineations of each zone (Figure 8.1).

Habitat 141° covers an area of some 14.5 million hectares. Specific projects occur across a range of land tenures. Areas in which connectivity projects should be focused have been defined for most of the zones. The portion of conservation land (public as well as private) within each zone varies from ~6% in the West Wimmera region to almost 60% in the Grampians–Gariwerd region (Table 8.1). Overall, some 23% of Habitat 141° is managed for conservation.

Table 8.1. Habitat 141° extent and conservation land

Zone	Habitat 141° zone	Area of zone (thousand ha)	Area of conservation land in each zone (thousand ha)	% Conservation land in each zone
1	Greater Glenelg	1450	102	7.0
2	West Wimmera	715	39	5.5
3	Wimmera Mallee-Tatiara	1400	194	13.9
4	Murray Mallee	4600	1681	36.5
5	Kanawinka Coast	480	41	8.5
6	Rangelands	3130	845	27.0
7	River Floodplain	890	126	14.2
8	Grampians-Gariwerd	310	180	58.1
9	Coorong	1525	139	9.1
	Total	**14 500**	**3347**	**23%**

Notes: (i) Area of conservation land is based on the Collaborative Australia Protected Area Database (CAPAD) 2008 (Australian Government 2012); (ii) The figures imply a level of precision that is not intended as boundaries of the zones are not intended to be rigid and fixed; (iii) Details of conservation activity on private land within each zone, other than what is in CAPAD, are not collated by Habitat 141° and thus not included in the above figures, though details are generally available from contributing organisations.

History

Habitat 141° was initiated by non-government environmental organisations (NGOs) that recognised the conservation values of the region and saw the potential to build connectivity between large areas of public land. This concept had been originally developed by the West Wimmera Tree Group and Hindmarsh Landcare Network (Poulio 2007; Sharp and Blackadder 2008) through their vision to link the Little Desert National Park and Wyperfeld National Park. Project Hindmarsh instigated thinking about corridors and linking areas within a landscape and proponents of Habitat 141° took this concept and extended it across multiple landscapes.

The Wilderness Society's (TWS) WildCountry program, which sought to promote conservation at large scales, highlighted a 'Tri-state initiative' with the Habitat 141° region as its focus (TWS 2005). In 2005 and 2006, TWS and Greening Australia Victoria (GAV) consulted people across the region and developed a broad stakeholder partnership of NGOs and regional natural resource management organisations. Parks Victoria joined the collaboration in 2007 and assisted the development of an institutional and organisational structure and governance. GAV were able to raise significant philanthropic funding to put towards establishing and planning Habitat 141°. Substantial collaborative conservation planning was resourced primarily by GAV and on-ground activity started in the Greater Glenelg and West Wimmera Zones. All of the key groups who have played significant roles in developing Habitat 141° are still largely involved in the initiative.

The original informal committee was formed in 2005, and included TWS, GAV and Mallee, Wimmera and Glenelg Hopkins Catchment Management Authorities. It was agreed at the time that a loose steering committee structure would be an appropriate starting point. In February 2007, Conservation Action Planning (CAP) was instigated across zones in the southern part of Habitat 141° and in July 2007 a science workshop was held (Koch 2009, 2011). Parks Victoria supported engagement (Vincent 2010) and hosted a stakeholder meeting of government and large NGOs in September 2008 to firm up the vision and objectives of Habitat 141° and to establish more formalised governance (Dunn and Jancauskas 2008). This led to the employment of the initial coordinator in October 2008 and to the formation of a formal steering committee and the development and signing of a memorandum of understanding between partner groups in 2009 (Habitat 141° 2009).

During 2009 and 2010 there was significant focus and effort put into deciding on and forming a governance mechanism for Habitat 141°. A meeting was held in March 2010 of community and NGO groups to determine their input to Habitat 141° (Bradey and Alexander 2010), and a governance workshop was held in July 2010 (Habitat 141° 2010), resulting in the formation of a Governance Working Group (GWG). An Interim Executive Committee tasked with developing a Habitat 141° legal entity was formed in April 2011 (Carr and Alexander 2011), following a recommendations report on future governance arrangements (Habitat 141° GWG 2010). From July until November 2011 (when funding for coordination ceased) the coordinator reported directly to the Interim Executive Committee and was hosted by GAV. The process and outcomes of forming governance has been documented in several reports that are available on Habitat 141° website (www.habitat141.org.au).

Structure

The Habitat 141° Interim Executive Committee currently has a broad overview role and is acting in an interim capacity until the Habitat 141° legal entity is formed when the inaugural executive will be elected. As Habitat 141° (as at the end of 2012) does not have any confirmed income, nor employs any staff, it is not driving conservation planning within zones, nor funding, coordinating or supporting on-ground action. Coordination and on-ground action is continued by existing groups and networks of organisations that plan and undertake projects. Some of these partner groups are developing agreements to improve coordination and they continue to contribute their time and some of their organisations resources to the support of Habitat 141°.

Habitat 141° is a large multi-sector and multi-partner initiative and this has proven to be a great strength but also a constraint, as the views and values of many sectors and organisations need to be sought and incorporated. Habitat 141° has evolved and developed its governance since 2008 and current arrangements are still in transition. Habitat 141° has a large number of supporting organisations and strong relationships with regional NRM organisations, state and Australian governments. However, it has not yet been able to build this support into an effective organisation structure to achieve the desired function and vision. Reasons for this are outlined later in this chapter.

Major successes

The planned focus of the Habitat 141° entity is on the 'gluing not the doing'. It would act in a supportive and facilitative role, leading in some areas and supporting member groups to achieve on-ground actions towards the Habitat 141° vision. There have been significant achievements over the past 7 years and though not all desired outcomes have been realised, major foundations for Habitat 141° are now in place. Habitat 141° has ongoing support from many different sectors and organisations and from the majority of NRM organisations who are active in the region. The initiative has also received substantial support from the Victorian and South Australian Governments, both directly and indirectly through in-kind contributions from Parks Victoria, South Australian Department of Environment, Water and Natural Resources and other government agencies.

Habitat 141° has extensive involvement and support from NGOs that have provided the majority of the direct financial support to date. The philanthropic and business sectors have also been supportive in providing direct and indirect support for on-ground work in Habitat 141°. Involvement of different sectors, types and sizes of organisations is a considerable advantage; however, it has also brought some challenges to its development. Habitat 141° through its partner groups has successfully undertaken Conservation Action Planning (CAP) using the framework developed by The Nature Conservancy (TNC 2007). This approach brings professionals together with the individuals, groups and organisations who are going to implement the plans. CAP planning uses a collaborative and participatory approach that increases understanding of the plan and drives its implementation. A CAP has been prepared for the Greater Glenelg zone and CAPs for two other zones are at draft stage (Koch 2010, 2011).

Habitat 141° has built the basic architecture of a multi-partner landscape-scale initiative including:

- definition of nine zones and recognition that some will be developed sooner or to a greater extent than others;
- conservation planning – undertaken in active zones;
- identifying spatially explicit areas of connectivity in which conservation investments should be targeted;
- engagement with stakeholder groups including discussion and communication with over 25 different organisations involved in on-ground delivery across three jurisdictions;

- a high degree of awareness of, and support for the initiative, its key players and directions across state governments in Victoria and South Australia as well as the Australian Government;
- using the zones as the operational level for collaborative planning and action;
- recognition by Habitat 141° partner groups of the potential for the carbon market to drive landscape-scale biodiverse restoration and carbon sequestration outcomes (see URS 2010); and
- development and documentation of a governance model and implementation of this as it continues to evolve.

There are many on-ground projects and programs that are directly contributing to the Habitat 141° vision. Some of these have occurred as a direct result of Habitat 141° action, while other initiatives were already underway and are now being refined or focused to achieve targeted actions identified through the conservation action planning process. Unfortunately, Habitat 141° has not consistently captured and recorded the on-ground outcomes achieved by partner groups. To date this has been undertaken at the individual zone level by some landscape zones. Habitat 141° is also conscious of not appropriating the outcomes achieved by on-ground implementation groups as Habitat 141°'s outcomes, without their explicit approval.

A wide range of outcomes have been achieved across the nine zones. Significant on-ground outcomes have been achieved within the West Wimmera Zone, including restoration of over 450 ha and management of 1850 ha of regionally significant and endangered vegetation communities. Collaborating groups and landholders have established at least 10 new conservation covenants and ongoing conservation work is underway on at least another 35 sites. There has been extensive engagement with volunteers in endangered species surveys and with business, environmental and conservation organisations in the on-ground works.

Major lessons

There are some important lessons from the development of Habitat 141° over the past 7 years.

It is apparent that the Habitat 141° vision exceeded the collective capacity of stakeholders to achieve that vision. There is a further gap between the capacity of people and organisations and the scale of investment needed to achieve the vision. The number of different sectors and groups involved significantly increased the complexity and length of decision-making processes. With the benefit of hindsight, a staged evolution starting with a smaller number of groups working at a smaller scale would have been more efficient and achievable. The desire to achieve Habitat 141°'s vision should be matched by increases in the capacity of the initiative and its supporting groups. An unintended consequence of extensive engagement is perhaps too much involvement without commensurate accountability and responsibility by supporting groups to achieve the structures and processes that are necessary to achieve the vision.

Governance of Habitat 141° was never going to be quick, easy or simple given the size of the initiative, the range of sectors involved and the number of organisations and interests represented. One factor influencing this was that some organisations were unclear about their interactions with Habitat 141° and expectations of how it operates, and what it could achieve. This made the process of establishing a Habitat 141° entity difficult. The reasons for, and functions of, governance have been carefully considered and documented by the

Habitat 141° GWG and Interim Executive Committee, which could provide valuable lessons for other large-scale collaborative initiatives that are investigating governance and forming governance structures.

The importance of adequate and clear communication by Habitat 141° has been recognised, but was not always demonstrated through the development process. There was not always sufficient or adequate communication with stakeholders during the process of developing governance. This was caused in part by the large number of people and organisations involved and capacity limitations arising from the scale of interactions and timing of engagement. The capacity for local groups to be engaged across such a big area was a major challenge. While many indicated that they wanted involvement – 'Habitat 141° is us!' was a comment from the community workshop – the resources to engage them were never sufficient. This led to some disengagement and lack of confidence in Habitat 141°'s ability to achieve its grand vision. This could have been resolved by initially focusing on a much smaller geographic area, such as one or two of the nine landscape zones, so that the available capacity matched the scale of the task until capacity was able to be increased.

Experience in forming and developing Habitat 141° indicates that collaboration is not always the normal way of doing business in conservation. Collaboration has to be learnt, developed and practised between all contributors. Although the need to work together is generally recognised, active collaboration is not always demonstrated and there is also competition between organisations for funding and recognition. Collaboration for landscape-scale conservation outcomes needs to be adequately supported and guided. Growing the capacity of Habitat 141° will involve growing the capacity of organisations to collaborate combined with growing the capacity of a central body to provide coordination and leadership.

Different governance structures were discussed with the consensus that Habitat 141° should form an independent legal entity whose role was to: (i) lead the alliance; (ii) communicate among and between stakeholders; (iii) facilitate planning and assist; and (iv) support groups achieve on-ground outcomes. Habitat 141°'s identity is distinct and different from any of its founding or funding organisations. Its identity reflects an agreed set of functions and values for the initiative that were decided while developing the agreed governance model (Habitat 141° GWG 2010). The functions and values of Habitat 141° were designed to make it distinct from any of the existing organisations. A key function was to provide leadership in a way that none of the existing organisations could. A new organisation was the most effective means of achieving this.

Applying the lessons

Habitat 141°'s powerful vision has attracted many different organisations. People's desire to understand, share and contribute to this vision is a significant opportunity for landscape-scale conservation. Organisations were encouraged to align and contribute their actions and to focus their on-ground efforts. However, this does not necessarily mean that all groups and organisations who want to be involved in achieving the vision need or should be involved in establishing and managing governance arrangements. Developing governance and capacity across different organisational perspectives, values and cultures and trying to achieve common understanding across more than 20 organisations has been difficult, time consuming and frustrating. Now that this work has been largely done it is essential that it is not wasted and is put into effect.

There was a lack of sufficient capacity to build communication and relationships and develop these into successful collaborative partnerships and governance arrangements. One of the contributing reasons for this is the large area and number of organisations involved.

Another factor has been the need to build partnerships across multiple scales including at the zone and whole of Habitat 141° scale simultaneously. An establishment period should have been more explicitly recognised including acknowledgement that Habitat 141° would not be able to undertake all of its proposed functions including planning, communications, partnership building, information sharing, and outcome documentation during its establishment. Given the scale and complexity of large collaborations such as Habitat 141°, a small but guaranteed source of funding to establish a small secretariat comprising of a director and communication manager for a 3 to 5 year period is needed during the crucial start-up period. There is a role for funding support from government, business and philanthropic sources. Support from a variety of sources would help balance year-to-year input from any one organisation or sector.

There needs to be greater recognition of the need for these 'gluing' functions that then allows other groups to focus on achieving on-ground outcomes. One difficulty is that NRM funding from governments for coordination and facilitation of landscape-scale projects has been limited except when directly tied to on-ground implementation. Without funding for 'gluing functions' the capacity to grow large-scale initiatives will be limited. Habitat 141°'s focus on achieving alignment through collaboration and not necessary through funding is a new role in natural resource management. The role that Habitat 141° should play in achieving and directing funding to on-ground works has been difficult to agree and define. Participating organisations generally will not access funding from the Habitat 141° entity but they will be able to achieve and demonstrate their alignment and outcomes through Habitat 141°. Organisations were wary of Habitat 141° becoming another funding body or another level in existing funding chains. Habitat 141°'s key advantage was seen in bringing in new funding streams that have not traditionally been available to NRM and also in significantly increasing the quantity of funding available.

Conclusion

Habitat 141° has realised that its powerful vision does not necessarily create the mechanisms needed to implement that vision. The proponents of Habitat 141° and all of its supporters did not realise how complex the task of building the mechanism and achieving governance would be and how long this would take.

The complexity of the institutional framework underpinning Habitat 141° arises from its scale, the range of jurisdictions and number and types of groups involved. Habitat 141° has engaged and involved the government, non-government, philanthropic and for-profit sectors. It is probably the largest and most diverse landscape-scale collaboration for biodiversity conservation outcomes in Australia in terms of the numbers and diversity of organisations involved. This extensive engagement, however, has made it difficult to reach agreement on what activities it should undertake and what governance will enable it to achieve its vision. The extensive work and governance undertaken to date will provide a sound basis for the future growth and development of Habitat 141° as well as providing some valuable learnings for the development of similar initiatives.

Acknowledgements

I'd like to acknowledge the work of the Governance Working Group and all members of the current Interim Executive Committee. Andrew Bradey, Ian Walker, and Karen Alexander have all provide their experience and assistance. Without the input of all individuals and groups that have readily given their time and expertise and, in some cases, funding, Habitat 141° would not have made the progress it has. However, this description and analysis of Habitat 141° are my views alone.

Biography

Ben Carr has worked in a variety of policy, program and project roles in government and non-government organisations. He has worked with Gondwana Link and Kosciuszko to Coast and was Coordinator of Habitat 141° from August 2010 to October 2011.

References

Australian Government (2012). Protected Area Information. Department of Sustainability, Environment, Water, Population and Communities, Canberra. Available: http://www.environment.gov.au/parks/nrs/science/capad/ [Accessed 12 March 2012].

Bradey A and Alexander K (2010). Report on community meeting of local groups in Habitat 141°. March 20–21, 2010, Little Desert Lodge. Unpublished report of outcome of a workshop organised by Greening Australia (Vic), The Wilderness Society, and the Victoria Naturally Alliance. Available: http://www.habitat141.org.au/wp-content/uploads/2011/06/Victoria-Naturally-Alliance_Mar2010_LittleDesertCommunityForumReport.pdf [Accessed 12 March 2012].

Carr B and Alexander K (2011). Report on stakeholder meeting deciding governance arrangements for 141° 'entity'. April 19–20, 2011, Little Desert Nature Lodge. Unpublished report compiled by Habitat 141°. Available: http://www.habitat141.org.au/wp-content/uploads/2011/07/Habitat141_April2011_Finalising-Governance-for-Entity.pdf [Accessed 12 March 2012].

Dunn J and Jancauskas E (2008). Habitat 141° Alliance Meeting, Dunkeld, 9–10 September 2008. Post Meeting Documents Version 2, 8 October 2008. Unpublished report produced by Now for Future Pty Ltd on behalf of the Habitat 141° Alliance. Available: http://www.nowforfuture.com/projects/74645/74645_resources/71615_booklet2.pdf [Accessed 12 March 2012].

Habitat 141° (2009). Habitat 141° Alliance Agreement Final March 2009. Available: http://www.nowforfuture.com/projects/89645/89645_resources/H141_Agreement.pdf [Accessed 12 March 2012].

Habitat 141° (2010). Outcomes of Habitat 141° Workshop, Sylvania Park Mohair Farm, Horsham 26–27 July 2010. Functions, Principles, Feedback on Governance Criteria. Unpublished reports compiled by Habitat 141°. Available: http://www.habitat141.org.au/resources/ [Accessed 12 March 2012].

Habitat 141° GWG (2010). Report and Recommendations of the Habitat 141° Governance Working Group to the Habitat 141° Alliance 2010. Unpublished report prepared by Ben Carr, Habitat 141° Coordinator, on behalf of the Governance Working Group. December 2010. Available: http://www.habitat141.org.au/wp-content/uploads/2011/03/habitat141-governance-report-recommendation-dec-2010.pdf [Accessed 12 March 2012].

Koch PJ (2009). Workshop Report: Conservation at Large Scales Ecological Research Workshop to inform Habitat 141° Planning. Available: http://www.habitat141.org.au/wp-content/uploads/2011/03/habitat141-science-workshop-report-final.pdf [Accessed 12 March 2012].

Koch PJ (2010). Habitat 141° Kanawinka Coast Zone 5 Conservation Action Plan, Unpublished Conservation Plan Version: 7/02/11. Available: http://conpro.tnc.org/1652/ [Accessed 12 March 2012].

Koch PJ (2011). Habitat 141° Zone 1 'Greater Glenelg' Conservation Action Planning, Summary Report, Feb 2011. Available: conpro.tnc.org/1447/Habitat_141_Zone_1_CAP_Report_Feb2011_v3.pdf/download_public [Accessed 12 March 2012].

Poulio M (2007). *Project Hindmarsh – Ten Years and Beyond: Growing the Future Together.* Hindmarsh Landcare Network, Nhill, Victoria.

Sharp A and Blackadder J (Eds) (2008). Making a Difference – A Celebration of Landcare. Department of Agriculture, Fisheries and Forestry, Canberra. Available: http://www.daff.gov.au/natural-resources/landcare/publications/making_a_difference_a_celebration_of_landcare [Accessed 12 March 2012].

TNC (2007). Conservation Action Planning Handbook: Developing Strategies, Taking Actions and Measuring Success at any Scale. The Nature Conservancy, Arlington, USA. Available: http://www.conservationgateway.org/sites/default/files/Cap%20Handbook_June2007.pdf [Accessed 12 March 2012].

TWS (2005). *WildCountry: A New Vision for Nature.* The Wilderness Society, Canberra. Available: http://www.wilderness.org.au/files/WildCountry-a_new_vision_for_nature2005.pdf [Accessed 12 March 2012].

URS (2010). A Natural Investment Project: A Regional Development Case Study. Final Report for Victoria Naturally Alliance. Prepared by URS Australia Pty Ltd. Available: http://www.habitat141.org.au/wp-content/uploads/2011/07/URS_2010_A-Natural-Investment-Project-Case-Study.pdf [Accessed 12 March 2012].

Vincent G (2010). Habitat 141° Stakeholder Interviews. Unpublished Report Prepared by GW Vincent and Associates for the Habitat 141° Interim Steering Committee. Available: http://www.nowforfuture.com/projects/89645/89645_resources/Interviews_and_Governance_Options.pdf [Accessed 12 March 2012].

9

Midlandscapes: matching actions to opportunities in landscape conservation in the Tasmanian Midlands

Stuart Cowell, Andrew Cameron, Daniel Sprod and Matt Appleby

Introduction

The Tasmanian Midlands (hereafter the Midlands) are arguably one of the most significant, yet least appreciated, landscapes in Australia. Typically, priority conservation landscapes in Tasmania are seen as those with tall wet forests or remote mountains and moraines. While these are important, the grassy ecosystems of the Midlands remain as one of the most under-protected landscapes in Australia with less than 4% of the bioregion protected.

There is a long history of landscape-scale and connectivity initiatives in the Midlands, each building on the work of its predecessors. Despite these efforts, there has been a concomitant decline in the extent and integrity of the grassy ecosystems, and the species that depend on those ecosystems. The Midlandscapes project is aiming to reverse that trend, and see conservation integrated into the predominantly agricultural focus of the region.

Midlandscapes is focused on the Tasmanian Midlands Biodiversity Hotspot, one of 15 biodiversity hotspots in Australia (Australian Government 2012), and encompasses the lowland plains and foothills of the Midlands up to an altitude of ~600 m between the Eastern Tiers, North Eastern Highlands and the Western Tiers and Central Plateau (Figure 9.1). The total area of the Midlandscapes project is 640 900 ha with a total of 120 000 ha of conservation assets, priority vegetation communities outside the current reserve system, identified by the project team through a Conservation Action Planning process. Midlandscapes aims to see 64 050 ha managed with a primary conservation focus by 2020.

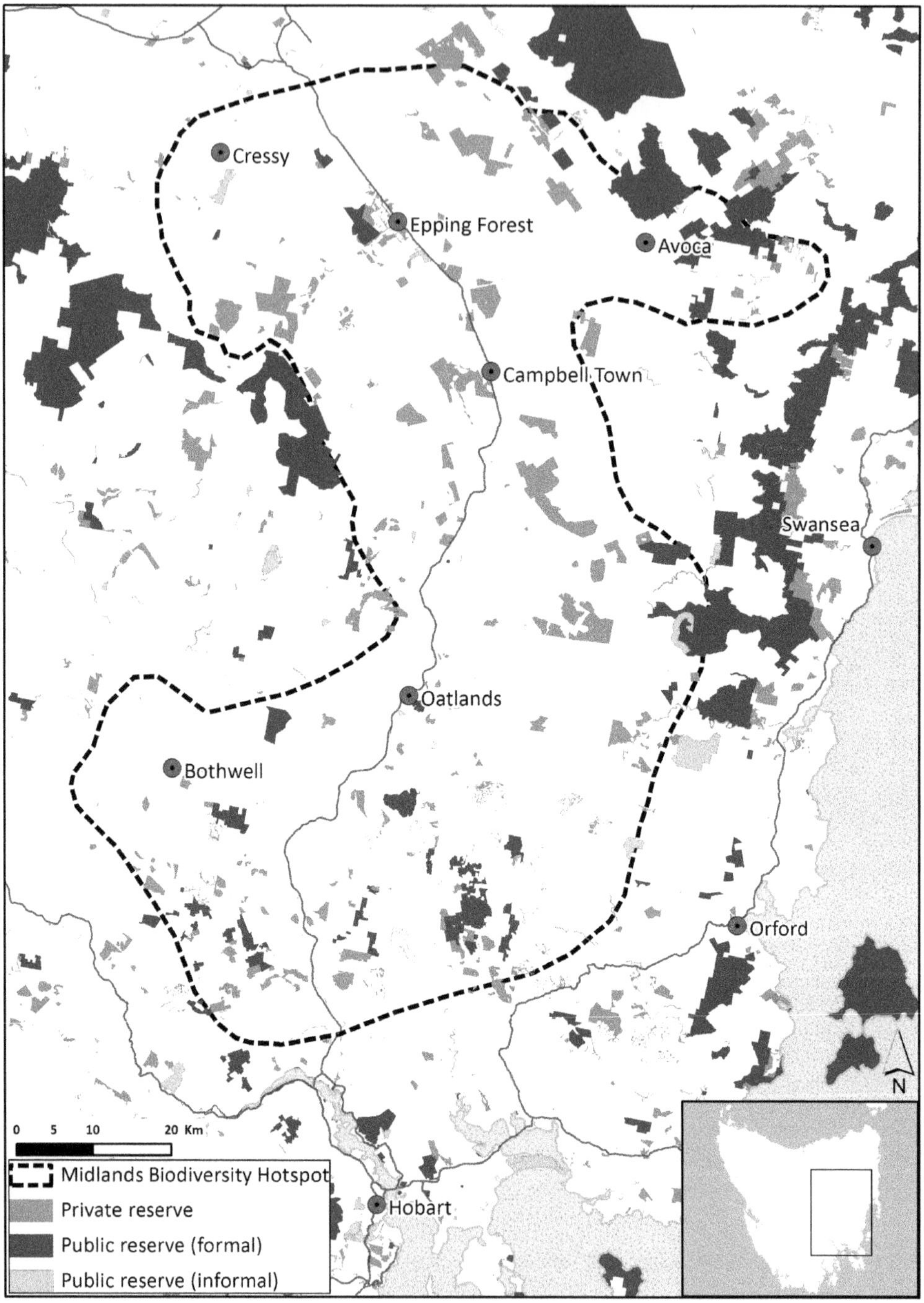

Figure 9.1 Location and conservation lands in the Tasmanian Midlands region.

The landscape of the Midlands is a mosaic of cleared land, forests, grassy woodlands, wetlands and native grasslands. The areas of remaining native vegetation are recognised as important contributors to the long-term functioning ecology of the region. The Midlands contain at least 12 endemic species, 32 nationally threatened species and more than 180 plants and animals listed as threatened at the state level – particularly daisies, lilies and orchids. Twelve wetlands are listed on the Directory of Important Wetlands in Australia and 10 wetlands are of regional significance.

The Midlands is home to one of the Australia's most endangered and least protected ecosystems: temperate grasslands (Kirkpatrick *et al.* 1995). In the Midlands these native grasslands principally occur on valley floors and have been reduced to less than 10% of their original extent. Those that remain support an astonishing array of plant diversity.

The Tasmanian Midlands is also an important cultural landscape. Aboriginal peoples managed the landscape with fire and maintained the Midlands as open woodland and native grassland (Kirkpatrick *et al.* 1995). Early colonists saw the region as ideal for pastoral and agricultural pursuits. The Tasmanian Midlands were among the earliest areas colonised in Australia and were almost entirely in private ownership by 1820. Land rarely changes hands outside of the families that originally settled the area and a strong sense of place and stewardship has developed over the last 200 years. Many landowners are interested in participating in conservation activities (see for example Gilfedder and Kirkpatrick 1997).

Established in 2008, the Midlandscapes project was developed to bring together several conservation 'actors' in the Midlands landscape to work towards a coordinated approach. This approach was influenced by analogous work in the Gondwana Link (Chapter 3) and Kosciusko to Coast projects (see Chapters 13 and 18), and was seen as a way of increasing the effectiveness of conservation efforts.

A landscape-scale approach to conservation is the most appropriate approach to protecting areas of high conservation value in the Midlands. The Midlands is a 98% privately owned 'working' landscape with diverse and unique conservation assets occurring throughout. These assets, for example temperate grasslands, native fauna, and grassy woodlands, are interdependent and cannot be dealt with without considering overall ecosystem function at landscape scale.

Midlandscapes is intended to (i) facilitate the creation of a conservation vision for the Midlands including a landscape-scale conservation plan; (ii) develop an income stream for conservation management including an investment fund and market based tools; and (iii) raise awareness of biodiversity values in the region.

The vision for Midlandscapes is: *Healthy natural ecosystems within the working landscapes of the Tasmanian Midlands.* The project objective for Midlandscapes is: *10% of the Tasmanian Midlands Biodiversity Hotspot managed primarily for biodiversity conservation by 2020, comprising 64,000 ha of six ecological communities and one fauna habitat which have been identified as the Key Conservation Assets.*

History and structure

As previously noted, there have been many precursor projects leading to the development of the Midlandscapes initiative. Key initiatives include:

- *Whole Farm Planning* demonstration farms (1980s) supported by Greening Australia and funded through the National Soil Conservation Program to show how native vegetation could enhance a farm's sustainability;
- *Midlands Habitat Corridor* (1990s) promoted by Greening Australia and funded by Save the Bush which focused on the long-term viability of native flora and fauna populations on private land by the establishment of a habitat corridor in the northern Midlands;

- *Upper South Esk Corridors and Green and Derwent Valley Corridors of Green* (also Greening Australia in the late 1990s) aimed to encourage regional-scale native vegetation management to achieve ecological and social benefits;
- *BushWeb* (late 1990s, early 2000s) funded by the Australian Government's Natural Heritage Trust and supported by local government (the Northern Midlands, Southern Midlands and Break O'Day Councils) to help landowners protect remnant vegetation and rehabilitate vegetation in poor condition while maximising benefits for wildlife; and
- *Non-forest Vegetation Program* (mid 2000s) also funded by the Natural Heritage Trust, involved the Tasmanian Government through the then Department of Primary Industries, Water and Environment, and the Tasmanian Farmers and Graziers Association. The Non-forest Vegetation Program aimed to protect threatened and under-reserved, non-forest native vegetation on private land, with a particular focus on native grasslands and working with landowners to provide benefits for conservation and primary production. The program secured over 20 000 ha under conservation covenant.
- In 2007 and 2008, the Maintaining Australia's Biodiversity Hotspots (MABH) Program (DEWR 2007) provided funding to the Tasmanian Land Conservancy (TLC) for delivery of the Midlands Biodiversity Hotspot Tender. The Tender provided an opportunity to road-test innovations developed under Midlandscapes.

Midlandscapes is led by a partnership of the TLC, Bush Heritage Australia (BHA) and the Tasmanian Department of Primary Industry, Parks, Water and Environment (DPIPWE). There is a formalised memorandum of understanding between TLC and BHA, with DPIPWE as an agreed partner. The partners have developed an agreed program of activities supported by the Midlands Conservation Action Plan and an accompanying Business Plan. Midlandscapes is directed by a Steering Committee comprising senior management staff from each of the partner organisations.

A Technical Working Group comprising NGO, state and research scientists provides advice to the Steering Committee on conservation prioritisation, landscape-scale mapping and identification of conservation assets, focal landscapes and potential landscape linkages. A coordinator oversees implementation and operations under direction from the Steering Committee.

The full-time coordinator is employed by TLC and jointly funded by TLC and BHA. The partner organisations make available administrative staff and specialist ecologists when required during stages of program delivery. These partner staff include operational officers for landowner liaison, on-ground site assessments, landscape ecological analysis (including site mapping), development of metrics to manage outcomes measurement, and conservation prioritisation.

Project skills are drawn from a pool of existing permanent or contract staff with working knowledge and experience with the Midlands and its community.

Major successes

The Midlandscapes project is, as with many landscape-scale conservation projects in Australia today, still developing and growing. Therefore, its successes relate to the achievement of those things that are markers along the path to overall project success.

In common with many landscape connectivity projects throughout Australia (see for example Worboys and Pulsford 2011), Midlandscapes is characterised by four key features that form the fundamental building blocks of any landscape project that is intent on success.

1. Collaboration not competition

Self-evident to many practitioners in this space, but still anathema to many policy makers and agencies in their self-serving pursuit of efficiency through competition, successful collaboration is essential to operating across multiple tenures, issues, stakeholders, resource needs and scales.

Successful collaboration requires patient investment and willingness to pursue the participation of others. Not to be confused with 'stakeholder consultation', where those with an interest no matter how passing or detrimental to the project goals are given an opportunity to influence the outcomes, collaboration is a focus on building mutual success (see Hoskins and Angelica 2005).

Midlandscapes has required that each of the core project partners (TLC, BHA, DPIPWE) become a successful collaborator, and significant effort has been made on relationship building, and mutual reinforcement of roles. Not without occasional tension, the collaborative nature of the partnership has allowed more open sharing of time and other resources, and increased project resilience by allowing for the fluctuating cycles of organisational capacity of the partners.

A competitive approach would not allow these opportunities.

2. A clear collective vision and action plan

Midlandscapes, as one of its early activities, invested in the development of a Conservation Action Plan and a Business Plan to guide implementation. Both were developed with active participation and input from a core group of landholders with an active interest in securing conservation investment into the region. Critically, the plans focused not just on the conservation assets and their 'protection', but also on developing a clear pathway for directing investment, and measuring the success of that investment.

The non-government partners in particular had previous experience with the Conservation Action Planning tools developed by The Nature Conservancy (e.g. Pasquini *et al.* 2011), and their absolute focus on pragmatic action and strategic resource allocation.

The process of developing the plan was then used to engage with landholders as a way of building a shared understanding and vision for the conservation goals of the project. The early collaborative development of the plan helped provide a common language for subsequent actions.

3. A big toolbox with willing 'tradies'

Allied particularly to the need for collaboration is the need to have a range of conservation tools to apply in the landscape. This is essential for three key reasons:

- i No tool is applicable in all situations (e.g. acquisition), and achieving the outcome at a site and landscape scale will require different approaches;
- ii Working at the landscape scale, particularly in a production landscape, will necessarily involve dealing with different property rights and other legal instruments, that will require sometimes novel responses (for example, over 90% of the Midlands is in private hands, with some farms still with the original granted families); and
- iii A variety of tools and options provides a greater opportunity to access a broader range of income sources, reducing the fragility of what will be complex long-term projects.

As an example, the Midlandscapes collaborators brought together market-based tender approaches through the Midlands Biodiversity Hotspot Tender and the Forest Conservation Fund,[i] offsets funding, for example through the Roaring 40s Eagles Nest Protection Program,[ii] property acquisition through both 'purchase and hold' and 'revolving fund' approaches, conservation covenants, and stewardship funding through the establishment of a private perpetual fund (the Midlands Conservation Fund), to support mid-term stewardship contracts through payment for ecosystem services (see Table 9.1 for a breakdown of conservation lands in the Tasmanian Midlands Hotspot).

4. Leadership

While collaboration is essential, there is a critical leadership role required to direct and drive the overall program. In the case of the Midlands, leadership has been provided by many individuals and organisations over time. The Midlandscapes partners formalised the role through the establishment of the Midlandscapes coordinator. In addition, informal leadership for innovative approaches in the Midlands continues to be provided by key individuals (landholders, policy makers and researchers).

It is insufficient for that role to be delivered by committee, or to be assumed, and without it a complex and difficult undertaking such as a landscape project will not be able to progress.

Table 9.1. Conservation mechanisms on private and public land in the Tasmanian Midlands hotspot

Conservation on private land by program	**Area**
Private Forest Reserves Program 1997–2006 (PFRP)[a]	10 571
Non Forest Vegetation Program (NFVP)[a]	6781
Forest Conservation Fund (FCF)[b]	7499
Forest Conservation Fund Direct Approach (FCFDA)[c]	3357
Midlands Biodiversity Hotspots Project (BHP)[d, 1]	1391
Midlands Biodiversity Hotspots Tender (MBHT)[c, 2]	6602
Protected Areas On Private Land (PAPL)	149
FCF Revolving Fund (RevFund)[c]	640
Roaring 40s Eagle Nest Protection Program (R40s)[c]	80
All conservation areas on private land	**37 070**
Conservation on private land by tenure	**Area**
Private land in perpetuity (on Tasmanian Reserve Estate TRE)	25 012
Private land variable term agreements (on TRE)[3]	10 715
TLC land in perpetuity (not on TRE)	580
TLC stewardship contracts (not on TRE)[4]	763
All private reserves in Hotspot	**37 070**
All public reserves in Hotspot	**32 704**
Total	***69 774***

Host organisation: (a) Tasmanian Department of Primary Industries, Parks, Water and Environment; (b) Australian Government Department of Water, Heritage and the Arts; (c) Tasmanian Land Conservancy; (d) Southern Midlands Council

Notes: (1) Includes input from NFVP & PAPL; (2) Includes funding from NFVP & FCF; (3) 5 to 48-year covenants or Vegetation Management Agreements (VMAs); (4) MBHT 6 &12-year contracts

Early identification of the need, and specific tasking of an individual to perform the role, should be one of the first steps taken by any team wishing to pursue these approaches.

An early and sustained investment by the Midlandscapes project in the specific role of coordinator, working in collaboration with other leaders within the Midlands, has maintained a strong and consistent focus.

Major lessons

Perhaps one of the strengths of those projects that appear – at least on the surface – as more resilient to changing funding and policy trends, is that they both survive and grow from the errors they inevitably make. Midlandscapes is no exception. Although to date the project has been fortunate to avoid truly threatening crises and errors, there are several things that could have saved both time and effort had they been more thoroughly considered.

1. We come in peace

It is not possible to put too much time into clarifying your objectives and intentions, and to continue to reiterate those. Several misunderstandings were precipitated, and with significant consequence, through what were misunderstandings of perspective or intent both between project partners and between partners and landholders. This is perhaps one of the most 'over learned' lessons in collaborative conservation projects.

2. It's going to take how long?

In project terms, a landscape approach is going to take some decades to generate tangible benefit that can be seen as truly sustainable. In building support for such projects, there is a strong temptation to 'bring forward' the benefits when discussing the investment timelines required to deliver those benefits. Ultimately this hampers future success by creating an impression of poor management.

The construction of a robust program logic, in the case of Midlandscapes through the Conservation Action Planning process, was critical for all parties to build confidence in the actual timelines required, and in mapping early progress towards the very distant goal.

3. Going under the bus

For some time, significant intellectual and relationship capital was resident in only one or two people involved in the project, notwithstanding broader participation by many parties. While it is inevitable that there will be some concentration of knowledge and relationships, this can create fragility. The transition of staff created significant and unnecessary relationship pauses because insufficient investment was made in ensuring the project was sustained by a more robust set of relationships.

4. Oh no, not again!

There is already a significant history of plans and planning in the Midlands. While these plans have certainly built goodwill within parts of the Midlands community, and may have slowed the rate of ecological decline (although this is untested), it is clear that the key conservation assets in the Midlands have continued to decline irrespective of this effort (see for example the advice for listing Lowland Native Grasslands of Tasmania under the *Environment Protection and Biodiversity Conservation Act 1999*).[iii] This inevitably generates planning fatigue for both project proponents and landholders, reinforced by a certain level of plan 'wallpapering', whereby a new plan is simply laid over the top of the old with perhaps cursory recognition of earlier work – although some effort was made in Midlandscapes to engage and include earlier leaders and planners to maintain some continuity.

Midlandscapes was different from previous approaches in the Midlands in the scale of its vision and private sector investment in defining the opportunity and seeking to resource that opportunity. Existing plans were indeed insufficient for that need. Nevertheless, the existing palimpsest of plans for the Midlands created an understandable reluctance to participate on the part of community members and landholders.

Applying the lessons

The Midlandscapes project itself, and the current focus of activity on the establishment of sustainable finance mechanisms through the Midlands Conservation Fund, reflect a response to the opportunities and issues identified above.

The initiating conceptual model of Midlandscapes was one of establishing a conservation reserve in the landscape using the traditional tool of securing control through acquisition, and then building relationships from that base (a strategy that has worked elsewhere, see Chapters 3, 7, 12; Pasquini *et al.* 2011). The ongoing failure of the acquisition model in the Midlands, despite significant available resources to purchase land and significant resources placed into pursuing land to purchase, precipitated a reconsideration of that core conservation model.

This was further influenced through discussions with key landholders about alternative approaches to achieve a conservation result in the landscape.

The Midlandscapes project has therefore shifted focus to more actively supporting conservation management as a service on private lands not managed primarily for conservation, initially through the Midlands Conservation Fund, a sustainable finance mechanism. More recently its focus is on considering such options as funding for carbon sequestration in combination with the Midlands Conservation Fund.

Considered in purely financial terms, management of land for conservation is seen by many as a liability. There are, of course, substantial benefits that accrue to society and nature as a whole, and those benefits are, at least intrinsically, valued. Private organisations, such as TLC and BHA, exist for the purpose of translating that intrinsic value into resourced actions undertaken for their own sake (Cowell and Williams 2006). They are a financial expression of a community desire to see a greater level of conservation in the landscape than exists in the public conservation estate.

Nevertheless, the social expectation of conservation management 'in perpetuity' places a substantial resource burden on the manager, whether the land is managed primarily for production or for conservation. The ongoing costs of production management are of course incorporated into the price structure of the goods produced, and these are able to vary over time. The same is not typically true of conservation management. While initial funds are typically provided to secure access to the conservation resource and to manage it into the future, these costs inevitably continue to grow beyond that provision, even in well-run and managed agencies. This can create both financial and ecological risk as resourcing fails to 'keep up' with management needs.

Private conservation organisations and their supporters explicitly or implicitly accept this risk as part of their business on their own lands, trading off the opportunity to increase the extent of conservation lands against the certainty of having resources to manage an existing estate. Other landholders within a broader landscape may not be willing to accept that risk themselves, or allow conservation organisations to, in effect, shift their own risk burden onto those landholders without compensation.

The critical issue facing landscape projects, and Midlandscapes is no exception, is to be able to sustainably resource, for all parties, the management of that risk.

Conclusions

Midlandscapes is not what might be thought of as a 'typical' landscape connectivity project in the current (at 2012) proclivity for 'corridors' and catchy alliterative titles 'from ... to ...' projects. Rather, the focus of Midlandscapes is on allowing significant biodiversity values to persist, and to be valued by landholders as part of a cultural and production landscape. Achieving this requires a focus at the landscape scale, and requires connectivity as a strategy.

Adaptation of strategy in the absence of a plan is simply 'making it up as you go along', responding to the push and pull of funding opportunities and circumstance. An explicit focus on developing a clear plan and vision has allowed the Midlandscapes project the opportunity for confident strategy revision in the face of changing circumstances while maintaining a focus on the end goal.

Without strong collaboration, a clear vision and intent, willingness and capacity to learn and adapt, and clear leadership, landscape projects such as Midlandscapes cannot succeed, even in their establishment phase.

Acknowledgements

This paper is based on the authors' extensive knowledge of the project, and a tectonic library of grey literature, such as emails, funding proposals and reports. We thank all those people who contributed to the creation of that library. As is the case with all successful landscape-scale projects, there is a very long list of dedicated and passionate people that keep these projects moving and maintain their vision. In particular we would like to acknowledge Louise Gilfedder for her untiring dedication to the Midlands conservation values, and the people that manage them. We also thank the many landholders who, with patience and good humour, have continued to help us see the rich tapestry of Midlands' values and needs.

Biographies

Stuart Cowell has worked in many natural resource management sectors, including pro duction, research, not-for-profit conservation and Indigenous land management over the last 25 years. He was a senior manager with Bush Heritage Australia from 2001–2010 where he led development of the Beyond the Boundaries program. He now works as a consultant supporting the development of landscape management plans and capacity.

Andrew Cameron is currently the coordinator of the Midlandscapes project and has been responsible for coordinating the implementation phase of the project since 2008–09. Since 1998 Andrew has worked as a negotiator and consultant on number of private land conservation programs including – the Private Forest Reserve Program, the Non Forest Vegetation Project and the Midlands Biodiversity Hotspots Tender. Andrew lives in the Midlands and maintains a close connection with its farming community through managing his family's sheep grazing and conservation property in the Nile Valley. He was a member of the Tasmanian board of Greening Australia from 1997 to 2007.

Daniel Sprod is Landscape Ecologist for the Tasmanian Land Conservancy with special interest in landscape-scale conservation planning and prioritisation. Current research topics are modelling woodland bird distribution, carbon methodologies and modelling connectivity for vulnerable mammals and birds. He has developed conservation management plans for biodiversity and sustainability, from the farm level, through regional, catchment and state scales.

Matt Appleby has a PhD and research background in the field of restoration ecology. Matt has worked on planning and implementation of both the Midlandscapes and Kosciuszko to Coast projects in addition to providing ecological input to management of Bush Heritage reserves across southern Australia. Matt is currently the Ecologist Coordinator for Bush Heritage Australia.

Endnotes

i. http://www.environment.gov.au/land/forestpolicy/fcf/index.html
ii. http://www.tasland.org.au/majorprogrammes/r40s
iii. http://www.environment.gov.au/biodiversity/threatened/communities/pubs/74-listing-advice.pdf

References

Australian Government (2012). Australia's 15 National Biodiversity Hotspots. Available: http://www.environment.gov.au/biodiversity/hotspots/national-hotspots.html [Accessed 20 July 2012].

Cowell S and Williams C (2006). Conservation through buyer diversity: a key role for not-for-profit land-holding organizations in Australia. *Ecological Management & Restoration* **7**, 5–20.

DEWR (2007). Maintaining Australia's Biodiversity Hotspots. Department of Environment and Water Resources, Canberra. Available: http://www.environment.gov.au/biodiversity/hotspots/publications/programme.html [Accessed 20 July 2012].

Gilfedder L and Kirkpatrick JB (1997). The influence of owners' intentions on remnant bush conservation in Tasmania. In: *Conservation Outside Nature Reserves.* (Eds P Hale and D Lamb) pp 217–225. Centre for Conservation Biology, University of Queensland, Brisbane.

Hoskins L and Angelica E (2005). *Forming Alliances: Working Together to Achieve Mutual Goals.* Fieldstone Alliance, Saint Paul, Minnesota.

Kirkpatrick JB, McDougall K, and Hyde M (1995). *Australia's Most Threatened Ecosystems: The Southeastern Lowland Native Grasslands.* WWF Australia, Sydney.

Pasquini L, Fitzsimons JA, Cowell S, Brandon K, and Wescott G (2011). The establishment of large private nature reserves by conservation NGOs: key factors for successful implementation. *Oryx* **45**, 373–380.

Worboys GL and Pulsford I (2011). 'Connectivity conservation in Australian landscapes'. Report prepared for the Australian Government Department of Sustainability, Environment, Water, Population and Communities on behalf of the State of the Environment 2011 Committee, Canberra.

10

Wedderburn Conservation Management Network: Kooyoora connections

Peter Morison and Wendy Murphy

Introduction

The Wedderburn Conservation Management Network (WCMN) encompasses low rises and surrounding plains along the northern slopes of Victoria around the township of Wedderburn, 80 km north-west of Bendigo (Figure 10.1). The area covers ~170 000 ha and consists of fragmented vegetation patches, most of which are on public land, surrounded by private farming land and some connecting vegetation on private land.

The area includes fragmented public land blocks, with most of the largest blocks being part of Kooyoora State Park, Wychitella Nature Conservation Reserve and Mt Korong Nature Conservation Reserve, as well as important freehold properties managed for conservation (Figure 10.1). Private land use is predominantly cropping and grazing, while eucalyptus oil production occurs on some public land and gold prospecting is a popular pursuit in the region.

The WCMN area straddles both the Goldfields and Victorian Riverina subregions (part of the Riverina and Victorian Midlands bioregions, respectively). The vegetation throughout is predominantly a combination of box and ironbark forests and mallee. These two main systems are intermingled producing a rich mix of flora and fauna (CFL 1988). These are surrounded by small pockets of plains woodlands and patches of hills woodlands. The predominant Ecological Vegetation Classes are Box-Ironbark Forests, Sandstone Ridge Shrubland and Hills Herb-rich Woodland, with smaller pockets of Hillcrest Herb-rich Woodland and Metamorphic Slopes Shrubby Woodland.

The formation of a Conservation Management Network (CMN) to work at the landscape level in the Wedderburn region was first recommended in 2001 by the Environment Conservation Council's (ECC) Box-Ironbark Forest and Woodlands Investigation (ECC 2001). The reason for selecting this area was to improve the connectivity and help secure the local population of the endangered Malleefowl (*Leipoa ocellata*). The choice of the CMN model was in recognition of both the high importance of the public land within the Wedderburn area, the linkages provided by vegetation on private land and that the local community needed to play a major role in any activities and decision making. A CMN structure would take a landscape view across all land tenures with a focus on conservation of biodiversity (ECC 2001).

Current composition

The WCMN comprises the Wychitella Nature Conservation Reserve, which includes seven separate vegetation blocks of between 300 ha and 1500 ha, most of which are poorly connected. There are also numerous smaller reserves between 10 and 200 ha scattered across the area. All these reserves are considered the key parts of the WCMN. The WCMN has taken the view that the Wychitella Nature Conservation Reserve is the cornerstone of biodiversity conservation in the region and has focused much attention on it. This includes improving connectivity, enhancing vegetation quality and increasing patch size.

The initial focus for the WCMN was working in and around the Wychitella Nature Conservation Reserve, where breeding Malleefowl are found. In this area the public land surrounds a cleared centre of private land and is referred to as the 'doughnut' (Figure 10.1).

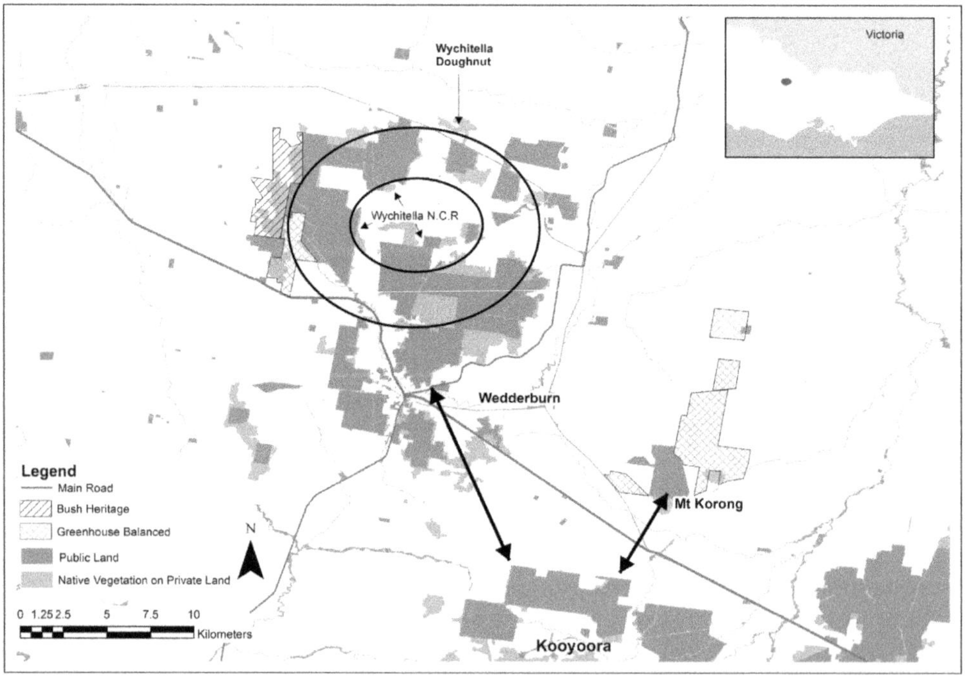

Figure 10.1 Location and components of the Wedderburn Conservation Management Network. Arrows indicate proposed future vegetation linkages.

The Malleefowl has long been the centre of local community action. In the 1960s a conservation battle commenced to have these public lands declared a conservation reserve with the plight of the Malleefowl the main focus. The battle was ultimately successful and the Malleefowl survives – for now. The Wedderburn Malleefowl population is the most south-easterly population in Australia and thus may have some greater significance in the face of climate change.

The main activities conducted are fox control on both public and private land to protect the Malleefowl and revegetation works on private land to connect or enlarge the habitat patches.

The WCMN area has since expanded its geographic scope, in response to funding opportunities provided by the Australian Government's 'Caring for our Country' program in 2009, and now includes connecting Mt Kooyoora, Mt Kerang and Mt Korong with contiguous vegetation (Figure 10.1). Over 20 private landholders have had revegetation works on their properties, with the non-government organisation, Bush Heritage Australia, and a private company, Greenhouse Balanced, now becoming major landholders in the WCMN area (Figure 10.1). Parks Victoria and the Department of Sustainability and Environment (DSE) continue to provide valuable support for works on public land. However, the WCMN does comprise other active players. These include, or have included, the Loddon Shire, North Central Catchment Management Authority, Mt Korong EcoWatch, Trust for Nature, Friends of Kooyoora, Wychitella District Landcare Group, as well as local community members and present and absentee landowners.

According to its Statement of Purpose, the WCMN '... works with the local community to protect and replenish the landscape by supporting the restoration of biodiversity and sustainable land management'. In addition, a recent WCMN workshop identified 'A healthy and resilient environment, an educated community and a fulfilled and enjoyable WCMN' as broad goals. There is a consensus that the continued survival of the Malleefowl is a core goal, as is establishing connectivity between Mt Kooyoora, Mt Kerang and Mt Korong that will provide greater opportunities for flora and fauna to move in response to climate change.

History and structure

While most groups are established in response to pressure from community members (i.e. 'bottom-up') this was not the case with the WCMN. It was established when the Victorian Government accepted recommendations by the ECC (2001) and subsequently funded a facilitator through DSE ('top-down') (see Chapter 21). This created initial difficulties in gaining support and interest from the community. Indeed when DSE first advertised in the local press for people who wanted to be on the initial committee it received only two responses. Subsequent canvassing of 'likely individuals', however, proved successful, and a committee of 11–13 was established, with a dynamic local, John Morieson, elected as committee chair.

DSE decided the original committee was to have key representatives from major relevant agencies but with a majority community representation. At the commencement, the committee meetings were more formal. However, as things evolved, the group became less formal and voting on issues was rare. What eventuated was a discussion format reaching some consensus being adopted. This original committee included representatives from Parks Victoria and the Loddon Shire through the Landcare facilitator. Over following years, representatives from the Department of Primary Industries, the North

Central Catchment Management Authority and Trust for Nature also participated. The WCMN was well funded by DSE with provision for a full-time facilitator and on-ground works.

Over time this committee structure fluctuated with some people leaving and new people joining. In December 2011, the WCMN changed back to a more formal structure with an elected committee of five which consists of local residents and landholders, as well as DSE and Parks Victoria representation. The wider WCMN membership includes representatives from organisations such as the Victorian Malleefowl Recovery Group, Bush Heritage Australia, Greenhouse Balanced, Trust for Nature, local Friends groups and Landcare groups as well as interested general public.

Major successes

In the early 2000s, the Department of Sustainability and Environment, purchased a key 186 ha property to link two major sections of the Wychitella Nature Conservation Reserve (Fitzsimons and Ashe 2003), in part influenced by the ECC's (2001) recommendation for greater connectivity conservation efforts in the region. Then, in 2004, an important 245 ha block of remnant vegetation adjacent to the Wychitella NCR came up for sale. The land had been well managed by the owner for many years and was in excellent condition. The WCMN facilitator approached Bush Heritage Australia (BHA) to seek their support for protection of the land. This proved fruitful when they purchased the land for conservation with donated funds. It then became known as the Judith Eardley Reserve (Fitzherbert 2004).

As a result of their interest and widely publicised works on pest plant and animal control on this property, neighbouring private land managers have approached Bush Heritage Australia to offer their land for sale. Two properties adjoining the Judith Eardley Reserve were subsequently purchased creating what is now the Nardoo Hill Reserve, comprising some 572 ha (BHA 2012).

In 2007, the private carbon emissions offset company Greenhouse Balanced was introduced to the area. The company director, Paul Dettman, was keen to work with a community group and so began a dramatic change in the region as large areas of land were revegetated in keeping with the WCMN targets and principles. To date Greenhouse Balanced have purchased and revegetated close to 700 ha of land in key connectivity areas around Mt Korong and south of BHA's Nardoo Hills Reserves. In total ~3000 ha of private land have been added to the conservation estate through conservation covenants.

Over 20 private landholders have had revegetation works conducted on their properties. Sizes have ranged from small areas of 0.4 ha to 2 ha to larger blocks of over 40 ha. The key criteria being that these sites link or extend remnant vegetation on public and private land. Some of this work has been conducted using voluntary labour, or by the use of contractors. Only some of these works are permanently protected with covenants.

To assist Malleefowl protection, the WCMN developed an action plan (Rowley 2004). From this plan a successful fox-baiting program has been in operation since 2004. Parks Victoria and Department of Primary Industries have complemented this baiting program with their own periods of baiting. This is a continuous program targeting all periods of the fox breeding season and offering protection for breeding Malleefowl and emerging chick survival. Official records from the area have been kept by the Victorian Malleefowl Recovery Group (VMRG) since 2008 showing that in 2008 there were three known active mounds, rising to four in 2009 and 2010. In 2011, only two mounds were active, but there have been numerous reports of single Malleefowl being seen in areas they have not been seen in for many years, including land revegetated by the WCMN. Most of these descriptions point towards young birds (see also Morison (2010) for observations on the Malleefowl recovery effort).

Major lessons

When it commenced, the WCMN Committee felt it was important to start quietly to avoid raising unrealistic expectations. It was also felt that it was better to first achieve some on-ground successes before any widespread promotion of the WCMN. Both of these strategies have since been justified. In working with a volunteer group it has proven important to support people who are passionate about a particular issue, even if it does not seem the highest priority.

Very early on the WCMN Committee decided that high kangaroo numbers were a concern for all landholders in the area. This proved a difficult issue for many but tackling it has built credibility in the community. By understanding, acknowledging and acting on community issues like this in an appropriate way, the WCMN has built credibility and ongoing support and acceptance.

The WCMN committee has not always been actively engaged in decision making in project selection. This has largely been done by the facilitator and the chair. This has led to some disenchantment within the group. Some felt they had no say in the group's direction and subsequently left. This approach may have been a product of the 'top-down' approach by DSE where the bureaucratic priorities of government may differ from those of the community. The danger of this is that members feel disenfranchised. The facilitator is now using a system that better involves the Committee without constraining negotiations with landholders.

Up until recently the WCMN had a loose committee structure, which did not suit some who wanted a more formal arrangement. Tensions also grew around the different workloads of members suggesting a need to provide a diverse range of activities to satisfy each one.

Some of the projects that have been conceived and funded have been large. Many committee members wanted to assist with the work, but were only able to do small projects at a time, which makes achieving set funding targets difficult. This necessitates the facilitator and ranger doing much of the work themselves. Hence the funded facilitator and ranger positions are vital for the successful running of the WCMN. Without them the regular flow of money and the ability to achieve targets would be very difficult.

While landscape-scale biodiversity enhancement activities are the focus of the WCMN, it has proved a little overwhelming for a relatively new group. One or two projects a year with a greater focus on community engagement may be more manageable. Education and community engagement are areas the group is very enthusiastic about.

Collation and sharing of information could be improved. Many monitoring programs are conducted under the WCMN banner, but much of the data is not collated nor analysed satisfactorily. Access to data is not always easy making it difficult to be drawn upon for project funding justification or comparison between years. Time constraints also limit the extraction and use of the data.

A recent WCMN workshop highlighted that two of the three broad goals were about community engagement and member enjoyment. This came as something of a surprise to the members and should have been stressed earlier.

Applying the lessons

In order to give the WCMN more autonomy and encourage members to become more active in project development, annual planning days have been established where attendees are asked to bring a wish list of projects they would like to see the WCMN undertake. An asset list has also been developed listing rare and threatened species in the WCMN area as well as the positive aspects such as large public land blocks and a dynamic group. This then enables the suggested projects to be matched with appropriate funding sources. If the group take on more active project development roles, less work is then loaded on to the facilitator and ranger, who can then act in a more supportive capacity.

As the committee is made up of local residents and landholders, they have the ability to extend the WCMN message to other community groups they might be involved in and draw new members or partnerships into the group and bring in new ideas. Also having a ranger who comes from the local area is a big advantage to the WCMN. Not only does the ranger know many of the landholders in the area but also has an understanding of their attitudes towards conservation activities. Many projects and works have been arranged in the local supermarket or at sporting events with the landholders often feeling more comfortable approaching the ranger in a less formal capacity.

Encouraging members to stay involved with the WCMN is often difficult as their interests change or sometimes, if satisfied with on-ground works, they do not feel the need to stay involved in the group. An adaptive group is important here – as new members come and old members leave there need to be people willing to take on new roles and the group must be willing to change their focus slightly. This is where promoting the group to adopt a more self-managing role and having a wide range of project ideas helps.

Delivering good on-ground works is important in encouraging landholders to do more and to talk to others about their activities. Many landholders are initially reluctant to undertake large-scale works on their properties because of the fear of failure or poor results. They often consider revegetation projects as locking away their land and need to feel as though they are getting something out of it. They will often offer small areas to be revegetated first to see the results, then if satisfied, they will likely offer larger areas.

Malleefowl habitat requirements and behaviour were not widely understood in the Wedderburn community, which has driven the WCMN to really focus on getting information out, especially to the younger people. The involvement of schools is encouraged and a Malleefowl Festival was initiated as part of the Loddon Shire's annual 'Naturally Loddon' Festival each September. This event is designed to inform the community about the Malleefowl and to give them more pride in knowing the Wedderburn area is quite unique in being home to this species and that the Wedderburn population is quite unique within the species.

Although the WCMN has a major focus on Malleefowl protection, funding opportunities for this species became limited and the focus had to alter and widen in scope. All subsequent projects can be related back to Malleefowl protection in some way, but the group really needed to think 'outside the box' to achieve this. Creative project wording on funding applications was needed so that the criteria for the funding could be met, but benefits for Malleefowl were also achieved.

Conclusion

Since its inception in 2003, the Wedderburn Conservation Management Network has come a long way in its development as a group, the partnerships it has bought in to the Wedderburn region and the area of private land managed for conservation. Being able to adapt from having a major focus on Malleefowl protection, to now having a focus area of over 170 000 ha covering a wide range of rare and endangered species and communities is a big achievement. Despite all the problems and issues, the WCMN has been the catalyst for landscape-scale change. In achieving this, the single biggest factor in this success has been having a funded facilitator and ranger. Without these positions most of the achievements would not have been possible or would have taken considerably longer. Much credit for this must go to DSE who fund both positions (at least in part). They facilitated the formation of an independent community group without guarantees of specified outcomes being delivered.

A strength of the group is its practice of supporting the endeavours and passions of individual WCMN members, despite them not always being top priority. This allows members to contribute in their own way and maintains the energy within the group. The group members have been active in the community, promoting the WCMN as a resource base for community education and conservation works. Some have worked hundreds of hours of volunteer time towards wonderful on-ground successes. It is anticipated that this will continue into the future.

It may take many small steps, but the overall goal of creating a contiguous vegetation link between Wychitella NCR, Kooyoora State Park and Mt Korong will be achieved. With the pressures of climate change, this will likely provide greater opportunities for the movement of Malleefowl and other fauna across the landscape.

The WCMN was established by DSE through a 'top-down' approach but one that required 'bottom-up' contributions from the community committee. This may not be the usual path towards the establishment of a community group but it has proved successful. This is in part the beauty of the WCMN model that it can be established in many differing forms depending on circumstances. The continued success of this group demonstrates that there is no single model that should be followed or that can be applied in all circumstances.

Acknowledgements

Thanks to Pam Clunie for her help with editing this chapter and John Morieson for leading the group into existence, Jill McFarlane and Michael Moore for their continued leadership roles as subsequent Chairs of the WCMN and Alison Jeavons for creating the map.

Biographies

Peter Morison has recently left the Victorian Department of Sustainability and Environment after working there for more than 20 years, mostly in private land conservation. He started at DSE as a Land for Wildlife Extension Officer and commenced as WCMN facilitator in 2003. More recently, he was working in community engagement with the DSE Land and Fire section. He is now working in a private capacity and on his conservation property at Pine Grove in northern Victoria.

Wendy Murphy (née Rose) grew up in Inglewood Victoria where she still lives today. After completing a Bachelor of Applied Science (Natural Resources Management) Degree at Deakin University in 1999, Wendy spent time working in vineyards around Inglewood and a winery in Echuca. In September 2004, Wendy was employed by the Department of Sustainability and Environment as the Wedderburn Conservation Management Network Ranger, a position she still holds today.

References

BHA (2012). Nardoo Hills Reserves. Bush Heritage Australia, Melbourne. Available: http://www.bushheritage.org.au/reserves_NardooHills [Accessed 25 April 2012].

CFL (1988). *Wychitella Flora and Fauna Reserve Resource Inventory*. Department of Conservation, Forests and Lands, Melbourne.

ECC (2001). *Box-Ironbark Forests and Woodlands Investigation Final Report*. Environment Conservation Council, East Melbourne.

Fitzherbert K (2004). Our first reserve in Victoria. *Bush Heritage News* **Summer**, 1–3.

Fitzsimons JA and Ashe C (2003). Some recent strategic additions to Victoria's protected area estate 1997–2002. *Victorian Naturalist* **120**, 98–108.

Morison P (2010). Landscape recovery – community resilience in the face of bureaucracy. *Transactions of the Royal Society of Victoria* **122**, lxxvi–lxxviii.

Rowley L (2004). 'Wedderburn Community Malleefowl Action Plan'. Report prepared for the Wedderburn Conservation Management Network.

11

Conservation Management Networks: The Gippsland Plains Story

Robyn Edwards and Trish Fox

Introduction

Agricultural landscapes in temperate Australia have endured major losses and degradation of their native vegetation, leading to severe decline in extent of many vegetation communities, flora and fauna (NLWRA 2002). Due to this severe habitat loss, vegetation became fragmented with remnant patches occurring predominantly on private land. Traditional methods of protecting biodiversity, such as reserving extensive areas of public land for conservation, were not applicable in this situation. In response, the late 1990s saw the development of Conservation Management Networks (CMN), providing new models for landscape-scale conservation.

A CMN is a biophysical network of remnant vegetation sites across a variety of tenures and a social network of managers, owners and interested people (Thiele and Prober 2000). A CMN brings together the social and biophysical networks in order to improve land management and biodiversity outcomes. The focus of a CMN is usually on a vegetation community or species that is threatened or under-represented in the reserve system and is usually located in fragmented landscapes. A CMN provides a conservation network model that complements the National Reserve System. Trust for Nature (TfN) played a major role in facilitating the establishment of the first CMN in Victoria and the second in Australia: the Gippsland Plains Conservation Management Network (GPCMN).

The GPCMN is located in south-eastern Victoria in an agricultural landscape (Figure 11.1). The landscape covers ~700 000 ha with less than 10% native vegetation remaining. The main focus of the GPCMN is the nationally threatened Gippsland Red Gum Grassy Woodlands and the associated plains vegetation communities. Within the

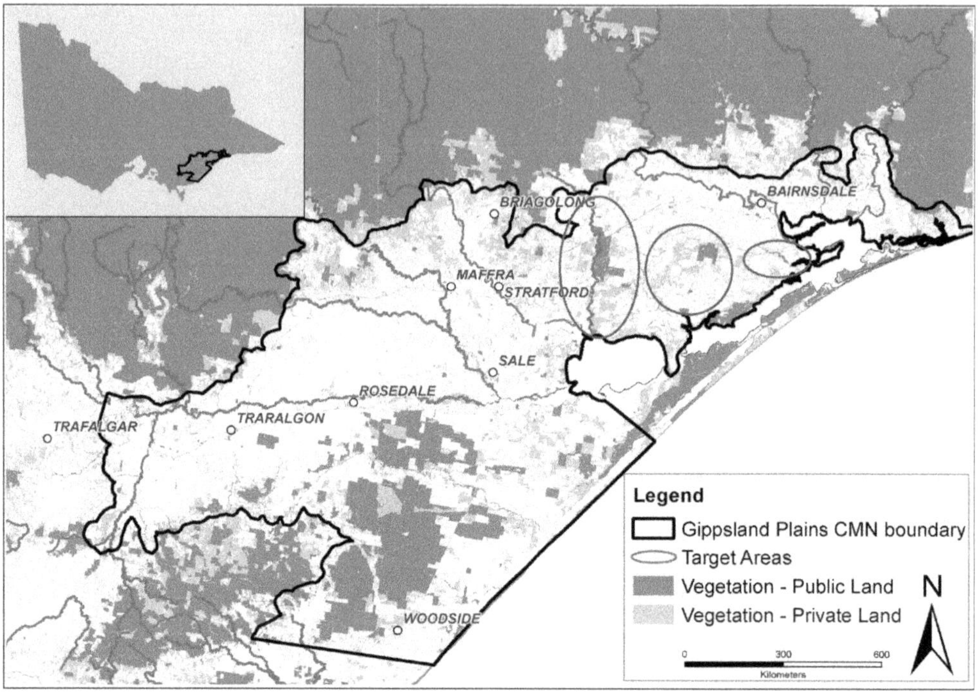

Figure 11.1 Outer boundary of the Gippsland Plains Conservation Management Network.

broad geographic area of the GPCMN three core areas have been identified as a priority for works and funding; (i) the Perry River Network; (ii) Moormurng Network; and (iii) the more recently established Forge Creek Network (Figure 11.2). Central to these networks are the core public land protected areas, Providence Ponds and Moormurng Flora and Fauna Reserves and the Forge Creek Reserve. Projects and programs are targeted to extend linkages and increase protected sites outward from these reserves across the landscape. Managers and owners of 61 sites are, at present, actively participating in on-ground works across 1365 ha (Table 11.1 shows a breakdown of hectares across tenure of the core sites of the GPCMN).

The GPCMN follows the three broad principles identified in the CMN model developed by Thiele and Prober (2000): (i) permanent legal protection; (ii) ecological management/restoration; and (iii) information sharing. These principles have been refined into a set of objectives in the Victorian CMN strategy (Context Pty Ltd 2008) and adopted by the GPCMN, in summary to: (a) increase the protection status of sites; (b) maintain, enhance and re-establish remnants across private and public land; (c) bring together owners and managers of vegetation remnants; (d) connect and buffer remnant patches; and (e) develop consistent and complementary management across sites.

History and structure

Trust for Nature is the main organisation in Victoria that provides permanent protection mechanisms on private land, and does so under the *Victorian Conservation Trust Act 1972*. This is achieved through its property purchase, revolving fund and conservation covenanting programs. Assistance with ongoing management of these private conservation lands occurs through a stewardship program. The Ecological Vegetation Class mapping

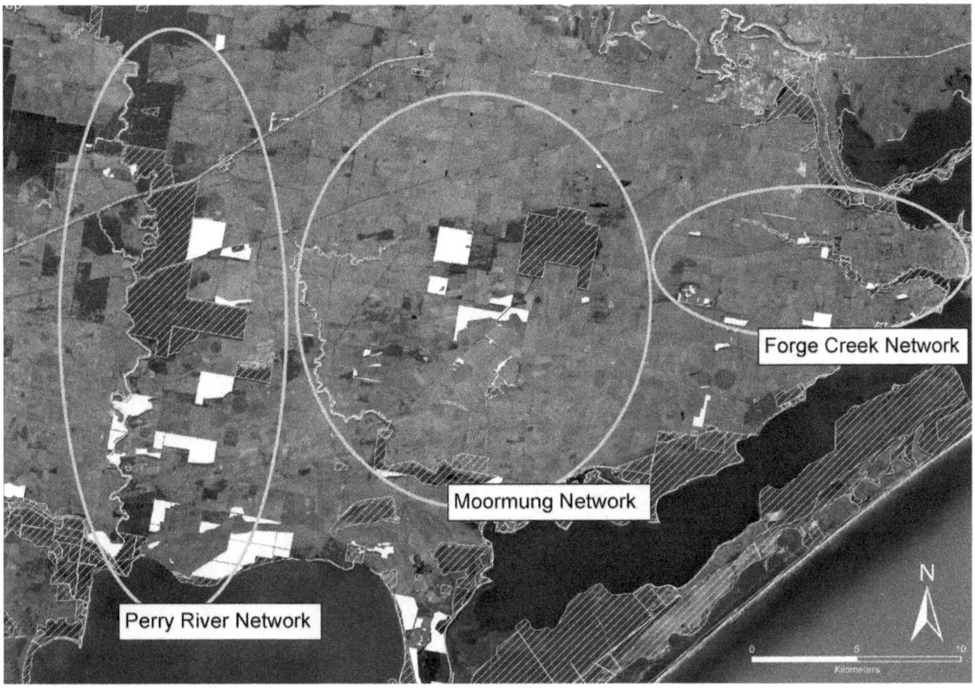

Figure 11.2 Priority areas within the Gippsland Plains CMN: Perry River Network, Moormurng Network and Forge Creek Network. Trust for Nature Reserves and covenanted properties are shown in white and Crown land reserves are hatched. (Map data @ 2010 MapData Sciences Pty Ltd, PSMA. Used with permission from MapData Sciences Pty Ltd (http://www.nowwhere.com.au/lic/NowWhereLic.htm) (www.mapds.com.au)).

Table 11.1 Core sites of the Gippsland Plains CMN

Sites that have been legally protected since 1999 are highlighted in bold.

Name	Tenure and/or land manager	Hectares
Moormurng and Providence Ponds Flora & Fauna Reserves	Crown land/Parks Victoria	3382
Forge Creek Reserve	Water Reserve & Water Frontage Crown land, Public Lands (DSE) & East Gippsland Catchment Management Authority (EGCMA)	45
Toms Creek Reserve	Water Frontage, Public Lands (DSE) & EGCMA	30
Rail Reserve under management agreement	Crown land, DSE, V/Line	10
Unused roads with conservation licence and Shire Parcels	Public Lands (DSE) and East Gippsland Shire	**35**
72 covenanted properties	Freehold	**4864**
5 Trust for Nature purchases	Freehold	**555**
2 DSE purchases facilitated by TfN (Swallow Lagoon & Bengworden Reserves)	Crown land/Parks Vic	**240**
26 fenced/unprotected sites	Freehold	101
	Total	8104

that was undertaken by the then Victorian Department of Natural Resources and Environment formed the basis of a TfN-targeted extension program on the Gippsland Plains in the mid 1990s. This mapping enabled identification of remnants on private land and landholders were subsequently contacted by TfN to encourage fencing and permanent protection of their grassy woodlands. However, this presented a challenge in how to secure the future of fragmented remnants across land tenure in an agricultural landscape. The national park system, the traditional model of conservation, did not apply in this situation as it was established for large tracts of intact habitat that were mainly found on public land. The TfN covenanting and stewardship programs addressed this gap in part by providing protection and management at the site level. However, the Trust's conservation program was limited at the time in that it did not include public land sites and there was no mechanism for bringing the sites and owners into a network. Binning and Young (1997) proposed the concept of Protected Area Networks (later known as Conservation Management Networks) as a way of improving landscape conservation and this model was considered highly applicable to the Gippsland Plains. In 1999, TfN secured funding from the Australian Government's National Reserve System program for the purchase of four properties in order to develop the Perry River Protected Area Network. This, combined with a targeted extension program and the establishment of nearby covenants, formed the foundation of the Perry River Network (Edwards 1999; Edwards and Traill 2002; Fitzsimons and Wescott 2005).

With the establishment of the Grassy Box Woodlands CMN in New South Wales (Chapter 12) and the holding of a national forum on CMNs in Canberra in 2001, the Thiele and Prober (2000) model of CMNs was gaining momentum. As a result of this forum, TfN and the Department of Natural Resources and Environment conducted workshops in Gippsland with other agencies and landholders and the Gippsland Plains Conservation Management Network was formalised. At the workshops agreement was reached: (i) to extend the Perry River Network to cover the extent of Gippsland Red Gum Woodlands (Figure 11.1); (ii) to change the name from the 'Perry River Protected Area Network' to 'Gippsland Plains Conservation Management Network'; (iii) to produce the newsletter *On the Plains*; and (iv) to form a steering committee of community representatives. Funding was also obtained by TfN to facilitate the GPCMN's development, to act as host organisation and to employ a ranger to undertake woodland restoration trials for the CMN.

In 2002, the GPCMN became an incorporated body enabling the CMN to apply directly for funding. A milestone in the development of the GPCMN was reached in 2005–06 when TfN's role as host organisation was no longer required. The CMN's independence from TfN was made possible through the recognition of the GPCMN as a partner by the East Gippsland Catchment Management Authority (EGCMA). This resulted in an increase in funding which enabled the facilitator position to increase to three days a week ensuring that all of the core facilitation duties of the CMN could be carried out.

The GPCMN has regular committee meetings and an Annual General Meeting to elect four office bearers and committee members. The current committee comprises Trust for Nature, East Gippsland Landcare Network, East Gippsland Shire Council, Wellington Shire Council, VicRoads, East Gippsland Rainforest CMN, Greening Australia, Department of Sustainability and Environment and seven private landholders (six of whom are farmers), with West Gippsland Catchment Management Authority as an associate member. In addition, the GPCMN has an editorial committee for *On the Plains & In the Rainforest* (a joint newsletter of the Gippsland Plains CMN and nearby East Gippsland

Rainforest CMN). The facilitator is engaged under contract as a consultant, minimising CMN employment responsibilities and liabilities and a bookkeeper is contracted to assist with financial management. The guiding documents of the GPCMN are the GPCMN Objectives; *GPCMN Cross-Property Biodiversity Conservation Plan: Red Gum Plains, 2010–2013,* and the *Strategic Plan for Conservation Management Networks in Victoria* (Context Pty Ltd 2008). Income is sourced predominantly from federal and state government grants through the EGCMA, with some private sponsorship for specific projects. Presently 80 paying members support the GPCMN while 300 individuals and organisations receive the newsletter.

Major successes

The most significant conservation achievements that have contributed towards meeting the GPCMN principles and objectives are outlined below. Sound leadership, committed members, passionate personal and strong partnerships have been paramount to these successes.

1. Increasing the protection status of sites

A priority of the GPCMN is to achieve the highest possible form of security over a site. The GPCMN promotes permanent protection programs and assists in engaging landholders to participate in these programs. Permanent protection is then facilitated by member organisations and has included conservation covenants, changing grazing licences to conservation licences on unused roads, land management agreements, land purchases by the Victorian Government in partnership with TfN, land purchases by TfN with funding support from the Australian Government's National Reserve System Program and, at present, working to change the reservation status of significant Crown land parcels. This has resulted in some 5576 ha having been legally protected since 1999. Permanently protected sites (on public and private land) are given priority for CMN funds and works ensuring that the investment of public dollars is secured in the long term.

2. Bringing together land managers and developing cross-tenure projects

The success of the social network of the GPCMN was documented by Fitzsimons and Wescott (2007), who found that 56% of GPCMN managers regularly corresponded with each other and 82% suggested that other sites in the network influenced their management decisions. The Forge Creek Network is one example demonstrating how effective the GPCMN is as a model for bringing people together and integrating management. The Forge Creek Network is a cross-tenure project consisting of water reserve, Crown land and freehold land (Figure 11.2). The diverse range of skills and funding brought together at one time at a site has resulted in a significant increase in conservation works. GPCMN has raised the profile of the importance of the creek and environs; supported the Landcare group's projects; secured additional funding; mobilised more of the community to action and facilitated the works between Landcare groups, individuals, EGCMA, East Gippsland Water and TfN. Achievements have included fencing Crown land frontage and adjoining scattered trees sites, the cessation of grazing licences, covenanting four properties, reinstatement of the chain of ponds and erosion control, reintroduction of threatened species, weed control, revegetation and woodland bird monitoring.

3. Improving approaches to ecological restoration

The early establishment of restoration trials enabled the development of local solutions to ecological problems, paved the way for works across sites and improved the approach to habitat restoration works. GPCMN members (TfN, DSE, and GPCMN ranger and facilitator) conducted trials which included researching the effect of spring and autumn burning on weeds and understorey establishment (Prober *et al.* 2009), exclusion fencing, Noisy Miner (*Manorina melanocephala*) removal and, more recently, Peregrine Falcon (*Falco peregrinus*) recovery actions.

The Peregrine Falcon is on the edge of extinction on the Gippsland Plains (Victor Hurley, head of the Victorian Peregrine Project, pers. comm.) but was not considered a high priority for action by the Victorian Government because it is not listed as threatened at a statewide level. Consequently, the GPCMN developed the Gippsland Plains Peregrine Falcon Project and joined forces with the Victorian Peregrine Falcon Group. The focus of the project has been on the establishment of nest boxes installed on disused power poles that are erected at key locations, as well as trialling the placement of a nest box in an old Gippsland Red Gum (*Eucalyptus tereticornis* subsp. *mediana*). Six nest boxes have been established with the first successful hatching in spring 2011 in the Gippsland Red Gum nest box (Figure 11.3).

Major lessons and applying the lessons

Many challenges have been overcome and there are still many ahead as the GPCMN continues to evolve. The following points stand out as being significant lessons that have been learned or which will lead to improvements in achieving the GPCMN's objectives.

Figure 11.3 Peregrine Falcon hatchlings spring 2011, the first trial of a nest box in a Gippsland Red Gum (Photo: Rohan Bilney).

1. Improving strategic design when connecting and buffering remnants

A feature of CMNs is the ability to *fill in the gaps* in biodiversity programs. This has been demonstrated through the GPCMN Scattered Tree Project. An important but previously overlooked habitat component of these grassy woodlands was the mature Gippsland Red Gum trees scattered across the agricultural landscape. These paddock trees contain the majority of tree hollows on the Plains but are in decline. In 2004, the GPCMN developed the Scattered Tree Project in order to improve their health and longevity. Initially, priority sites were neither identified nor targeted, resulting in single isolated trees being fenced with only small biodiversity gains. Using recent research on both the proximity of scattered trees to remnants and the effects of tree density on bats (Lumsden and Bennett 2005), the facilitator developed new guidelines for the scattered tree project including: (i) targeting groups of trees and sites close to remnants; (ii) using scattered trees as a basis for connecting remnants; and (iii) targeting scattered trees sites adjacent to remnants as the basis for restoration works to buffer and re-establish grassy woodlands. These sites are more likely to be ecologically resilient by containing relictual grassland species, regenerating eucalypts, a stored seed bank and a seed source from the adjoining remnant.

2. Development of partnership agreements and integration of funding at priority sites

Another key feature of CMNs is the integration of land managers and owners – 'the social network'. Early biodiversity projects tended to be delivered in isolation from one another resulting in less efficient use of resources, often at low priority sites and in an *ad hoc* fashion across the landscape. In response to this an *integrated partnership approach* was developed between the GPCMN and its members TfN, East Gippsland Landcare Network and Greening Australia. Project funding and activities were combined at agreed priority sites which substantially improved on-ground results. For instance, at one property over 2 years the combined funding achieved: 20 ha of remnant vegetation fenced, 26 ha of creek and unused road fenced and planted, 16 ha of scattered tree sites fenced and direct seeded, 12 ha of corridors direct seeded or planted, restoration of ground covers through exclusion fencing, burning and seeding, and all freehold areas were covenanted.

However, there were still large amounts of resources that had been or were being invested in particular properties by different partners. The gain on the ground was impressive, but this investment had not been protected in any way, and there were instances where years of work had been destroyed through a change of ownership or management. The GPCMN looked to improve the situation by developing a *partnership agreement.* This agreement is in the form of a Memorandum of Understanding (MOU) that is signed by each partner. The MOU outlines actions and funding to be allocated by each partner which are combined to invest in the biodiversity works 'wish list' of the owner. What was new to this partnership approach was that funds were conditional on covenanting significant remnants, placing water frontage agreements over creek frontages and partners agreeing not to undertake any further works on the property unless these conditions were met. This approach is presently being negotiated with a farmer who has participated in a range of biodiversity, land and creek projects over the years and if successful this will ensure that the sites where public dollars have been invested are protected on title.

3. Group governance: administration

With an increase in the number of projects and associated income, the GPCMN has rapidly grown. The internal processes established during the commencement phase were no longer adequate. One response to this was to commission a bookkeeper to assist with financial management.

An annual business plan has been identified as a way to improve administration. Once developed, it will include grant income, targets and resource allocation, financial processes and approval requirements. In addition, the development of a Gippsland Plains CMN strategy that builds on the cross-property landscape plan would improve planning and monitoring. Ideally the strategy would identify measurable 5-yearly targets and establish a system to monitor GCPMN's progress towards achieving CMN objectives; however, such planning activities are not usually funded through grants.

4. Increasing participation in the social and biophysical networks

In some instances engaging with land managers of remnant vegetation continues to be a challenge. There are sites not currently part of the network where there is either a lack of management, management for purposes other than conservation, or where resources and skills are inadequate. The GPCMN has made attempts at the local level to support the management of priority Crown land sites but has faced difficulties because many ecologically significant reserves such as public recreation reserves attract no biodiversity conservation management funds from the Public Lands Services branch of the Department of Sustainability and Environment. A shift in engagement and commitment has evolved slowly but has a greater chance of occurring when there is: (i) an ongoing partnership through the long-term continuation of the facilitator or ranger role; (ii) ongoing funding that can be directed to the site; (iii) biodiversity management becoming part of the local culture; and (iv) a respect for the group (CMN) which is often developed through the group's achievements and representatives.

Obtaining sufficient funding to increase the facilitator role to a full-time position and to re-engage a ranger would allow more opportunities to strengthen existing partnerships and form new ones. Additional ongoing funding for on-ground works and incentives is required to gain further commitment to protect and restore high value sites. Applying for, managing and reporting on small grants is time consuming and often not the best use of a facilitator's skill set. Ideally, state and federal government grants need to be for 3 to 5-year periods.

Conclusions

The Gippsland Plains CMN model features permanent protection, cross-tenure management and the integration of land managers and owners. It is not centrally managed by one organisation but is the sum total of members and partners, bringing everyone equally together. The CMN model has the broad aim of recovering a particular threatened or fragmented vegetation community (or communities) and this aim does not change over time. While methods to achieve this vary considerably across CMNs, the end goal and the broad objectives remain consistent regardless of changes to membership, managers, owners and funding targets.

GPCMN is making ongoing and consistent progress in meeting the broad CMN objectives of increasing permanent protection, ecological management/restoration, integration of land managers and owners and information sharing. In summary, the main lessons and achievements of the GPCMN include:

1 improving connectivity and buffering remnants by using the presence of scattered trees to influence the design of links and buffers;

2 developing partnership agreements for integrating funding at targeted sites that are conditional on increasing protection status;
3 continuing to improve administration and governance as the group evolves and grows;
4 developing measurable targets and monitoring systems;
5 recognising that additional resources are needed to increase participation by key stakeholders;
6 appreciating that sound leadership skills, a passionate and dedicated facilitator, a ranger to support the program and a committed CMN committee are paramount to the success of a CMN.

Acknowledgements

We would like to acknowledge past and present GPCMN committee members and the community of Gippsland for their ongoing support; Brett Mills, Susan Taylor and Suzanne Prober for their expertise and participation in the research projects; Trust for Nature staff; Bill Peel; the East Gippsland Catchment Management Authority; the West Gippsland Catchment Management Authority; the East Gippsland Landcare Network; and sponsors: Australian Geographic Society, SP Ausnet, Alinta, and the Police & Publicans Association.

Biographies

Robyn Edwards has been working with Trust for Nature (Victoria) for the last 17 years as Regional Manager, East Gippsland. She facilitated the establishment of the first Conservation Management Network (GPCMN) in Victoria and, in conjunction with the East Gippsland Catchment Management Authority, facilitated the establishment of the East Gippsland Rainforest Conservation Management Network. She is a committee member and treasurer of the GPCMN. Over her time with Trust for Nature she has facilitated the permanent protection of over 6000 ha in Gippsland through covenant or purchase mechanisms.

Trish Fox came to Australia in 1998 to undertake a PhD at the University of Western Australia, after 10 years as a journalist in Northern Ireland. She studied landscape preferences and ecology of remnant salmon gum woodlands in the Western Australia wheatbelt and in 2001 moved to Victoria. She has been involved with the Gippsland Plains CMN since its inception.

References

Binning C and Young M (1997). *Motivating people: Using Management Agreements to Conserve Remnant Vegetation.* Paper 1/97 National Research and Development Program on Rehabilitation, Management and Conservation of Remnant Vegetation. Environment Australia, Canberra.

Context Pty Ltd (2008). *Strategic Plan for Conservation Management Networks in Victoria: Working Together to Protect Biodiversity.* Context Pty Ltd, Brunswick, Victoria.

Edwards R (1999). Saving the red gum woodlands of the Gippsland Plains. *Conservation Bulletin* **Summer**, 3–7.

Edwards R and Traill B (2002). Getting beyond field days: targeting extension to protect threatened ecosystems on private land. *Ecological Management & Restoration* **3**, 229–231.

Fitzsimons JA and Wescott G (2005). History and attributes of selected Australian multi-tenure reserve networks. *Australian Geographer* **36**, 75–93.

Fitzsimons JA and Wescott G (2007). Perceptions and attitudes of land managers in multi-tenure reserve networks and the implications for conservation. *Journal of Environmental Management* **84**, 38–48.

Lumsden LF and Bennett AF (2005). Scattered trees in rural landscapes: foraging habitat for insectivorous bats in south-eastern Australia. *Biological Conservation* **122**, 205–222.

NLWRA (2002). *Australian Terrestrial Biodiversity Assessment.* National Land and Water Resources Audit, Canberra.

Prober S, Taylor S, Edwards R, and Mills B (2009). Effectiveness of repeated autumn and spring fires for understorey restoration in weed-invaded temperate eucalypt woodlands. *Applied Vegetation Science* **12**, 440–450.

Thiele K and Prober S (2000). Conservation Management Networks – a model for coordinated protection and management of remnant vegetation. In: *Balancing Conservation and Production in Grassy Landscapes, Proceedings of the Bushcare Grassy Landscapes Conference, Clare, South Australia 19–21 August 1999.* (Eds T Barlow and R Thornburn) pp. 58–63. Environment Australia, Canberra.

12

Conservation Management Networks for grassy ecosystems in New South Wales

Toni McLeish, Rainer Rehwinkel and Lorraine Oliver

Introduction

The Conservation Management Network (CMN) model was first conceived and promoted by CSIRO scientists Suzanne Prober and Kevin Thiele in the late 1990s. It was a way to provide for the conservation of widely dispersed remnants of rare vegetation communities occurring in fragmented agricultural landscapes (Thiele and Prober 1999, 2000; Prober *et al.* 2001).

The first Conservation Management Network, the Grassy Box Woodlands CMN, was established in 1998 in New South Wales (NSW), and focused on the Box-Gum Grassy Woodlands of the western slopes and tablelands (see Figure 12.1). These woodlands encompass ecological communities that are threatened at a national level (*White Box–Yellow Box–Blakely's Red Gum Grassy Woodland and Derived Native Grassland* listed on the federal *Environment Protection and Biodiversity Conservation Act 1999* [EPBC Act]) and state level (*White Box–Yellow Box–Blakely's Red Gum Woodland* listed on the NSW *Threatened Species Conservation Act 2005* [TSC Act]). The predominant feature of these woodlands is the presence of White Box (*Eucalyptus albens*), Yellow Box (*E. melliodora*) and/or Blakely's Red Gum (*E. blakelyi*), along with several other tree species, and a grassy groundlayer supporting a diversity of grasses, flowering herbs and sparsely scattered shrubs. These ecosystems offered accessible and relatively fertile grazing and cropping opportunities to the earliest settlers and were systematically cleared, leaving widely dispersed Box-Gum Grassy Woodland remnants across a predominately agricultural landscape.

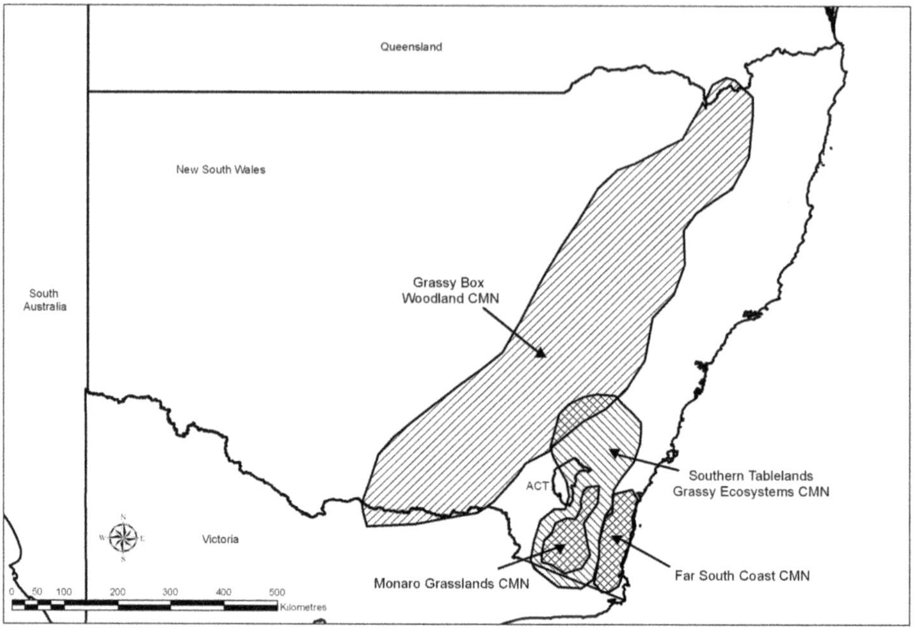

Figure 12.1 Location of Conservation Management Networks in New South Wales

The Grassy Box Woodlands CMN (GBWCMN) was followed by the establishment of the Southern Tablelands Grassy Ecosystems CMN in 2002 (Rehwinkel 2002) and Monaro Grasslands CMN in 2003 (Eddy 2007). Both CMNs focused on the threatened grassy ecosystems of south-eastern NSW (including the Box-Gum Grassy Woodlands mentioned above, the EPBC Act-listed *Natural Temperate Grassland of the Southern Tablelands of NSW and the ACT* and the TSC Act-listed *Tablelands Snow Gum, Black Sallee, Candlebark and Ribbon Gum Grassy Woodland*).

The Far South Coast CMN was established in 2007 and focused on the conservation of all vegetation types in the highly cleared Bega Valley. There is only minor discussion of this CMN in this chapter.

The GBWCMN services land managers and owners of Box-Gum Grassy Woodlands in NSW (Prober *et al.* 2001; Fitzsimons and Wescott 2005). It is a network of public and private landowners, and other interested individuals, who manage Box-Gum Grassy Woodlands for biodiversity outcomes. By focusing on one type of endangered ecological community, the GBWCMN is able to deliver specific and targeted management information to interested land managers and to focus on emerging issues that threaten the ecosystem at scales larger than any one local government or catchment management authority area. For example, lobbying by the GBWCMN was integral to the listing of the invasive species, Coolatai Grass (*Hyparrhenia hirta*), as a noxious weed in 31 shires in NSW.

NSW legislation has established a Priority Action Statement for all NSW threatened entities (OEH 2012a) and the Australian Government has commissioned national recovery plans (e.g. Environment ACT 2005; DECCW 2010); these documents identify CMNs and their various functions as necessary for the conservation of threatened grassy ecosystems.

The recovery actions described in these documents guide CMN activities, including the delivery of: (i) best practice information; (ii) the promotion and establishment of conservation agreements; (iii) the participation of the community in conservation initiatives; and (iv) enhancing the ability of governments and non-government organisations (NGOs) to incorporate conservation of the specific threatened communities into their planning and management processes.

The NSW CMN model predominantly focuses on a specific vegetation community or set of related communities, as the management needs of such remnants are generally similar across the landscape. Importantly, CMNs communicate with and support the owners or managers of the vegetation remnants across all tenures, and other interested or influential stakeholders. Membership of a CMN is free.

The GBWCMN currently communicates with 1504 members, including landholders, local government staff (25 councils), catchment management authority staff (seven CMAs), NSW government agency staff (Office of Environment and Heritage including the National Parks and Wildlife Service, and the Department of Primary Industries including the Land and Property Management Agency), NGOs (the Nature Conservation Trust and various Landcare organisations, including NSW Landcare, regional Landcare organisations and individual Landcare groups), Livestock Health and Pest Authorities (managers of the travelling stock reserve network), educators (Department of Communities and Education, including the Environmental Education Centre Network) and conservation groups (including 'friends of' groups, birdwatchers and field naturalists).

The Monaro Grasslands CMN ceased to operate in late 2008 due to lack of funding. At the time of cessation the Monaro Grasslands CMN was servicing 189 members. The Southern Tablelands Grassy Ecosystems CMN merged with the GBWCMN in 2010, also due to lack of funding. Its members continue to receive information via the GBWCMN.

History and structure

History and evolution of the CMNs

The continuing operation of CMNs in NSW has not been straightforward, with the model initially developed by Thiele and Prober (1999, 2000) having evolved, largely through the innovation of CMN coordinators, to marry the specific needs of each ecological community with their CMN membership-base and also with the changing focus of funding streams.

The GBWCMN began with four part-time extension positions. However, since 2003, it has operated with only one part-time coordinator. This coordinator relies heavily on the support of local OEH staff and facilities and on the formation of partnerships with other organisations, developed through individually funded projects, to provide additional resources. The Far South Coast CMN continues to be funded by the Southern Rivers Catchment Management Authority and Bega Valley Shire Council. As discussed earlier, the Monaro Grasslands CMN and the Southern Tablelands Grassy Ecosystems CMN are now dormant, having both been unable to secure ongoing funding.

A CMN coordinator provides technical support and resource material to members and facilitates information-exchange between educators, researchers, policy-makers and practitioners. The coordinator organises training for extension staff and land managers, field events, forums, conferences and workshops, maintains the membership mailing lists and performs other administrative functions. The key means of communication among members is currently via newsletters and the website.

The CMN coordinator also has a role to play in maintaining broader community and government understanding of current and emerging issues that impact on the threatened ecological communities. The coordinator is effectively the ecological community's spokesperson and supports individuals, groups and organisations to implement recovery actions for the ecological community.

Thiele and Prober (1999, 2000) saw the negotiation of legal protection of sites as a critical aim of their CMN model. Legal protection remains a priority for the GBWCMN and its coordinator regularly promotes the various conservation mechanisms including:

- conservation agreements applied as in perpetuity covenants through the Nature Conservation Trust and NSW OEH;
- formal stewardship agreements, up to 15 years, applied to sites through the Australian Government; and
- property vegetation plans, applied as either incentives or as offsets by catchment management authorities.

While the GBWCMN maintains a membership database, an important part of the Thiele and Prober model was to develop site management databases. The Southern Tablelands Grassy Ecosystems CMN developed the *Grassy Ecosystems Database*, which holds information, including the location, flora species lists, habitat and condition data and brief management histories from hundreds of sites across eastern NSW, including the region covered by the GBWCMN. These sites are not all necessarily part of the CMN or managed by a CMN member, however, the information adds to the understanding of the ecological community and is promoted as an information tool by the CMN. The Grassy Ecosystems Database information is being transferred to the NSW Government online platform (OEH 2012b).

The database has been used for a range of purposes including:

- production of maps of Endangered Ecological Communities (EECs) for inclusion in recovery plans (e.g. figure 2 in Environment ACT 2005);
- development of grassy ecosystem models using remote-sensing techniques (e.g. ERIC 2001); and
- development of spatial landscape plans (e.g. Fallding 2002).

Current governance arrangements/organisational structure

The GBWCMN is managed by an externally funded coordinator, who is hosted by NSW OEH. The position of coordinator is not secure, with current funding sources varied, competitive and short term. Two years is the longest appointment that the GBWCMN coordinator has experienced in the last 10 years, with more common funding periods being of 12 months' duration.

Funding for the CMN program has, in the main, been provided through Australian Government initiatives, with office space and transport costs provided by the NSW OEH. The in-kind support of individual NSW OEH staff based in Queanbeyan and Dubbo has provided technical, strong advocacy and administrative support, through the development of applications and management of externally funded projects on behalf of CMNs. The GBWCMN operates under its hosting agency's institutional arrangements and its success is currently measured by its longevity or ability to raise funding, and membership numbers. By better engaging with managers of threatened vegetation remnants, it is thought that such remnants will improve in condition, although condition improvements are hard to measure and no actual estimates in improvement in condition have yet been made.

Feedback from members currently drives the direction of the GBWCMN. Members are surveyed biannually to determine their needs, in a mail-out that accompanies the network's newsletter *Woodland Wanderings*. More recently, network members have been surveyed to determine if their current site-management includes any recommended best-practice. This feedback has to be correlated with the National Recovery Plan (DECCW 2010) and NSW Priority Actions for endangered ecological communities (OEH 2012a) that guide the operation of the CMN. The coordinator also has to keep abreast of government policy and direction, influencing this where possible, and keeping the members updated.

Success in accessing funding programs to maintain the GBWCMN also means the focus of the network changes, depending on the particular priority of the funding program. For example, the recent funding from the Australian Government's Caring for our Country program required the development of partnerships and landscape approaches to conservation of woodlands. Earlier funding from the Lachlan CMA aimed to collect information on the conservation values of travelling stock reserves and routes, while the original funding from the former Natural Heritage Trust centred on building network membership. None of the priorities are incongruous with the aims of the CMN; they just mean the focus constantly changes.

Major successes of the CMN model in NSW

Long-term programs develop profiles in the community, gaining respect and credibility. At the time of writing, the GBWCMN has been operating for 14 years, currently mailing out to more than 1500 members. As outlined below, the GBWCMN has demonstrated a wide array of innovations in its years of operation.

The GBWCMN has established and continues to develop a range of web-based tools, with resources that include grassy woodland education material and extensive resources such as videos, a grassy ecosystem website with management advice and reference lists and the NSW Grassy Ecosystems photostream (http://www.flickr.com/photos/nswgrassyecosystems/). There are increasing opportunities for using social networking to link the GBWCMN members.

The GBWCMN, in partnership with the NSW OEH, CMAs and the Livestock Health and Pest Management Authority, has for several years raised the profile and documented the important biodiversity assets of travelling stock reserves among the community and government agencies, commissioning surveys focusing on vegetation and woodland birds at over 800 sites and covering over 30 000 ha.

In early 2008, the GBWCMN coordinator was approached by an Australian Government agency in regard to the development of an environmental stewardship program. The agency sought the support of the network and its members for a stewardship initiative focusing on Box-Gum Grassy Woodlands. Network members had been discussing the need for such an initiative for many years. The GBWCMN coordinator became a member of the Australian Government Consultative Committee for the Environmental Stewardship Program. This initiative has been hugely successful in Box-Gum Grassy Woodlands across the network's area and beyond, providing managers of the threatened community with resources to implement best management practice, while encouraging 'in-perpetuity' conservation covenants. Several such covenants have now been signed between stewardship site's owners and the NSW OEH.

Network members have often raised the issue of receiving conflicting management messages. In response, in late 2009, the GBWCMN, in partnership with OEH, established a partnership of government and non-government groups that resulted in a Landcare

NSW-led project titled 'Communities in Landscapes: Working together to integrate conservation and production across Box-Gum Woodlands'. Alongside Landcare NSW, OEH and NSW Department of Primary Industries, partners included CSIRO, University of Sydney and Native Grasses Association Inc. (STIPA). The bringing together of these partners provided an opportunity to integrate management advice for production and conservation across agencies, science institutions and NGOs. The partners achieved many on-ground outcomes across the project area, including (i) the production of a series of educational videos and trialling of a *Virtual Woodland Excursion* attended by over 200 rural students; and (ii) the Cross Property Biodiversity Conservation Planning initiative, in which groups of neighbouring landholders in six targeted landscapes participated in identification and connectivity conservation management workshops and then developed and began implementation of property-focused, and landscape-scale connectivity conservation plans.

Major lessons

There is little doubt that the size of the network's geographic area affects how the coordinator and members interact. Focusing on a smaller geographic area allows for greater personal interaction between coordinators and members. This was an important function of the Monaro Grassland CMN.

The much broader geographic coverage of the GBWCMN means the coordinator may only interact remotely with some members via newsletters and websites, when current funding demands focus in a particular geographic region. This is observed by Fitzsimons and Wescott (2007, 2008), who found that there was less communication between managers of sites within the GBWCMN, compared with other multi-tenure conservation networks in south-eastern Australia, and suggested this was likely to be due to the relatively wide geographical separation between those sites. Yet the broad geographic coverage also means the network has a broader landscape perspective in addressing landscape threats and management issues, while also providing a landscape perspective of how the ecological community is recovering by communicating with a diverse and large membership.

CMNs in NSW have relied on discontinuous sources of temporary project funding. The CMN model and its successes always excite government policy-makers and community leaders; however, this has not translated into secure and regular funding.

While the GBWCMN has been funded continuously for 14 years, this has been from at least five different project funding sources, which means the CMN's functions have constantly had to adapt to meet the criteria of particular funding bodies. Limited funding for natural resource management has created competition among practitioners where there should be collaboration and partnerships. Partnered projects initiated and facilitated by the network to achieve part of a recovery action have provided much-needed resources. Key agencies and organisations seek to collaborate and partner with a CMN in order to access their membership base but some of these relationships have weakened once contact with members has been made. Public acknowledgement of CMN contributions pivotal to many projects may determine future funding opportunities for the network

There is often no acknowledgement from other NRM groups and agencies that landholders and managers can benefit by being part of a CMN. It has been suggested to the Australian Government that Environmental Stewardship holders, particularly those with little knowledge about managing their grassy ecosystem sites for conservation, would benefit from the management advice provided by the CMN. However, membership information (including the fact that membership is free) has never been forwarded to stewardship holders. The reasons for this are unclear.

Applying the lessons

The CMN model relies on having a strong, clever and effective coordinator who can sense opportunities when they arise. Successful coordinators have some connection or understanding of the agricultural landscape and a passion for the grassy ecosystems in which they are working. These qualities are not easy to write into a job description.

The adaptive capacity of CMNs should be encouraged as a means of meeting network members' needs and funding priorities. It is important, however, that the broader community gets a consistent message. CMNs are networks of remnants managed for conservation and their land managers who share social and biodiversity conservation objectives.

A strength of the CMN is its willingness to work with everyone and share all its information, expertise and contacts without reservation, but this also poses challenges. There needs to be greater and more formal recognition of the CMN's contributions to partner projects than just the inclusion of a network logo. Formal acknowledgement and public recognition must be achieved.

The need for partnerships and collaboration has been the mantra of the GBWCMN, making possible many funding opportunities that would otherwise not have occurred. The development of relationships that share decision making (partnerships) takes an exorbitant amount of time and effort, and this is not currently considered in funding applications. Funding bodies need to acknowledge this and allow for inclusion of these costs in future programs.

The challenge of a CMN covering a large geographic area has been met in the past by the GBWCMN in partnership projects that have employed additional regionally based staff to facilitate field events and site visits. This model requires more funding than is usually available, but offers the advantage that these 'woodland officers' often continue to promote the CMN and its aims well after they gain new NRM roles in their local area.

In future, equivalent resources and support extended to woodland officers could be provided by the GBWCMN coordinator to a range of other community network facilitators working with groups in smaller geographic areas (e.g. Landcare and other NRM facilitators) that fall within the network's boundaries. The coordinator could continue working with educational and research institutions to broker knowledge and build facilitator capacity to deliver coordinated on-ground recovery actions.

To address the issue of its large geographic area, the GBWCMN proposes to take a coordination role in capacity-building of facilitators from the growing number of smaller geographically based community groups (including other CMNs) across the extent of Box-Gum Grassy Woodlands. This will ensure that consistent and up-to-date best management practice advice for Box-Gum Grassy Woodlands is effectively delivered by local people in local landscapes. The GBWCMN proposes to provide regular 'train the trainer' opportunities to facilitators and other NRM extension officers (e.g. CMAs and local governments) and to improve opportunities to connect practitioners and facilitators via social and mobile media.

If CMNs are to continue their success and growth, which to date have been reliant on Australian Government funding, there is a need for a national coordinator to represent

CMNs at various levels of government and to provide guidance and support to new networks as they emerge. A national coordinator could further streamline NRM advice, reporting and tracking of the recovery of ecological communities nationally, providing economies of scale for ecological communities that cross state jurisdictions, or for nationally similar communities, such as the fragmented grassy woodlands and grasslands in the eastern and western states.

Conclusions

Conservation Management Networks in NSW have shown they are effective in providing for the conservation of widely dispersed remnants of rare vegetation communities occurring in fragmented agricultural landscapes, as Thiele and Prober (1999, 2000) originally proposed. Making CMNs a permanent fixture within the landscape is now the greatest hurdle, particularly given that the successes of CMNs are often low-key and behind the scenes. Heralding the achievements of CMNs and constant promotion must be a core part of the coordinator's job.

To achieve coordinated landscape benefits for listed ecological communities, CMNs need long-term, consistent financial support. CMNs in NSW are the champions of ecological communities and their managers. Governments acknowledge this by the inclusion of CMNs in national recovery plans and state priority action statements, yet there is no commensurate long-term funding. CMNs suffer from the same funding blockages that affect all natural resource management organisations that cannot currently collect payment for their output. Without consistent funding there is limited point in building a membership-base of land managers interested in updated best practice information. Ideally CMN coordinator positions would become permanent positions with local, state or Australian Governments, which have legal obligations with regard to protecting endangered ecological communities.

Finally, a national coordinator of CMNs could represent CMNs at multiple levels of government and provide guidance and support to new networks as they emerge. A national coordinator could streamline NRM advice, reporting and tracking the recovery of ecological communities, while ensuring CMN activities are given due recognition and acknowledgement.

Biographies

Toni McLeish is the coordinator of the Grassy Box Woodland CMN, a position she has held since 2003. Formerly, while living on her farm at Manilla, NSW, Toni was a Woodland Officer in the early stages of this CMN.

Rainer Rehwinkel is a Senior Threatened Species Officer with the Biodiversity Conservation Section of the Office of the Environment and Heritage in Queanbeyan, NSW. He is responsible for the conservation of grassy ecosystems and their associated flora and fauna. A co-author of a popular field guide for grassland flora, he initiated the Southern Tablelands Grassy Ecosystems CMN.

Lorraine Oliver is a Threatened Species Officer with the Biodiversity Conservation Section of the Office of the Environment and Heritage in Queanbeyan and she has a lead role in supporting CMNs.

References

DECCW (2010). *National Recovery Plan for White Box–Yellow Box–Blakely's Red Gum Grassy Woodland and Derived Native Grassland. A critically endangered ecological community.* Draft for Public Comment. Department of Environment, Climate Change and Water NSW, Sydney. Available: http://www.reec.nsw.edu.au/geo/woodland/text/wood1e.doc [Accessed 16 July 2012].

Eddy D (2007). The Monaro Grassland Conservation Management Network: reconnecting the sward. *Ecological Management & Restoration* **8**, 165–176.

Environment ACT (2005). *National Recovery Plan for Natural Temperate Grassland of the Southern Tablelands (NSW and ACT): An Endangered Ecological Community.* Environment ACT, Canberra.

ERIC (2001). *Remote Sensing Detection of Native Grasslands using Multi-Image Spectral Analysis – South Eastern Highlands of NSW.* Prepared for NSW National Parks and Wildlife Service by Environmental Research and Information Consortium Pty Ltd.

Fallding M (2002). A *Planning Framework for Natural Ecosystems of the ACT and NSW Southern Tablelands.* Natural Heritage Trust, NSW National Parks and Wildlife Service and Land & Environment Planning.

Fitzsimons JA and Wescott G (2005). History and attributes of selected Australian multi-tenure reserve networks. *Australian Geographer* **36**, 75–93.

Fitzsimons JA and Wescott G (2007). Perceptions and attitudes of land managers in multi-tenure reserve networks and the implications for conservation. *Journal of Environmental Management* **84**, 38–48.

Fitzsimons JA and Wescott G (2008). The role of multi-tenure reserve networks in improving reserve design and connectivity. *Landscape and Urban Planning* **85**, 163–173.

OEH (2012a). About the NSW Priorities Action Statement. Office of Environment and Heritage. Available: http://www.threatenedspecies.environment.nsw.gov.au/tsprofile/home_PAS_new.aspx [Accessed 29 June 2012].

OEH (2012b). NSW BioNet. Office of Environment and Heritage. Available: http://www.bionet.nsw.gov.au/ [Accessed 29 June 2012].

Prober SM, Thiele KR, and Higginson E (2001). The Grassy Box Woodlands Conservation Management Network: picking up the pieces in fragmented woodlands. *Ecological Management & Restoration* **2**, 179–188.

Rehwinkel R (2002). Our heritage – grasslands and grassy woodlands. *Austral Bugle. Newsletter of the Southern Tablelands Conservation Management Network in association with the Grassy Box Woodland Conservation Management Network* **1**(1), 1.

Thiele KR and Prober SM (1999). Reserve concepts and conceptual reserves: options for the protection of fragmented ecosystems. In: *Temperate Eucalypt Woodlands in Australia: Biology, Conservation, Restoration and Management.* (Eds RJ Hobbs and CJ Yates) pp. 351–358. Surrey Beatty and Sons, Chipping Norton.

Thiele K and Prober S (2000). Conservation Management Networks – a model for coordinated protection and management of remnant vegetation. In: *Balancing Conservation and Production in Grassy Landscapes, Proceedings of the Bushcare Grassy Landscapes Conference, Clare, South Australia 19–21 August 1999.* (Eds T Barlow and R Thornburn) pp. 58–63. Environment Australia, Canberra.

13

The Great Eastern Ranges Initiative: a continental-scale lifeline connecting people and nature

Ian Pulsford, Gary Howling, Rob Dunn and Rosemary Crane

Introduction

The Great Eastern Ranges Initiative is a continental scale connectivity corridor of essentially natural vegetation that extends along the eastern seaboard of Australia from the Grampians in Victoria, through eastern New South Wales (NSW) and the Australian Capital Territory (ACT), to the Atherton Tablelands and beyond in far North Queensland. These rugged uplifted lands form the watershed of the Great Dividing Range, tablelands and the Great Escarpment of Eastern Australia (Oilier 1982). They extend for 3600 km, or more than 28 degrees of latitude. These ancient geological features range in elevation from 100 m on parts of the escarpment to over 2228 m above sea level on the Great Dividing Range at Mount Kosciuszko.

Australia is one of Earth's 17 mega diverse nations (Mittermeier and Mittermeier 1997), with 22 000 flora and 6794 vertebrate fauna species, including some 1350 endemic terrestrial vertebrates, which is the highest number for any nation (Chapman 2005). The greatest concentration of this biodiversity is found along the eastern side of the continent and comprises a substantial part of Conservation International's 35th Global Biodiversity Hotspot, the 'Forests of East Australia' (Williams *et al.* 2010).

Context

The Great Eastern Ranges Initiative (GER) has evolved in several stages (Worboys 1996; Pulsford *et al.* 2004, 2012). Australia's first continental-scale corridor concept was announced on 27 February 2007 by The Hon. Bob Debus MP, NSW Minister for the

Environment, as a strategic adaptive response to a range of threats to biodiversity including biome shifts caused by climate change (Woodford 2007). It aims to help conserve one of Australia's richest concentrations of wildlife (Mackey *et al.* 2010), including the highest concentrations of bird and mammal species (NLWRA 2002), and the most intact ecosystems remaining along the eastern seaboard of the continent (Worboys 1996; Pulsford *et al.* 2004, 2012). This can be achieved by protecting existing connectivity and re-establishing linkages. The announcement recognised the need for habitat conservation and management of lands beyond protected areas. These ranges contain catchments that provide clean water to most of eastern Australia's population.

The current composition of the Great Eastern Ranges Initiative

Embedded in a matrix of mostly interconnected forests, woodlands and shrub lands is an archipelago of hundreds of protected areas, wilderness areas and wetlands, including three World Heritage areas. Surrounding or interconnecting the protected areas are large areas of state forest, other public lands, leasehold and private land. Often there are natural linkages inland and to the coast. Much of the private land is used for agricultural production activities including grazing and horticulture; production forestry is common on both public and private lands. Mining (primarily coal and gravel extraction) is well established in several landscapes (notably the Hunter Valley in central eastern NSW), and expansion of peri-urban settlements continues around most population centres.

During the first 4-year establishment phase, on-ground conservation activities largely focused in five priority regional connectivity 'gaps' and 'buffer' areas in NSW: (i) Border Ranges (north-east NSW/south-east Queensland see Chapter 15); (ii) Hunter Valley; (iii) Southern Highlands; (iv) South Coast (Kosciuszko2Coast), and (v) South-west Slopes (Slopes to Summit; see Chapter 14). A new partnership – the 'Jaliigir Alliance' was established in May 2012 (Figure 13.1).

Long-term vision for the corridor

With the passage of time and with increased community understanding and feedback, the corridor vision which was originally limited to the Australian Alps to Atherton (DECC 2007) has evolved to encompass a corridor extending over the full length of the Great Dividing Range: *To conserve and manage a 3,600km 'continental lifeline' of habitats, landscapes and people, that will support the continued survival of native plants and animals along the Great Eastern Ranges from the Grampians in Victoria to far north Queensland and maintain the natural processes on which they depend* (OEH 2012).

History and evolution of the initiative

The GER Initiative was established by the NSW Department of Environment and Climate Change (now the Office of Environment and Heritage [OEH]) in mid 2007 with the formation of a core team of staff. The NSW Government allocated $6.7 million through the NSW Environmental Trust. During this period four respected non-government conservation organisations (Bush Heritage Australia, OzGREEN, Greening Australia NSW, Nature Conservation Trust of NSW) and the Department of Environment and Climate Change, established five regional partnerships in priority connectivity gaps (Figure 13.1). By mid 2011, some 120 organisations were variously involved as members of these regional partnerships, participating in conservation activities ranging from community education and advocacy, to the negotiation of voluntary habitat protection instruments and habitat

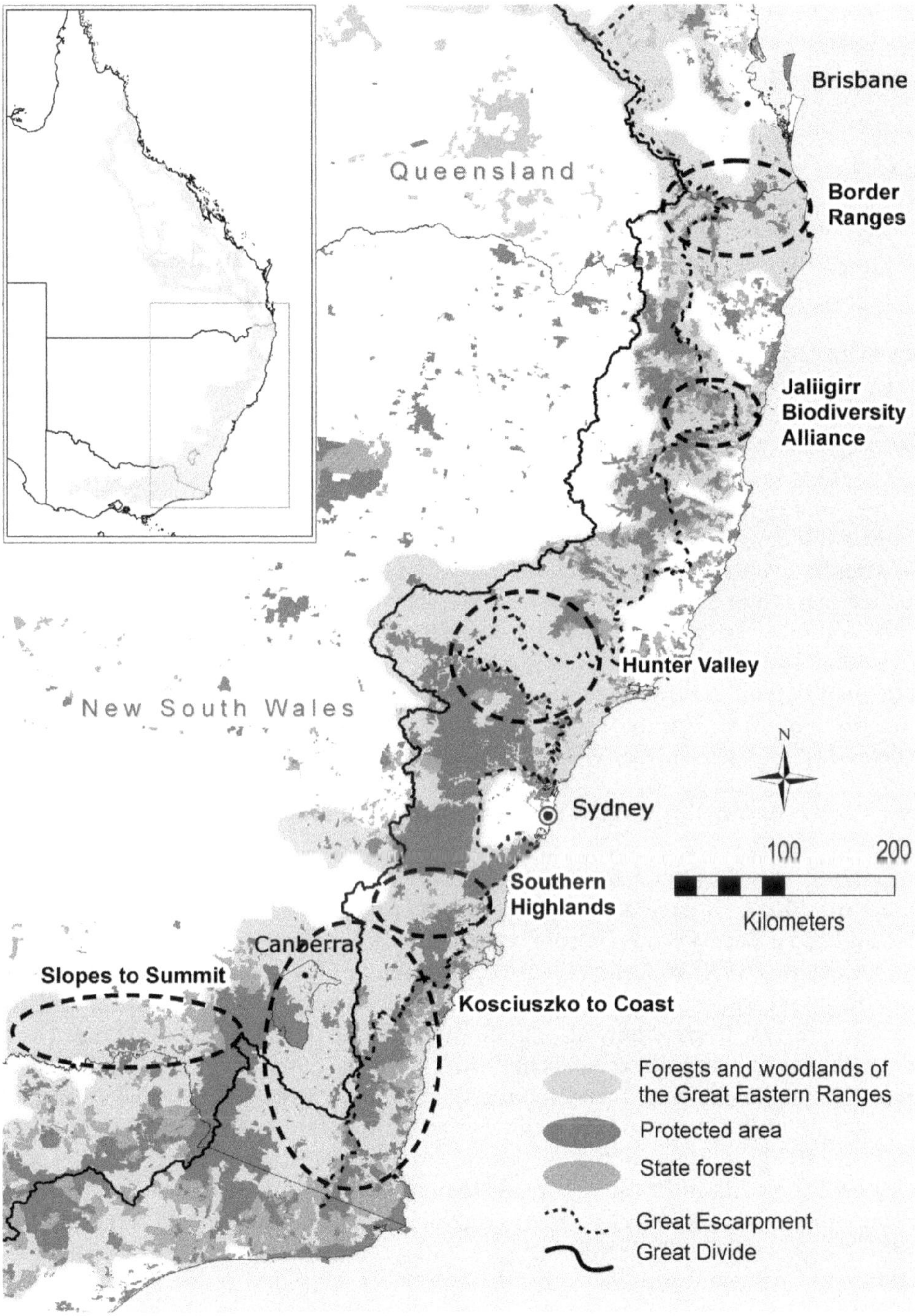

Figure 13.1 The Great Eastern Ranges corridor extends from the Grampians in western Victoria, through NSW and the Australian Capital Territory (ACT) to far north Queensland. Dashed circles represent priority regions for establishment of connectivity conservation partnerships. Parts of the corridor at the southern terminus (e.g. the Grampians) are somewhat isolated remnants of the Great Dividing Range that emerge from the surrounding plains.

rehabilitation. These organisations included non-government conservation organisations, public agencies, catchment management authorities, Aboriginal groups, local government, industry (mining and agriculture) and Landcare.

In December 2011, the NSW Government committed a further $4.4 million to consolidate the transition to a community-led initiative. The funds were allocated to: (i) existing partnership areas; (ii) promote the GER vision for a national corridor; and (iii) establish new regional partnership areas, which included linking the New England Tablelands with the NSW north-east coast (the 'Jaliigirr Biodiversity Alliance').

During 2012, the Australian Government implemented two world-leading initiatives to support large landscape-scale habitat restoration and connectivity conservation projects: (i) introduction of the Clean Energy Futures Initiative (including establishment of the $946 million Biodiversity Fund); and (ii) development of a *National Wildlife Corridors Plan* (see Chapter 17). In April 2012, the Australian Government Environment Minister, The Hon. Tony Burke MP, announced the allocation of funding to 317 projects Australia-wide, including over $25 million over 6 years for several GER partnership projects. This included funding for a link from the Greater Blue Mountains World Heritage Area to the inland rivers of the Murray–Darling Basin (the Kanangra Boyd to Wyangala Link).

While the focus to date has been on NSW, the momentum of the GER Initiative has generated increasing interest from national and state non-government conservation organisations and other existing partnerships and networks operating across the full extent of the corridor. These organisations and partnerships are being invited to become 'affiliates' and to identify their existing programs as contributing to the GER vision. This is expanding the reach of the GER Initiative considerably and is fostering increasing involvement of partners in NSW, Victoria, the ACT and Queensland.

Governance arrangements and organisational structure

During the first 3 years, a non-government group of lead partners and Department of Environment and Climate Change worked to built trust in one another. This took considerable time and investment. This led to governance being transitioned from the department to a Lead Partners Group consisting of Greening Australia NSW, National Parks Association of NSW, Nature Conservation Trust of NSW, OzGREEN and the Office of Environment and Heritage (Figure 13.2). This arrangement was formalised by a Memorandum of Understanding signed in July 2010. Greening Australia, as an incorporated body, administers funds on behalf of the partners. In 2010, a Chief Executive Officer (CEO) was appointed who reports to the Lead Partners Group and coordinates day-to-day operations. A regional facilitator was funded or in some cases co-funded in each of the established partnership areas, to provide local leadership in partnership activities. The CEO and Lead Partners provide overall coordination. Facilitators provide the essential link between regional partners (e.g. Slopes to Summit (Chapter 14) and Border Ranges (Chapter 15)) and the CEO (Figure 13.2). A business plan prioritises activities, guides expenditure and provides a framework for reporting to investors (DECC 2007; OEH 2012).

The Initiative encourages landholders, industry groups, Traditional Owners, government agencies, non-government organisations and others via newsletters, a website and field days to join the partnership or collaborate in delivering their respective skills and programs. The combined effort serves to create alignment of conservation activities in

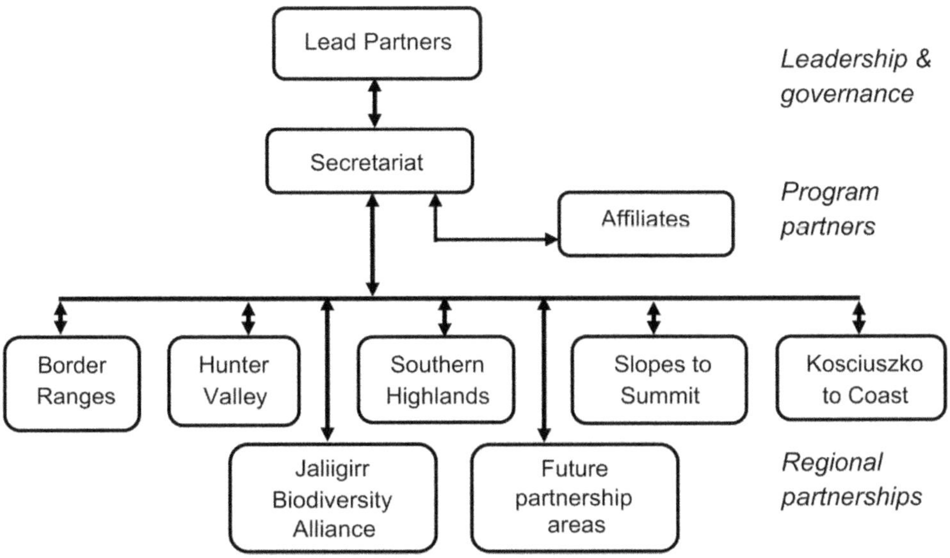

Figure 13.2 Great Eastern Ranges Initiative governance structure. Lead partners include non-government organisations Greening Australia, OzGREEN, Nature Conservation Trust (NSW), National Parks Association (NSW), and the NSW Government's Office of Environment and Heritage.

identified priority areas (gaps or hotspots), which in turn combine to influence connectivity on a wider scale.

Major successes

During the 4-year start-up and establishment phase the following key factors were judged to have been important to the success of the Initiative.

1. A bold vision

A bold connectivity conservation vision is a critical element that can provide the context for local individual actions. It generates impetus for collective local planning for parts of the corridor. This Initiative implements a polycentric style management approach where 'big picture' biodiversity conservation outcomes are achieved thanks to a clear vision, quality local planning guidance and integration of multiple individual and organisational efforts (Worboys *et al.* 2010; Worboys and Pulsford 2011). The timely achievement of on-ground successes is vital for sustaining community interest and involvement. This social capital is easily lost as governments frequently change or terminate programs after only a few years causing 'program fatigue', apathy, frustration and cynicism.

2. A small secretariat can catalyse major co-investment

A small and multi-skilled secretariat working with regional facilitators to coordinate implementation and communicate the corridor vision has proven essential and effective. Five separate regional partnerships were established in connectivity gaps identified by analyses conducted by the former Department of Environment and Climate Change. These analyses identified intact forests, woodlands and shrublands and included three World Heritage Areas, large numbers of protected areas, wetlands of international importance and three nationally recognised biodiversity hotspot areas (DSEWPC 2012a) (see Figure 13.1).

3. Offer an integrated suite of conservation instruments

The best outcomes are achieved if a suitable mix and range of conservation instruments are made available to landholders and managers in priority target areas. The suitability and availability of this mix varies in each region. Instruments range from land purchase for conservation protection and restoration, voluntary in-perpetuity conservation covenants, wildlife refuges, non-competitive and competitive grants and many others. This mix enables landholders to choose what best suits them in differing circumstances and increases the likelihood of engagement. Conservation planning to select target areas must use the best available data and prioritisation techniques (Meir *et al.* 2004; Barrett *et al.* 2010).

Over 4 years, 156 voluntary agreements were established with landholders, to improve the protection and management of 16 195 ha of habitat in the priority connectivity areas identified in regional partnership Conservation Action Plans (Table 13.1). Projects successfully completed with partners delivered $3.56 million in Environmental Trust funds to landholders. Just under $16 million was leveraged from partners as co-contributions. This included $12.32 million of in-kind contributions to match Environmental Trust investment in a ratio of 4:1. New cash contributions from partners matched $3.55 million of Environmental Trust funds in a ratio of 1:1 (OEH 2011).

Table 13.1. Number and area of voluntary agreements funded by the Great Eastern Ranges Initiative from 2007–2011 (Source: OEH 2011)

Type	Number	Area (hectares)
Property purchase (Bush Heritage Australia, Nature Conservation Trust of NSW – revolving/covenanting fund)	6	4989
In-perpetuity conservation agreements (*National Parks and Wildlife Act* and *Nature Conservation Trust Act*)	86	9526
Long-term (15-year property vegetation plan, OEH wildlife refuge)	34	1220
Other (whole-of-paddock restoration, Landcare term agreements)	30	460
TOTAL	156	16 195

4. Supported by the best available science

A landmark report on the scientific principles for connectivity conservation and the Great Eastern Ranges was published in 2010 (Mackey *et al.* 2010). It provides a strong scientific foundation for landscape-scale connectivity conservation of Australia's unique ecosystems. Many of these principles have been incorporated into *Australia's Biodiversity Conservation Strategy 2010–2030* (NRMMC 2010), draft *NSW Biodiversity Strategy* (DECCW 2010) and the *National Wildlife Corridors Plan* (DSEWPC 2012b).

Major lessons

1. Engage champions

The NSW Environmental Trust appointed a subcommittee to oversee the development of business plans (DECC 2007), as well as monitor progress and reporting in the GER. The subcommittee of 10 members included representation from catchment management authorities, local government, conservation organisations, farmers' associations, government agencies (environment and primary industry), NSW Aboriginal Land Council and a university. In hindsight, the addition of a representative from the philanthropic and business sectors to assist in securing non-government funding may have added value. The appointment of an independent patron to promote the vision may also have opened doors to business, philanthropic and political leaders to secure more funding and support.

2. Increase investment in building capacity

Building relationships takes considerable time skill and care. Success in achieving on-ground delivery has depended significantly on a small secretariat to guide the efforts of local facilitators to harness the capacity and willingness of partner organisations. Financially, for the Environmental Trust it was not possible to maintain a multi-disciplinary departmental-based secretariat team for longer than the first 3 years. During this time governance was transferred from the OEH to the mainly non-government Lead Partners Group outlined above. This group retains a secretariat of one Environmental Trust-funded centrally-based CEO, supported by a part-time OEH-based technical officer, part-time contract communications manager, program support officer and regional facilitators. More extensive training in project management, communication, media, facilitation and relationship management would be highly beneficial for increasing the skills and capacity of the facilitators (see Figure 13.3). Appropriate milestones and metrics for this critical type

Figure 13.3 Lauren Van Dyke, Kosciuszko2Coast Regional Facilitator, welcoming participants at the Kosciuszko2Coast community open day event at Bredbo in May 2008. Over 300 people from the surrounding region attended. The open day featured presentations, displays, information booths, and inspections of conservation activities, equipment, food and entertainment. (Photo: Ian Pulsford)

of investment (and other investments in building social capital) should have been included in the early business plans and thus investment in this critical work was not appreciated by decision makers.

3. Recognise the needs of the development phase of projects of this scale and measure against appropriate milestones

This program aimed to gain the support of private landholders as well as already established local conservation and industry organisations. It relied on the team's capacity to communicate the bold vision, establish trust and the understanding that the Initiative is scientifically credible and achievable.

The work during this phase was the foundation for all that followed. The 'measures for success' during this phase must be appropriate for this task. While early on-ground work is an important part of managing community expectations and building support for the idea, it should not be the primary marker for success in the development phase. This period is by definition about establishing the science, identifying priorities, creating capacity and establishing relationships for the long term. The time needed to develop trust-based relationships in particular, are generally underestimated and undervalued in the assessment of the progress in the development stage. A sufficient start-up period, measured against phase-appropriate milestones with a surety of funding to see it through to well beyond the start-up phase, will allow for the most efficient process in the long term. These milestones should be included in the Business and Operations Plans.

4. A program for the future

The GER Initiative established a successful government–private sector partnership for the delivery of cross-regional, whole of landscape conservation in NSW. The requirements for this approach to be successful are diverse and complex. For example, the development of systems and processes that enable partners to have confidence that their efforts would have positive impacts are required. Engagement of regional delivery partners proved to be a good model for realigning existing programs with those needed to achieve the shared vision. It also required the rapid transfer of leadership of the Initiative from NSW Government to a partnership with community-based organisations.

5. Expanding the Great Eastern Ranges Initiative and its governance to a national scale

Expansion of the GER Initiative beyond NSW to include the Victorian, ACT and Queensland sections is a key objective. Between 2007 and 2010, there was a particularly rapid and frequent change in Environment Ministers in NSW and other jurisdictions. This 'loss of corporate/cultural memory' and change in government priorities inhibited effective trans-boundary and integrated management of the GER corridor (Pulsford *et al.* 2012). 'Loss of memory' and momentum also occurred within the program delivery partners when there was rapid turnover of staff, potentially resulting in lack of awareness, indifference or cynicism. This loss of social capital caused loss of momentum and reduced effectiveness in program delivery.

Applying the lessons

From the lessons and challenges outlined above, the next steps to overcome roadblocks are likely to include the following.

1. Expanding to national scale

Recent changes in Australian Government policy are providing a new impetus and leadership required to achieve a continental-scale GER corridor (see Chapter 17). Trans-state boundary governance of large-scale corridors is difficult when there are different rates of implementation and support for the concept by different governments. The *National Wildlife Corridors Plan* and *Australia's Biodiversity Conservation Strategy* (NRMMC 2010) provide a much needed national framework for efforts to secure trans-boundary agreements at Environment Minister level and between non-government organisations. A priority is to consolidate negotiations that have been underway for several years, with government and non-government organisations in Victoria, the ACT and Queensland. In the long term, participation by governments in each jurisdiction at every opportunity must be taken.

2. Reform of legislation, policy and instruments

Further policy reform is needed at state and national levels to provide national-scale adaptation approaches for the conservation of ecosystems and biodiversity as climate changes. For example, there are several policies, support frameworks and instruments available which would otherwise help to facilitate implementation of this corridor on the ground (Pulsford *et al.* 2012). These include better use of conservation offsets policies in order to optimise landscape conservation outcomes. However, this situation is changing rapidly. Recent developments include changes in the Australian Government's policies on biodiversity and climate change adaptation such as the introduction of a tax on carbon pollution, the establishment of the Biodiversity Fund, and the *National Wildlife Corridors Plan* is providing a much needed national policy and science framework for corridor establishment and management (DSEWPC 2012b; Chapter 17).

3. Need for ongoing government support

It is a basic responsibility of the Australian Government to fulfil its international obligations in relation to biodiversity conservation. The emergence of large-scale biological corridors is now a recognised means of achieving this. For this approach to be effective there is a need for basic financial support from governments for a small governance entity and supporting secretariat to drive and coordinate action in the right places. This is because opportunities for securing philanthropic support in Australia are less available than in (for example) the USA (Pulsford *et al.* 2012). In addition, further efforts are required to involve the philanthropic sector in the GER. For GER, one solution may be to invite one or more philanthropic partners or business leaders to join the Lead Partners Group or become affiliates.

4. Resolve data licensing, standards and exchange arrangements

An absence of (i) standardised GER corridor-wide baseline condition data; (ii) a capacity to track change in condition; and (iii) a coordinated research effort for GER are significant constraints to effective planning and management. This inevitably means that achievements may not be adequately recognised. For this data to be assembled requires a considerable level of cooperation between catchment management authorities, state and federal agencies and private conservation bodies. Much work has already been undertaken, however, significant barriers remain. Privacy legislation is often cited as a barrier to data exchange between organisations. Increasingly these data will be required to provide meaningful reporting of progress as a requirement for funding for programs carried out under national and state funding bodies. This issue needs to be resolved urgently by all parties starting with leadership at National Ministerial Council and soon to be established National Wildlife Corridors Council level.

Conclusions

The Great Eastern Ranges Initiative has been effective in laying the foundations for one of Australia's largest connectivity conservation programs. Large-scale corridors are not a panacea to save all our biodiversity as climate change biome shifts occur. There will be many species and ecosystems lost or altered as conditions change. However, the GER corridor provides a unique opportunity to protect ecosystems that extend over the greatest range of elevation and climatic conditions anywhere in Australia. The corridor's natural advantages will provide more opportunities for more species and ecosystems to adapt or move than in any other part of Australia. This is because it contains Australia's richest concentrations of species (Mackey *et al.* 2010); including one of our greatest concentrations of endemic species (Chapman 2005) and large numbers of climatic refugia (Mackey *et al.* 2012). In this sense the GER corridor is 'an Ark' that has a unique role in the conservation of biodiversity and ecological processes in Australia as climate changes.

Achieving this bold vision will take more than a lifetime. In moving towards this goal the GER Initiative's greatest achievements during the first 4 years include: (i) successfully navigating the difficult start-up phase; (ii) building widespread community awareness and support; and (iii) commencing on-ground implementation.

The great challenges for the future include: (i) the need to strengthen governance and leadership; (ii) the expansion of partnerships beyond NSW; (iii) securing permanent funding for a small secretariat to provide leadership, governance and coordinate program delivery; (iv) securing long-term funding for on-ground connectivity conservation activities; (v) further development of effective communications infrastructure to maintain partner and community involvement; and (vii) the development of an effective monitoring, evaluation and reporting system.

Acknowledgements

We acknowledge the visionary leadership provided by the Lead Partners Group (Greening Australia NSW, National Parks Association of NSW, Nature Conservation Trust of NSW, OzGREEN and OEH), the Hon. Bob Debus MP (former NSW Minster for the Environment), NSW Environmental Trust, NSW Environment Trust GER sub-committee, secretariat staff, regional facilitators, partners and affiliates whose involvement and commitment have been instrumental in its progress to date.

Biographies

Ian Pulsford is a specialist in protected areas and linking landscapes, with over 32 years' experience in conservation policy and practice with the New South Wales Government including selection, design and management of protected areas. He was a divisional manager with the NSW Department of Environment and Climate Change and then the founding manager of the Great Eastern Ranges Initiative from 2007 to 2010. He is a member of the IUCN World Commission on Protected Areas and has served on various government committees, including as an independent expert on a panel advising the Australian Government on the draft *National Wildlife Corridors Plan.*

Gary Howling is a senior policy officer with the New South Wales Government. Since early 2008 Gary has worked as principal conservation analyst with the Great Eastern Ranges Initiative. In this role, he has provided government agencies, non-government organisations, and private conservation partners with specialist scientific, technical and conservation assessment advice to guide the development of a wide-ranging portfolio of projects. Gary has a wide background in regional planning and landscape-scale conservation programs, biodiversity and native vegetation conservation policy, community engagement and conservation science brokering.

Rob Dunn is the CEO of the Great Eastern Ranges Initiative. Prior to this position Rob was the CEO of the Nature Conservation Trust of NSW for 4 years, an organisation that protects high conservation value private land under in-perpetuity conservation agreements. From 2000 to 2004, Rob was the Company Secretary of Landcare Australia. The role involved the negotiation of sponsorship agreements with large corporates to support this national community-based movement. Rob is a chartered accountant and has gained broad commercial experience over 20 years before working in the not for profit sector. He gained a Masters in Environmental Management in 2005 from Charles Sturt University. He is a current board member of BirdLife Australia.

Rosemary Crane is a specialist in strategic communications and is also a documentary and video producer. Rosemary was responsible for community engagement and communications for the Great Eastern Ranges Initiative during its formative period and was ultimately responsible for developing and managing the partnership process to take the Initiative beyond the initial funded period. Rosemary's background includes the development of communications strategies for government and business organisations as well as work in community engagement. Rosemary manages a production business with a substantial track record in corporate solutions.

References

Barrett T, Drielsma M, Love J, Howling G, and Anderson G (2010). *Modelling of Fauna Habitat and Connectivity Values to Support Conservation Planning Decisions in the Hunter Region of NSW as Part of the Great Eastern Ranges Initiative.* Department of Environment, Climate Change and Water, Sydney.

Chapman AD (2005). *Numbers of Living Species in Australia and the World.* Department of Environment and Heritage, Canberra.

DECC (2007). *Alps to Atherton Initiative: NSW Business Plan 2007–2010.* Department of Environment and Climate Change NSW, Sydney.

DECCW (2010). *Draft New South Wales Biodiversity Strategy 2010–2015.* Department of Environment, Climate Change and Water NSW, Sydney.

DSEWPC (2012a). Australia's 15 National Biodiversity Hotspots. Available: http://www.environment.gov.au/biodiversity/hotspots/national-hotspots.html [Accessed 1 July 2012].

DSEWPC (2012b). *National Wildlife Corridors Plan: A Framework for Landscape-scale Conservation.* Department of Sustainability, Environment, Water, Population and Communities, Canberra.

Mackey B, Watson J, and Worboys GL (2010). 'Connectivity Conservation and the Great Eastern Ranges Corridor'. ANU Enterprises Pty Ltd. Report to the Department of Environment, Climate Change and Water, Sydney.

Mackey B, Berry S, Hugh S, Ferrier S, Harwood TD, and Williams KJ (2012). Ecosystem greenspots: identifying potential drought, fire, and climate-change micro-refuges. *Ecological Applications* **22**, 1852–1864.

Meir E, Andelman S, and Possingham HP (2004). Does conservation planning matter in a dynamic and uncertain world? *Ecology Letters* 7, 615–622.

Mittermeier RA and Mittermeier CG (1997). *Megadiversity: Earth's Biologically Wealthiest Nations*. CEMEX, Mexico City.

NLWRA (2002). *Australian Terrestrial Assessement*. National Land and Water Resources Assessment, National Heritage Trust, Canberra.

NRMMC (2010). *Australia's Biodiversity Conservation Strategy 2010–2030*. Natural Resource Management Ministerial Council, Commonwealth of Australia, Canberra.

OEH (2011). 'Great Eastern Ranges Initiative'. A Report to the NSW Environmental Trust describing funded activities from 2007 to 2011, Sydney.

OEH (2012). *Great Eastern Ranges – Business Plan 2012–15*. Office of Environment and Heritage, Sydney.

Ollier CD (1982). The Great Escarpment of Eastern Australia: tectonic and geomorphic significance. *Journal of the Geological Society of Australia* **29**, 13–23.

Pulsford I, Worboys GL, Gough J, and Shepherd T (2004). The Australian Alps and the Great Escarpment of Eastern Australia Conservation Corridors. In: *Managing Mountain Protected Areas: Challenges and Responses for the 21st Century*. (Eds D Harmon and GL Worboys) pp. 106–114. Andromeda Editrice, Italy.

Pulsford I, Worboys GL, Howling G, and Barrett T (2012). Australia's Great Eastern Ranges Corridor. In: *Climate and Conservation: Landscape and Seascape Science, Planning, and Action* (Eds JA Hilty, CC Chester and MS Cross) pp. 96–105. Island Press, Washington DC.

Williams KJ, Ford A, Rosauer DF, De Silva N, Mittermeier R, Bruce C, Larsen FW, and Margules C (2010). Forests of East Australia: The 35th Biodiversity Hotspot. In: *Biodiversity Hotspots: Distribution and Protection of Conservation Priority Areas*. (Eds FE Zachos and JC Habel) pp. 295–310. Springer, Heidelberg.

Woodford J (2007). 2800km coastal route for wildlife to escape warming. *Sydney Morning Herald* **27 February 2007**. Available: http://www.smh.com.au/news/environment/route-for-wildlife-to-escape-warming/2007/02/23/1171734021093.html [Accessed 18 July 2012].

Worboys GL (1996). Conservation Corridors and the NSW Section of the Great Escarpment of Eastern Australia. Paper presented to the IUCN World Conservation Congress. Montreal, Canada.

Worboys GL and Pulsford I (2011). 'Connectivity conservation in Australian landscapes'. Report prepared for the Australian Government Department of Sustainability, Environment, Water, Population and Communities on behalf of the State of the Environment Committee 2011, Canberra.

Worboys GL, Francis W, and Lockwood M (Eds) (2010). *Connectivity Conservation Management: A Global Guide*. Earthscan, London.

14

Slopes to Summit: focusing on bite-size landscapes that matter

Peter G. Spooner, Sam Niedra, Veronica A.J. Doerr and Christine Cumming

Introduction

Slopes to Summit (S2S) is one of six regional partnerships of the Great Eastern Ranges Initiative (GER), one of the world's largest connectivity conservation programs (Chapter 13). It was established in 2007 by the former New South Wales (NSW) Department of Environment and Climate Change (DECC, now NSW Office of Environment and Heritage (OEH)) and the Nature Conservation Trust of NSW (NCT). S2S is a regional partnership of land management and scientific research organisations, that aims to achieve connectivity conservation outcomes in the NSW South Western Slopes Bioregion – a high priority area for biodiversity conservation (e.g. Benson 2008).

The S2S study area encompasses eastern sections of the upper Murray River and Billabong Creek catchments, from areas west of Culcairn to Albury, to Khancoban and Kosciuszko National Park in the east (Figure 14.1). The project is aligned to the eastern domain of the Murray Catchment Management Authority (MCMA) implementation area, and principle population centres include Albury, Holbrook, Culcairn, and Tumbarumba. The region has a predominantly cool temperate climate, where topography mainly consists of low undulating hills and flat riverine plains in the west, to larger mountainous blocks in the east, with occasional granitic outcrops scattered throughout the region.

The S2S connectivity corridor consists of a patchwork of large blocks and remnant patches of native vegetation, some of which are interconnected, situated in a mosaic of rural land uses (e.g. cropping, grazing and pine forestry). In less fertile country, there are several protected areas that extend in a partly interconnected chain from the high country

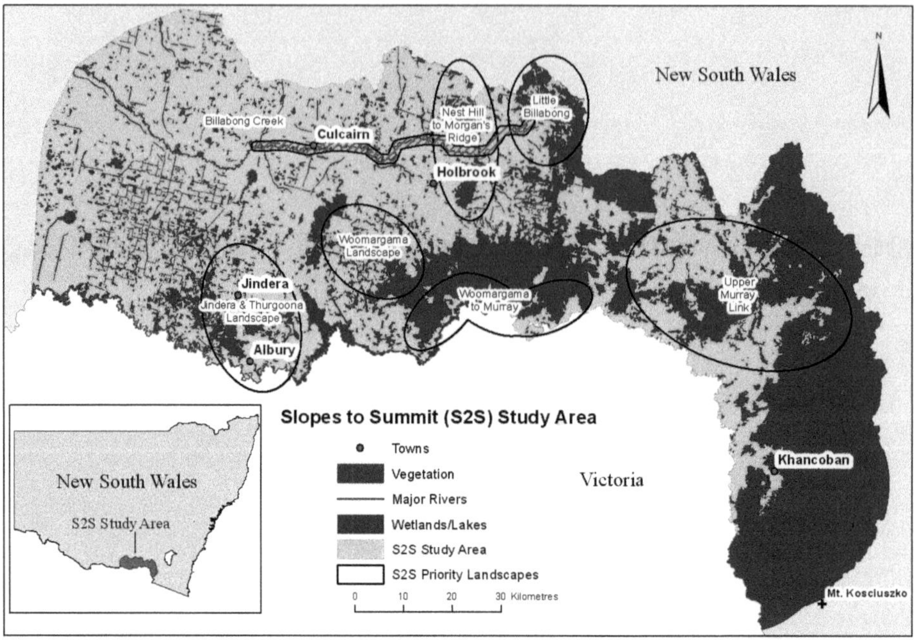

Figure 14.1 The Slopes to Summit (S2S) landscape, part of the Great Eastern Ranges Initiative, situated primarily in the NSW South West Slopes bioregion. Major areas of remnant vegetation are shown, and priority local landscapes indicated.

of the Kosciuszko National Park, which is located on the Great Dividing Range, to Woomargama National Park on the south-west slopes. There are many other smaller patches of remnant vegetation on the surrounding private land (Figure 14.2). In more fertile country in the west or on valley floors, much of the native vegetation has been

Figure 14.2 The Woomargama priority local landscape, showing remnant woodland vegetation and scattered trees in an agricultural matrix. (Photo: Scott Hartvigsen)

cleared and what remains is highly fragmented. Here riparian areas, roadsides and Travelling Stock Reserves and Routes (TSRs) provide important refuges for endangered species and ecosystems (Spooner and Smallbone 2009; Lentini *et al.* 2011).

Owing to a strong altitudinal gradient, and variety of landforms and human land uses, there exists a diversity of species and remnant ecosystems within the S2S region. Key communities of conservation importance include endangered White Box–Yellow Box–Blakely's Red Gum Woodlands, Grey Box Woodlands and Native Temperate Grasslands. The region has over 58 threatened fauna and flora species, which include the Booroolong Frog (*Litoria booroolongensis*), Striped Legless Lizard (*Delma impar*), Brown Treecreeper (*Climacteris picumnus victoriae*), and the Squirrel Glider (*Petaurus norfolcensis*) (Figure 14.3), the latter being the focus of recent on-ground connectivity conservation activities as part of S2S (OEH 2012).

The S2S region was selected by DECC as a target for connectivity conservation activities, in recognition of its significant ecological values and range of physical characteristics described above. It builds upon previous community-based initiatives in the region, such as the *Upper Billabong Land and Water Management Plan* (UBLWMPG 1999), *Riverina Highlands Regional Vegetation Management Plan* (NSWDLWC 2003), the 2005 Holbrook Heartlands Initiative (CSIRO 2012), and other programs by the MCMA. The S2S initiative is governed by a steering committee, whose present representatives are: NCT, OEH, CSIRO, Charles Sturt University, the Australian National University, the MCMA, Holbrook Landcare Network, Parklands Albury-Wodonga, and Albury Conservation Co. The vision of S2S is: *To achieve a healthy Upper Murray and Billabong Creek catchment with resilient, connected and functional ecosystems, and to maintain and restore ecological processes to provide greater security for faunal movement and dispersal in response to climate change.*

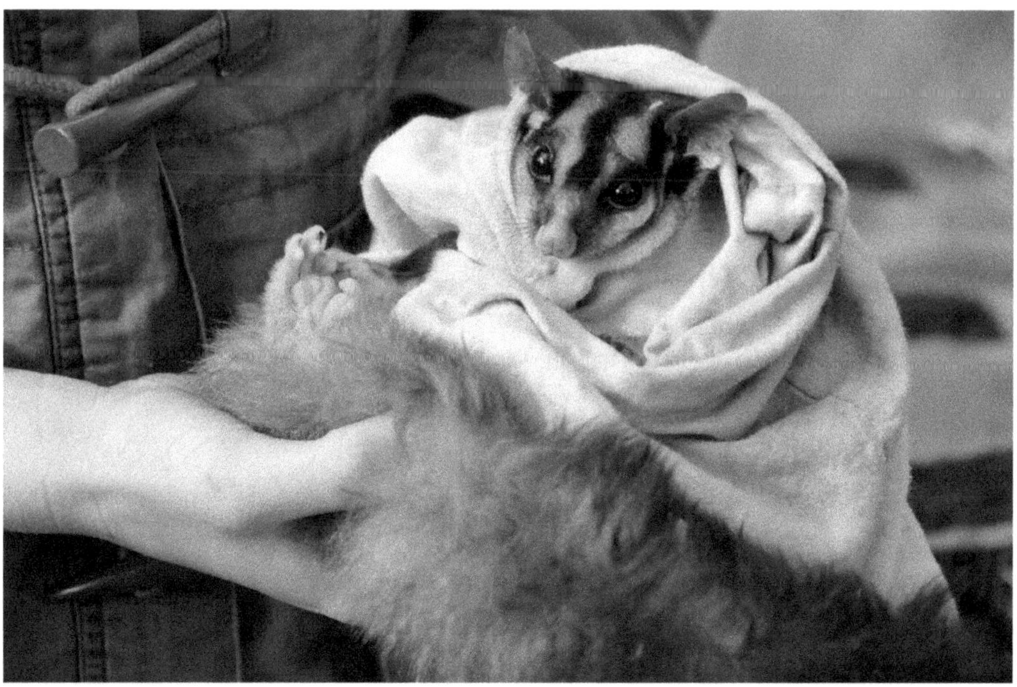

Figure 14.3 Squirrel Glider (*Petaurus norfolcensis*), a focus of on-ground connectivity conservation activities in the Slopes to Summit landscape. (Photo: Alex Bonazzi)

History

In August 2007, the NCT conducted a three-day workshop in Albury to launch GER. A Conservation Action Planning (CAP) process was conducted with regional stakeholders to determine major biodiversity assets, their threats, and to identify priority areas for investment. The workshop was a success in terms of attendance, where over 30 key natural resource management (NRM) stakeholders participated, and nine organisations later became regional partners.

A dedicated S2S Partnership Facilitator was provided by NCT (funded by NSW DECC from the NSW Environmental Trust Fund in 2008–09); a position that continues to provide a key leadership role. In 2008, NCT and DECC continued to drive the development of the S2S Strategic Plan, which was revised to articulate a vision, improve key goals, and identify complementary mechanisms for implementation. These revisions were completed in 2009. The CAP was reviewed and amended in late 2010, building on new 'key habitats and corridors' analysis provided by NSW DECC. These data, combined with further steering committee work, were used to define several focus (local) landscapes (Figure 14.1).

In 2010–11, S2S experienced a period of financial uncertainty, and faced the potential loss of its dedicated facilitator. This was because NSW DECC had not committed funds to continue the project at this point, and NCT funding was uncertain. In addition, applications to the Australian Government's 'Caring for our Country program' by S2S were unsuccessful. So in 2010, the S2S steering committee prepared a strategy to ensure the long-term future of the partnership, which included applications for grant funding from a variety of sources. Essentially S2S moved from a fully-funded NSW Government program, with a commensurate degree of resources, to a more independently funded and governed model. However, in December 2011, the NSW Government committed $4.4 million of funding over 4 years to GER, including funds for the S2S facilitator and on-ground works.

Present governance of S2S is mostly by consensus of the steering committee, based on agreed priorities and scientific expertise, where the S2S facilitator role is critical in maintaining momentum by coordinating S2S group meetings, and maintaining regular group correspondence. When the need arises, working groups are established to undertake matters of particular importance, such as funding applications. The group continues to seek financial sustainability via a combination of federal and state government, philanthropic and corporate sponsorship grants to implement on-ground works.

Major successes

The success of the S2S initiative can be measured against its aims by considering progress in terms of landscape-scale planning, on-ground activities, and funding for additional S2S-scale activities which are summarised as follows:

Landscape-scale planning

Although many partner organisations of S2S already had spatially explicit landscape plans in place for the South West Slopes region, the S2S initiative was able to provide greater information through the 'key habitats and corridors' analysis, where in 2010, maps were provided by the former NSW Department of Environment, Climate Change & Water (DECCW) on how to provide landscape connectivity for native species. This approach has now been used by other GER partnerships and many other landscape-planning groups in Australia, and has provided S2S with a stronger scientific basis with which to make decisions about on-ground actions specifically at landscape scales. Presently NCT is working with Charles Sturt University to further develop spatial mapping and prioritisation tools.

On-ground actions

Since 2007, NCT has implemented a range of conservation management incentives in southern New South Wales. To date, over 900 ha of high value conservation lands across 11 properties has either been protected by in-perpetuity conservation covenants under the *Nature Conservation Trust Act 2001* (NSW) or purchased under a revolving fund and covenanted when onsold, most of which are located in the S2S region. Although program outputs are modest in relation to the physical scale of S2S, these areas will be managed in perpetuity (covenant on title), and represent important building blocks for future programs. In addition, the MCMA has recently moved to align its management agreement and other incentive programs with S2S priorities where practical.

Partner organisations have also made significant efforts to coordinate their community education activities. To develop connectivity conservation partnerships such as S2S requires a large investment of time to work with key stakeholders and landholders, to develop trust and ownership of the project. In conjunction with continued on-ground liaison activities, S2S has successfully held five major community events, including the Biodiversity on Farms Field Day in May 2010. These events were attended by local farmers, townspeople and businesses, where key stakeholders assisted with on-ground and educational activities (NCT 2009, 2010, 2011).

Additional funding for S2S-scale activities

As OEH and NCT funding support was withdrawn in 2010, attempts were made to address funding shortfalls via grants applications, and to urgently seek out new and diverse partnership opportunities. To this end, the group has been successful in securing $171 400 to support S2S operations beyond 2012 from the following organisations: (i) *Caring for our Country (Community Action Grant) – Woolshed Thurgoona Landcare Group*, for revegetation works, hosting of community field days, and a community fauna census event; (ii) *Albury Conservation Co. (ACC) – Thurgoona*, where NCT, on behalf of S2S, has been contracted for 2 years to assist with the implementation of the ACC's 5-year Strategic Plan, which includes on-ground works, and community education activities aimed at conserving threatened species in the Jindera–Thurgoona S2S priority area (e.g. Squirrel Glider); (iii) *Atlas of Living Australia – Thurgoona biodiversity pilot project,* a biodiversity information sharing project; and (iv) *Bjarne K Dahl Trust Eucalyptus Grant,* to produce a documentary showcasing the biodiversity values of *Eucalyptus* species in the region ('Snow Gums to River Reds – a story of Eucalypts from Mount Kosciuszko to the Murray River').

In mid-2012, S2S, with Holbrook Landcare Network as the lead proponent, was successful in obtaining an Australian Government Clean Energy Future (Biodiversity Fund) grant for $2 297 000 over 6 years, which will help fund revegetation activities for carbon retention and connectivity aims, in S2S priority landscapes.

Major lessons

Method to launch a connectivity conservation group

S2S was launched in 2007 by inviting key stakeholders to a workshop, and engage in a Conservation Action Planning (CAP) process (see The Nature Conservancy 2012). Unfortunately the participants had little input on decisions regarding the boundaries of the region, or rationale on the use of the CAP planning tool; rather, they were facilitated though a CAP process by the NCT. Although useful, this approach caused immediate problems in terms of application, relevance and future governance of the working group, as it became clear that the CAP process in itself was not sufficient to provide the starting foundation for a connectivity conservation group.

Participation in the CAP process was mainly questioned by some participants due to its inability to provide the appropriate tools or framework to identify critical landscape linkages or corridors. This is because the CAP process was originally designed for site-scale rather than landscape-scale planning. As the new group experienced, the CAP is not designed to be a spatial prioritisation tool. The lessons learnt from this process (in hindsight) were to have spatial analysis conducted before focus meetings, to help participants make more informed decisions on priorities.

Secondly, and as highlighted by Whitten *et al.* (2011), the CAP is very much a generic project management tool, which relies on the expertise of participants to identify and prioritise key strategies and actions. However, much of this knowledge was already available in present planning instruments for the S2S region, to which most participants had already contributed much time and energy (e.g. NSWDLWC 2003). Therefore the use of a CAP was also questioned by participants in terms of its appropriateness in kick-starting a connectivity program in this particular region.

Importantly, a new connectivity conservation group should build upon existing planning instruments, both to capitalise on existing knowledge and recognise the prior work of all partners involved, which helps to create the necessary cooperative and trusting atmosphere for multi-partner conservation groups.

Governance – who is really steering this ship?

The CAP process facilitated by S2S should have clearly identified which partners had which responsibilities, definitions of success, and what would be monitored by whom to gauge success (see The Nature Conservancy 2012), but this did not clearly emerge from initial workshops or later meetings. Thus, an oversight of S2S during its development was its inability to clearly identify who was implementing the strategic actions. Much was often assumed. Initially the NCT provided strong leadership to the S2S partnership, and was responsible for conducting 'connectivity conservation activities' (NCT 2009). However, over time NCT was limited in the amount of resources it could direct to S2S. By the end of 2010, it was made clear to the steering committee by the Great Eastern Ranges Initiative and NCT, that to remain viable, the group had to find a more self-sustainable funding model.

As a result of this situation, key partners who originally joined S2S under a previous funding and governance framework (as 'participants' in 2007), found themselves in more of a stand-alone 'partnership' (via a steering committee in 2011), with a commensurate degree of responsibility to implement S2S actions. This situation was problematic, as many of the organisations represented by the S2S steering committee also lacked the resources to implement on-ground works at the scale desired by GER.

Fortunately, new funding provided by the NSW Government in 2012 to GER (from the Environmental Trust Fund) should help maintain desired momentum by funding the facilitator position. However, this financial change may realise new governance issues, as it represents a shift back towards a government-driven model. This issue highlights the pitfalls of implementing a large-scale conservation program without the long-term backing of a well-funded, well-resourced lead organisation(s).

Applying the lessons

Becoming self-sufficient

To date, S2S has admittedly struggled to implement many of the envisaged on-ground actions, while some of the activities have not matched key objectives. The main reasons for this situation are shown below.

- *A lack of funding for on-ground actions*, and where funding has been received, this *has not matched the scale of initiative visions and objectives to date*. Delivering the principles of connectivity conservation, as advocated by GER, is highly problematic on such a large scale when funding is so limited. As a result, current on-ground works are still fundamentally designed around the needs of site-scale conservation rather than landscape-scale conservation.
- *Lack of ongoing technical support*: Aerial photos and satellite imagery are the main decision-making tools used by S2S, because existing connectivity analysis demonstrated by OEH resides with OEH, and is not readily available. This situation has prevented the group from exploring GIS analysis at finer scales to guide on-ground actions. Even with GIS access, there are few resources to undertake appropriate analysis at the initiative scale.
- *Reliance on partners to do their 'bit'*. S2S relies largely on a model where the steering committee provides a forum to exchange information, explore collaboration activities and address S2S priorities. However, existing partners still operate within their own guidelines as – without specific S2S project funding – most on-ground actions in the region are largely business-as-usual. As Whitten *et al.* (2011) describe, a reliance on partners to achieve on-ground outcomes is fraught with risk, as partners can only undertake actions that fulfil their existing organisational goals. While some partners have moved to align their programs with S2S priorities (e.g. MCMA), all are somewhat limited in their abilities to do this.

By the end of 2011, the group realised that to remain a viable proposition, it needed to: (i) broaden its consideration of funding and partnership opportunities (irrespective of new government funding); and (ii) scale down activities to 'bite-size', local landscapes (see below).

Continental-scale project – a local landscape approach

Given that realised levels of funding were significantly below what was required to achieve conservation at a large scale, a certain degree of strategic planning was required to target on-ground activities towards smaller, more defined priority areas (approx. 15–20 km). In order to do this, S2S identified seven priority 'local landscapes' to target implementation works, by building on the CAP process, 'key habitats and corridors' analysis, but largely by expert opinions of participants (Figure 14.1). Within these local landscapes, the aim is to increase the size and quality of patches of native vegetation communities as well as connectivity between them.

In order to assess connectivity values when defining priority local landscapes, the group first needed to address a key question – *what is connectivity?* As described in the literature, landscape connectivity is the degree to which the landscape facilitates or impedes movement of a particular species among patches (Taylor *et al.* 1993). This concept can be divided into two constructs: (i) *structural connectivity* – the landscape features that physically 'link' discrete areas of habitat; and (ii) *functional connectivity* – the degree to which organisms actually move between discrete habitat patches, disperse, and cause gene flow (e.g. Crooks and Sanjayan 2006; Hilty *et al.* 2006), where a habitat patch is a discrete area of habitat occupied by a species, surrounded by areas of more unsuitable habitat described as the 'matrix' (Doerr *et al.* 2010).

Guided by research expertise within the group, S2S quickly moved from a structural interpretation of connectivity, to a more functional approach concerning keystone species. However, as data on functional connectivity for species are limited, S2S first used available 'key habitats and corridors' analysis at a relatively coarse level – to identify local landscapes in which targeted efforts (particularly matrix management) would likely produce landscapes with 'connected' patches of native communities (Figures 14.1 and 14.2). Building on the CAP process, the group then used expert advice (e.g. Doerr *et al.* 2011) to inform finer-scale on-ground decisions. Key priorities/approaches follow several management guidelines recommended for connectivity conservation projects (Mackey *et al.* 2010), namely to:

- utilise various instruments and mechanisms, and promote conservation management activities designed to minimise or eliminate threats in the matrix, in order to maximise ecological connectivity and permeability; and
- identify and develop a system of local and extensive corridors (e.g. patches to act as 'stepping stones', riparian corridors, road and stock route elements) for focus species.

S2S is presently developing a range of new and exciting connectivity conservation initiatives in the priority local landscapes, using research to help guide and evaluate on-ground decisions. In 2012–13, S2S will focus on-ground works in the peri-urban landscape of Thurgoona, an area experiencing rapid residential development, north of Albury. Here, S2S has been highly successful in facilitating partnerships between a diverse array of community groups, illustrating that connectivity conservation is not just about vegetative links, but also about connecting people around a common vision for the future of the landscape.

Conclusions

The S2S partnership has achieved some important on-ground success, and enabled the development of several major collaborative funding applications, integrating science and NRM planning expertise. In order to do this, S2S and GER have had to be successful in raising its profile in the community. This success is evidenced in recent funding success and present NRM approaches, where awareness of S2S goals has influenced the prioritisation of on-ground activities by our major partners – particularly in key priority landscapes. Despite some lingering governance problems, a strong regional partnership has developed, which provides an appropriate framework to attract new funding and achieve desired outcomes at local and regional scales.

The intent of S2S is to build on what existing partners are doing, with a focus on connectivity conservation. The problem was, when S2S had no funding, it was trying to call on partners' time and resources to obtain funds to continue. This was detracting and distracting from what partners were doing themselves, rather than adding value to it. Given the voluntary nature of partner commitments, an ongoing lack of funds (until recently) has made it very challenging to maintain the interest of the S2S partners. However, with new funding from the NSW Government (2012–15), the Australian Government's Clean Energy Future Biodiversity Fund (2012–17), and further potential support from philanthropic, business and other interests, considerable progress will be made towards the vision of S2S.

An established partnership is now in place to implement, and evaluate, via university and CSIRO partners, connectivity conservation programs in the region. Future funding will be directed to priority local landscapes, to achieve significant on-ground outcomes. That is when the real value of the corridor initiative will be seen – *with further measureable outcomes.* Otherwise the initiative could become largely symbolic. Climate change predictions underline the urgency for the broader community to continue to support this bold initiative.

Acknowledgements

The authors would like to thank the contributions of previous S2S facilitators Nicky Bruce and Alison Skinner, who compiled previous NCT reports which assisted in writing this chapter, Nigel Jones for his thoughtful suggestions to improve an earlier draft, NCT and Charles Sturt University for in-kind support, and Liam Grimmett for assistance with Figure 14.1.

Biographies

Peter Spooner is a senior lecturer at Charles Sturt University, and has 12 years' experience in vegetation and restoration ecology, with specific research interests in disturbance ecology of woodland ecosystems, connectivity and road ecology. On the latter, Dr Spooner is an expert in biodiversity conservation and environmental history of Travelling Stock Routes and Reserves (TSRs). He has published over 25 peer-reviewed journal articles, and has extensive experience with fencing incentive programs, vegetation condition assessments, and liaison work with farmers and government agencies.

Sam Niedra has a Degree in Applied Science (NRM) and 12 years' experience in the NRM industry in the not-for-profit and local government sectors. Speciality areas include native vegetation management and restoration, and community engagement and capacity building. These skills have been developed in rural and urban landscapes, across a variety of roles including native seed collector, Landcare coordinator, council NRM Officer, and S2S partnership facilitator since July 2010 (employed by NCT).

Veronica Doerr has been researching animal dispersal for 16 years, using her knowledge to ensure landscape-scale conservation initiatives really will allow native species to move through landscapes where and when they need to. As a Senior Research Scientist with CSIRO, Dr Doerr's work involves a blend of radio-tracking dispersing individuals, modelling landscape processes, translating results into useful information for land managers, and assisting with the development of government policies for landscape conservation.

Christine Cumming has a B.Sc (Hons), and has worked in public and private land management as both a scientist, government officer and farmer. Chris is currently the Executive Officer of Holbrook Landcare Network, a Not-for-Profit organisation delivering a range of environmental and agricultural projects and services in the S2S region. The focus of Chris's work over the last 4 years is to engage, educate, inspire and support private land managers to be part of a resilient farming community (economically, environmentally and socially), with a strong culture of care for the land.

References

Benson JS (2008). New South Wales vegetation classification and assessment: Part 2. Plant communities of the NSW South-western Slopes Bioregion and update of NSW Western Plains plant communities, Version 2 of the NSWVCA database. *Cunninghamia* **10**, 599–673.

Crooks KR and Sanjayan M (Eds) (2006). *Connectivity Conservation*. Cambridge University Press, Cambridge, UK.

CSIRO (2012). Improving Murray Darling Health – Heartlands. Available: http://www.csiro.au/partnerships/psc8#a2 [Accessed 30 January 2012].

Doerr VAJ, Doerr ED, and Davies MJ (2010). *Systematic Review #44: Does Structural Connectivity Facilitate Dispersal of Native Species in Australia's Fragmented Terrestrial Landscapes?* Collaboration for Environmental Evidence, Bangor, UK.

Doerr VAJ, Doerr ED, and Davies MJ (2011). Dispersal behaviour of brown treecreepers predicts functional connectivity for several other woodland birds. *Emu* **111**, 71–83.

Hilty JA, Lidicker WZ, and Merenlender AM (2006). *Corridor Ecology: The Science and Practice of Linking Landscapes for Biodiversity Conservation*. Island Press, Washington, DC.

Lentini PE, Fischer J, Gibbons P, Hanspach J, and Martin TG (2011). Value of large-scale linear networks for bird conservation: a case study from travelling stock routes, Australia. *Agriculture Ecosystems & Environment* **141**, 302–309.

Mackey BG, Watson J, and Worboys GL (2010). 'Connectivity conservation and the Great Eastern Ranges corridor: An independent report to the Interstate Agency Working Group (Alps to Atherton Connectivity Conservation Working Group)'. Environment Heritage and Protection Council/ Natural Resource Management Ministerial Council, ANU Enterprises Pty Ltd, Canberra ACT.

NCT (2009). *Slopes to Summit Program (GER) Partnership Facilitation: Final Project Report*. Nature Conservation Trust of NSW, Albury, NSW.

NCT (2010). *Slopes to Summit Program (GER) Partnership Facilitation: Final Project Report*. Nature Conservation Trust of NSW, Albury, NSW.

NCT (2011). *Great Eastern Ranges Initiative Final Report. Slopes to Summit (S2S) Partnership Facilitation*. Nature Conservation Trust of NSW Albury, NSW.

NSWDLWC (2003). *Riverina Highlands Regional Vegetation Management Plan*. New South Wales Department of Land and Water Conservation, Sydney.

OEH (2012). Threatened Species, Populations and Ecological Communities of NSW. NSW Office of Environment and Heritage, Sydney. Available: http://www.threatenedspecies.environment.nsw.gov.au/tsprofile/index.aspx [Accessed 30 January 2012].

Spooner PG and Smallbone L (2009). Effects of road age on the structure of roadside vegetation in south-eastern Australia. *Agriculture, Ecosystems & Environment* **129**, 57–64.

Taylor PD, Fahrig L, Henein K, and Merriam G (1993). Connectivity is a vital element of landscape structure. *Oikos* **68**, 571–573.

The Nature Conservancy (2012). *Conservation Action Planning*. Available: http://www.conservationgateway.org/topic/conservation-action-planning [Accessed 30 January 2012].

UBLWMPG (1999). 'Upper Billabong Land & Water Management Group plan'. Unpublished report for Holbrook Landcare, Albury NSW.

Whitten SM, Freudenberger D, Wyborn C, Doerr VAJ, and Doerr ED (2011). 'A Compendium of Existing and Planned Australian Wildlife Corridor Projects and Initiatives, and Case Study Analysis of Operational Experience'. A report for the Australian Government Department of Sustainability, Environment, Water, Population and Communities. CSIRO Ecosystem Sciences, Canberra, ACT.

15

The Border Ranges Alliance: a Great Eastern Ranges regional partnership

Georgia Beyer and Lynn Baker

Introduction

The Border Ranges partnership region spans the far north-east of New South Wales and south-east corner of Queensland. The region aligns with the area covered by the *Border Ranges Rainforest Biodiversity Management Plan* (DECCW 2010) and extends along the coast from Beenleigh in the north to Evans Head in the south and as far west as Killarney (Figure 15.1). The region occupies over 1.5 million hectares (DECCW 2010).

The Border Ranges region is one of the most biologically diverse landscapes in the world, supporting a large number of rare and endemic species. This outstanding biodiversity is recognised by the Australian Government's identification of the Border Ranges North and South (Queensland and New South Wales) Biodiversity Hotspot (DSEWPC 2012) and its World Heritage status. The 'Forests of East Australia', of which the Border Ranges is part, has recently been identified as the 35th Global Biodiversity Hotspot (Williams *et al.* 2011). The only other listed global hotspot in Australia is south-west Western Australia. The Border Ranges region is also one of eight National Landscapes recognised and promoted by Tourism Australia (2012).

The national parks and state forests of the Border Ranges region mostly occur in higher elevation areas of over 300 m (Figure 15.1). Substantial areas of the slopes and plains have been cleared for agriculture and much of the forest on the mid-slopes and ranges has been harvested for timber (DECCW 2010). The high population growth and associated clearing for urban development along the coast and the lower valleys and flats is a major cause of habitat loss and fragmentation (DSEWPC 2012). However, the system of national parks is linked by a network of corridors of vegetation on public and private land that are generally

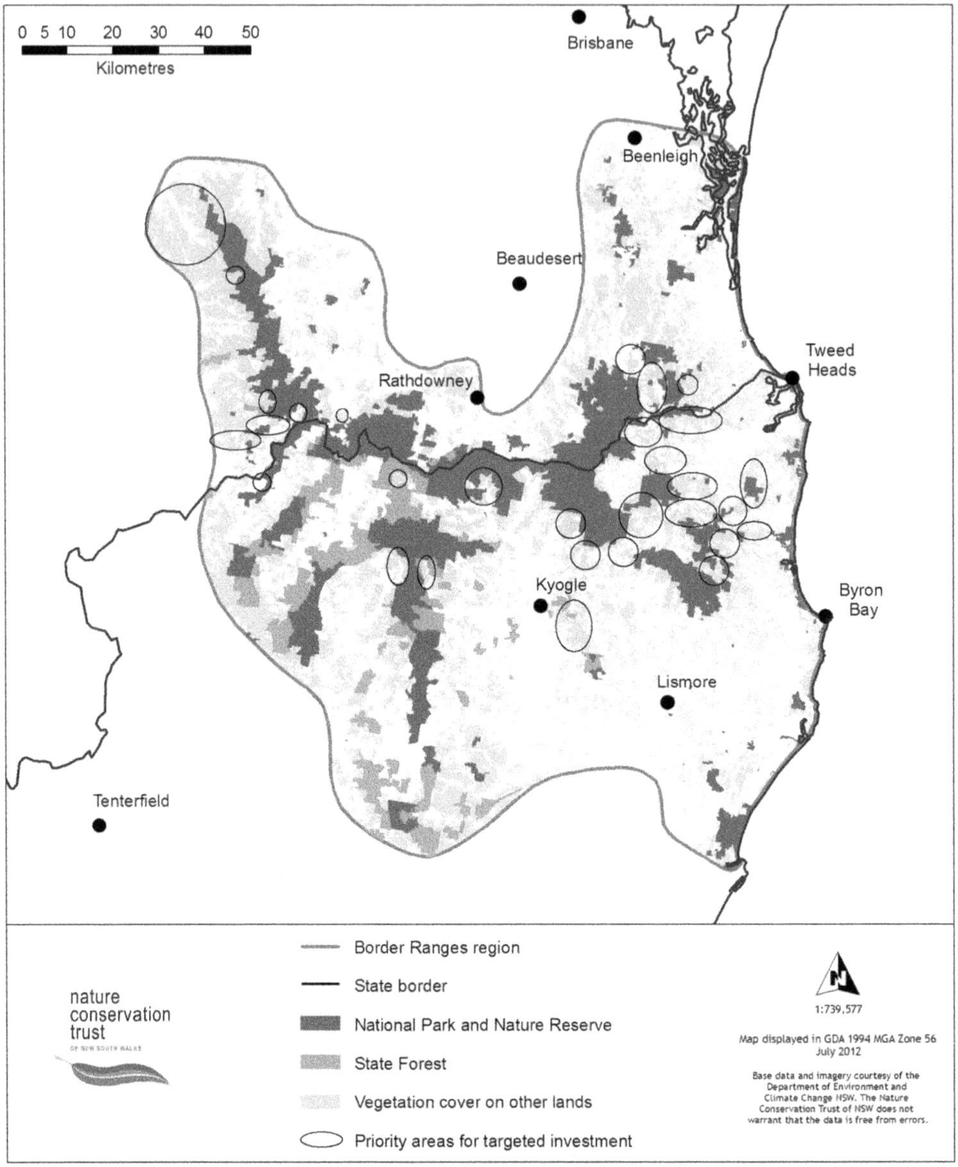

Figure 15.1 The extent of the Border Ranges Alliance region in north-east New South Wales and south-east Queensland.

well connected, but in some places contain areas that have cleared or degraded vegetation. Private lands make up 76% of the region, 15% is public protected area and 5% is state forest. Around half of the national park estate is part of the Gondwana Rainforests of Australia World Heritage Area. There are numerous legally binding and non-legally binding conservation agreements that protect native vegetation on private land in the region (Halliday *et al.* 2012). In 2010, 0.5% of the region was protected via legally binding private land conservation agreements (DECCW 2010). Native vegetation covers 49% of the region of which subtropical rainforest is the most extensive, followed by wet sclerophyll forest (DECCW 2010). Weed invasion is a major threat to conservation and connectivity.

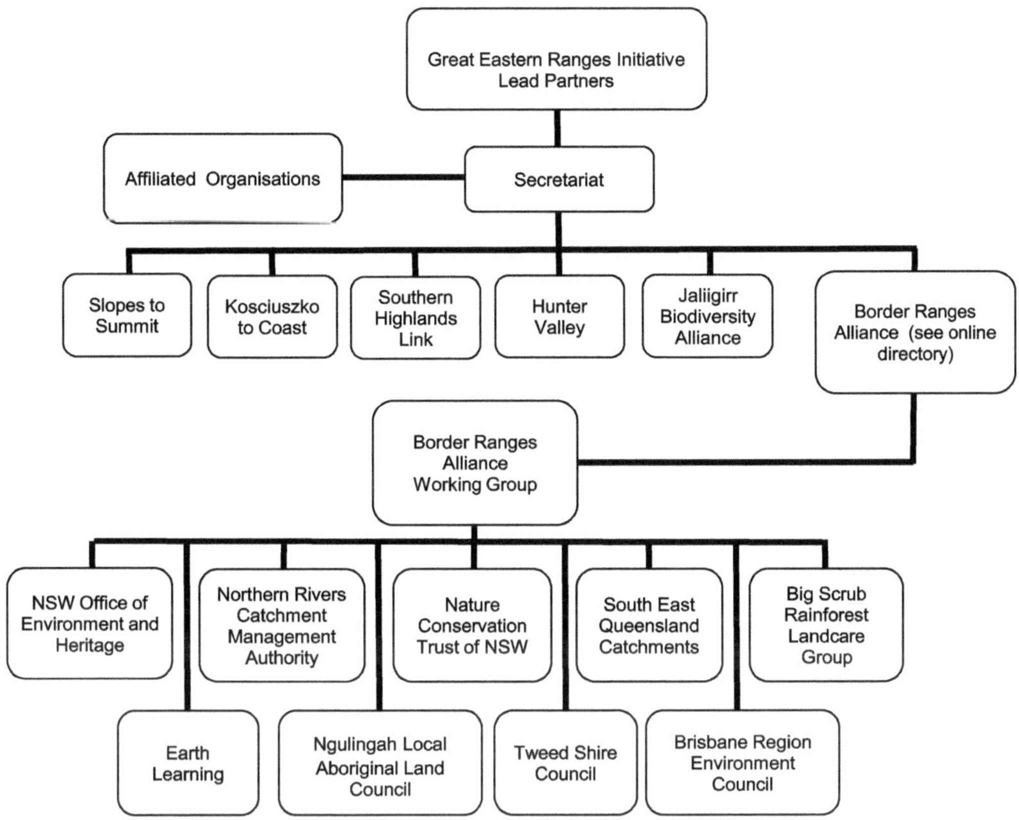

Figure 15.2 Organisational relationship between the Border Ranges Alliance as part of the Great Eastern Ranges Initiative.

The Border Ranges Alliance (the Alliance) is one of six regional partnerships formed as part of the Great Eastern Ranges (GER), an initiative to create a landscape-scale conservation corridor along the Great Dividing Range and Great Escarpment of eastern Australia from Victoria to north Queensland (Chapter 13). The Alliance is made up of a diversity of groups from north-east New South Wales and south-east Queensland. It includes over 40 representatives of federal, state and local government organisations, natural resource management (NRM) organisations, community, environmental, Aboriginal and Landcare groups (see Figure 15.2).

In addition to the overarching goals of the Great Eastern Ranges, the Border Ranges Alliance has the following aims.

1 To develop and maintain relationships with and between Border Ranges Alliance partners.
2 To provide information (and coordination where relevant) on regional activities, conservation strategies and research that is relevant to the goals of the Great Eastern Ranges Initiative.
3 To develop a collective understanding about how individual and collaborative efforts contribute to the vision of the Great Eastern Ranges Initiative.
4 To agree on a package of activities that will provide a focus for efforts and that will attract industry, community and government support.

5 To investigate and seek additional resources for activities in the Border Ranges.
6 To identify opportunities to invite industry and education groups to participate in the Alliance.

History and structure

A meeting was held in early 2008 to discuss the beginnings of a Great Eastern Ranges partnership in the Border Ranges. Attended by representatives of several natural resource management groups in the region, there was broad support for a partnership and it was decided that the Nature Conservation Trust of NSW would become the host of a Border Ranges Regional Facilitator position for the following 12 months. Funding was provided by the NSW Environmental Trust through the then NSW Department of Environment, Climate Change and Water. The GER Initiative and the Nature Conservation Trust each provided 50% in-kind contribution to the facilitator position. The Nature Conservation Trust continues to host the Border Ranges Facilitator.

A second forum of partners was held in mid-2009 where the nature and structure of the partnership were further developed and Terms of Reference drafted. The group decided the partnership was more aptly called an 'alliance' and the Border Ranges Alliance was formed.

The Border Ranges Alliance is not a formal or incorporated body and there is no formal membership. Instead, the groups considered to be members are those willing to be listed in a 'directory' of Border Ranges Alliance Groups on the Great Eastern Ranges website. At present there are over 40 listed groups. As per the Terms of Reference, membership is voluntary and is open to any organisation, community group, individual, agency, public authority, business, industry group, education institution or non-government organisation that has an active interest in the health of the environment within the Border Ranges.

The Terms of Reference that outline the nature of the partnership, its aims and how it operates, were developed by the members of the Alliance, placed on the GER website and adopted at the annual forum (Great Eastern Ranges Initiative 2012). A Working Group was established which generally meets every two or three months. The Working Group has cross-sector representation including state and local government, Landcare, non-government, NRM and Aboriginal organisations and has representatives from both Queensland and New South Wales (Figure 15.2).

The Alliance meets annually, and in recent years has helped determine the work plan of the facilitator. The Working Group oversees implementation of the work plan to define specific tasks and to track progress, which is then reported back to the Alliance. This ensures the facilitator is working to achieve regionally relevant outcomes that best assist the Alliance groups in their connectivity works.

The *Border Ranges Rainforest Biodiversity Management Plan* is a federally funded regional recovery plan established under the *Environment Protection and Biodiversity Conservation Act 1999* that covers the Border Ranges partnership area (DECCW 2010). It uses a Biodiversity Forecasting Tool (for spatial analysis) provided by the NSW Office of Environment and Heritage and expert and community knowledge to identify priority areas for repair and conservation of native vegetation. The Plan also contains several maps including connectivity corridors that identify strategic areas for conservation efforts to aid adaptation to climate change (DECC 2007), endemic fauna refugia and key habitat for forest fauna (Scotts 2003).

The Biodiversity Forecasting Tool and additional mapping was used to identify priority connectivity areas within the region specifically for the Great Eastern Ranges Initiative (Figure 15.1). The Alliance uses the Management Plan and the priority connectivity areas to guide its on-ground works and implementation. Some member groups also have their own strategic documents that relate to connectivity conservation, such as the *Northern Rivers Catchment Action Plan* (NRCMA 2005), *South East Queensland Catchments Strategic Investment Plan 2009–2014* (SEQC 2010) and *Byron Biodiversity Conservation Strategy* (Byron Shire Council 2004). The Alliance activities contribute to delivering targets within these strategic documents.

Major successes

The Great Eastern Ranges Initiative provides an overarching vision, a set of goals and resources, and at the same time encourages flexibility in implementation within the regional partnerships. This has allowed the Alliance to identify what the needs are within the Border Ranges, how to be most regionally relevant, and how best to add to what is already occurring.

Within the Border Ranges there are many well established non-government and community groups, and NRM community support officers providing support for smaller groups and landholders. The focus of the Border Ranges facilitator has been to act as a liaison to integrate the efforts of these groups and provide support to those who are already engaged with landholders in conservation works and improving landholder capacity. This has created a multi-tiered partnership that contains a wealth and diversity of knowledge, skills and experience. The Alliance provides an opportunity for partners to combine their knowledge, skills and experience to create integrated projects, reduce project overheads and duplication of effort.

The Alliance takes advantage of existing strategic plans, primarily the *Border Ranges Rainforest Biodiversity Management Plan*, so that the development of a new conservation plan has not been required as it has in other Great Eastern Ranges corridor regions.

Along with funding for a regional facilitator, the Great Eastern Ranges Initiative has provided funds for on-ground works. Funding has been distributed through small grant programs, overseen by the working group, to which partner groups can submit applications. This has provided direct benefits to groups that receive grants, and has helped to maintain the involvement of partner groups in the Alliance. Specific projects have included weed control and bush regeneration to enhance connectivity in mapped corridors between protected areas, development of landholder agreements and community engagement events that promote connectivity conservation.

Aboriginal communities have been actively engaged in developing biodiversity and cultural values management plans for community-held freehold lands, as well as fire plans through the Rural Fire Service's Hotspots Fire Project (fire planning for protection of life, property and biodiversity). The communities have used these plans to successfully seek NRM funding to resource on-ground management activities, training and capacity building of community youth and employment of Aboriginal Green Teams. Aboriginal Green Teams and NRM teams are promoted as preferred contractors by the Alliance Working Group.

A *Cultural Connections to Wahlubal Country in the Great Eastern Ranges* multi-media resource kit was developed in conjunction with community elders and local primary schools to share cultural biodiversity knowledge with schoolchildren in the region. There is also a *Cultural Connections to Nimbin Rocks* multi-media resource kit that demonstrates natural resource management being implemented by the Ngulingah management team on community-owned land at Nimbin Rocks. In addition, several communities have declared

Indigenous Protected Areas over their lands, expanding the National Reserve System within the Border Ranges.

A partnership with the Atlas of Living Australia (2012) will provide an online system to record and collate NRM activity data so that the contribution of Alliance projects towards improvements in connectivity can be more easily measured and communicated. The Atlas of Living Australia (ALA) pilot project has adapted the ALA Citizen Science tool for recording species observations to allow recording of activity data. For example, instead of submitting a record of observing 10 honeyeaters, a project manager from an Alliance group will be able to submit a record of 10 trees being planted. The total number of trees planted across all Alliance projects will then be able to be calculated. Similarly, other activities such as weed control and education events will be able to be recorded into the ALA.

Major lessons

With so many competent groups already working to achieve biodiversity and connectivity conservation in the region, it has been important to make sure that the Great Eastern Ranges Initiative adds value to those works and provides additional outcomes. One of the main obstacles to achieving those outcomes is a lack of funding. Seeking funds for on-ground works and capacity building events has been challenging for the Alliance. Despite a clear need for more conservation and connectivity work, applications by the Alliance for large competitive grants have so far been unsuccessful.

In addition, accessing private or corporate funds has been a challenge. There are over 320 000 not-for-profit groups in Australia competing for a limited amount of philanthropic funding and most charities would get much less than 10% of their total funding from philanthropic sources (Philanthropy Australia Inc. 2012). The Alliance will need to seek partnerships with or services from groups or individuals with marketing or business expertise and connections to increase opportunities to access corporate or philanthropic resources.

The Border Ranges Alliance covers a large geographic area requiring substantial travel time for members to meet. It was originally intended that the Alliance would meet twice yearly, but with many groups stretched for time or relying on volunteers, annual forums are held, and there has been a focus on the use of the Internet and email for communications. There have been advantages from this, such as the inclusion of profiles on the GER website for groups that do not have their own websites and a collection of on-line resources. However, the valuable interactions gained from group meetings have been limited. Working Group representatives are also widely spread and meetings are held by phone conferences.

Operating across the New South Wales/Queensland border has been effective. Many of the issues that conservation groups face are similar on both sides of the border as they share the same biological landscape with similar values and threats. Prior to the Alliance there were generally limited opportunities to work cooperatively or to share knowledge and experiences. The cross-border partnership has created some challenges relating to different policy and funding sources, but the partnership has allowed a better understanding of those differences to help overcome them and better achieve coordinated landscape outcomes. The sharing of approaches and methods for on-ground works and capacity building ensures new innovations or improved management practices from either side of the border are more quickly learnt and adopted throughout the region. An example is the rapid uptake of a new method of lantana control, the 'splatter gun' method, that has cut the costs of lantana control and increased the size of areas able to be treated.

Applying the lessons

Seeking philanthropic support for connectivity works will be a high priority for the Alliance. Taking advantage of this as yet untapped potential may add to the on-ground implementation of connectivity works in the region and lessen the dependence on government grants. A prospectus for investment has been prepared but the Alliance needs to acquire the appropriate expertise to make the most of this opportunity. There are many iconic projects that are lacking funding and would make suitable projects with tangible outcomes to partner with private supporters.

Continuing to apply for large government grants for collaborative projects will remain a priority. Despite some unsuccessful applications, securing funding for large-scale projects will be a good way to achieve strategic and coordinated landscape-scale improvements in connectivity. The process of developing multi-partner projects for funding applications has in itself strengthened the knowledge between groups of their respective aims and their projects.

The Alliance will continue to improve its communications through the Internet by updating the Great Eastern Ranges website and using the new ALA Border Ranges NRM activity recording site. Monitoring and evaluation of Alliance projects will become more efficient through the ALA and the success of the Alliance in achieving improved connectivity will be more easily measured.

Conclusions

The Border Ranges is a rich landscape with outstanding biodiversity and a large number of at-risk species. However, there are ongoing threats, many of them relating to the fragmentation and lack of connectivity of native habitats. The region is also rich with groups and individuals who are committed to protecting biodiversity and have excellent capacity to achieve conservation outcomes. Prior to the establishment of the Border Ranges Alliance there had already been significant achievements. In addition to the significant prior investment by governments that established an extensive protected area network and World Heritage declaration, NRM groups had been actively working to improve the condition of native habitats in the region for many decades. The Great Eastern Ranges has added a new framework within which Alliance groups can cooperate more effectively and has increased connectivity outcomes in line with up-to-date scientifically identified priorities.

The Great Eastern Ranges Initiative provides an overarching framework and goals, and the Alliance has developed its own governance and aims within that framework. The ability for all Alliance members to shape and direct the type of partnership and its priority actions has ensured that their activities are relevant to the region and will improve on-ground outcomes. This involvement has given broad ownership and commitment to the partnership and GER.

The Great Eastern Ranges Initiative has been a useful conservation model for the region by putting resources directly into on-ground projects in key connectivity areas and value-adding to the work of individual groups. At the same time it is investing in long-term outcomes through facilitating the partnership that helps the Alliance achieve its collective conservation goals.

Acknowledgements

We thank all the people and groups in the Border Ranges Alliance who have contributed their time, energy and enthusiasm to maintain the partnership, and for their commitment to improving conservation connectivity in the Border Ranges region for the benefit of our ecosystems and communities into the future.

Biographies

Georgia Beyer is the facilitator for the Border Ranges Alliance, Great Eastern Ranges Initiative. Georgia is also Senior Ecologist with the Nature Conservation Trust of NSW where she develops conservation agreements for private land. She has been working in biodiversity conservation for 20 years.

Lynn Baker has worked in biodiversity research, conservation and land management around Australia over 33 years, including 14 years with the NSW Government promoting biodiversity conservation in north-east NSW. One of the key outcomes was the development of the Border Ranges Rainforest Biodiversity Management Plan with the biodiversity team and cross-regional partners. Lynn continues to foster partnerships and opportunities to promote connectivity conservation and improved Aboriginal cultural biodiversity outcomes across the region.

References

Atlas of Living Australia (2012). About the Atlas. Available: http://www.ala.org.au/about-the-atlas/ [Accessed 21 May 2012].

Byron Shire Council (2004). *Byron Biodiversity Conservation Strategy*. Byron Shire Council, Mullumbimby, NSW.

DECC (2007). 'Key Altitudinal, Latitudinal and Coastal Corridors for response to Climate Change: A report for the Northern Rivers Catchment Management Authority'. Department of Environment and Climate Change NSW, Coffs Harbour.

DECCW (2010). *Border Ranges Rainforest Biodiversity Management Plan – NSW & Queensland*. Department of Environment, Climate Change and Water NSW, Sydney.

DSEWPC (2012). Australia's 15 national biodiversity hotspots, Border Ranges north and south (Queensland and New South Wales). Department of Sustainability, Environment, Water, Population and Communities, Canberra. Available: http://www.environment.gov.au/biodiversity/hotspots/national-hotspots.html#hotspot3 [Accessed 10 February 2012].

Great Eastern Ranges Initiative (2012). Border Ranges Alliance Terms of Reference. Available at: http://www.greateasternranges.org.au/border-ranges/alliance/alliance [Accessed 18 May 2012].

Halliday L, Castley JG, Fitzsimons JA, Tran C, and Warnken J (2012). Fire management on private conservation lands: Knowledge, perceptions and actions of landholders in eastern Australia. *International Journal of Wildland Fire* **21**, 197–209.

NRCMA (2005). *Northern Rivers Catchment Action Plan*. Northern Rivers Catchment Management Authority, Grafton, NSW.

Philanthropy Australia Inc (2012). Establishing a charity. Available: http://www.philanthropy.org.au/seekfund/charity.html [Accessed 13 February 2012].

Scotts D (2003). *Key Habitats and Corridors for Forest Fauna*. Occasional Paper 32. NSW National Parks and Wildlife Service, Coffs Harbour.

SEQC (2010). *South East Queensland Catchments Strategic Investment Plan 2009–2014*. South East Queensland Catchments, Qld.

Tourism Australia (2012). Australia's National Landscapes: Australia's Green Cauldron. Available at: http://www.australia.com/campaigns/nationallandscapes/AustraliasGreenCauldron.htm [Accessed 13 February 2012].

Williams KJ, Ford A, Rosauer DF, De Silver N, Mittermeier R, Bruce C, Larsen FW, and Margules C (2011). Forests of East Australia: The 35th Biodiversity Hotspot. In: *Biodiversity Hotspots: Distribution and Protection of Conservation Priority Areas*. (Eds FE Zachos and JC Habel) pp. 295–310. Springer, Berlin.

16

The Bunya Biolink: an application of Greening Australia's strategic approach to large-scale conservation

David Freudenberger, Lainie Grigg and Ralf Regeer

Introduction

The Bunya Biolink initiative, located in south-east Queensland, aims to link significant biodiversity assets from Barakula State Forest to the dry rainforests of the iconic Bunya Mountains, the eucalypt woodlands of Crows Nest National Park, and onto the rainforests of the Conondale Ranges (Figure 16.1). The Bunya Biolink aims to create diverse habitat corridors encompassing several environmental gradients (e.g. altitude), thereby increasing landscape resilience to climate change. The aims of the Bunya Biolink include improving the condition of the upper catchment of the Brisbane River that flows into Lake Wivenhoe, which supplies drinking water to 3 million people in south-east Queensland. This catch ment is also a major source of sediment and nutrient pollution that flows into Moreton Bay, an internationally significant, Ramsar-listed wetland.

Within a regional context, the Bunya Biolink is a potential component of the Great Eastern Ranges initiative (Chapter 13) where the native vegetation of the Eastern Ranges is at its narrowest in Queensland (Mackey *et al.* 2010). The Bunya Biolink initiative is also within the Birdsville-to-Bay climate change corridor concept that aims to protect and restore the vast and diverse environmental gradient stretching from the mulga woodlands and spinifex grasslands of Queensland's semi-arid south-west to the subtropical east coast of rainforests and mangroves (Regeer 2009).

The Bunya Biolink is a result of implementing Greening Australia's ambitious national strategy called Transforming Our Landscapes that was established in 2007. This strategy

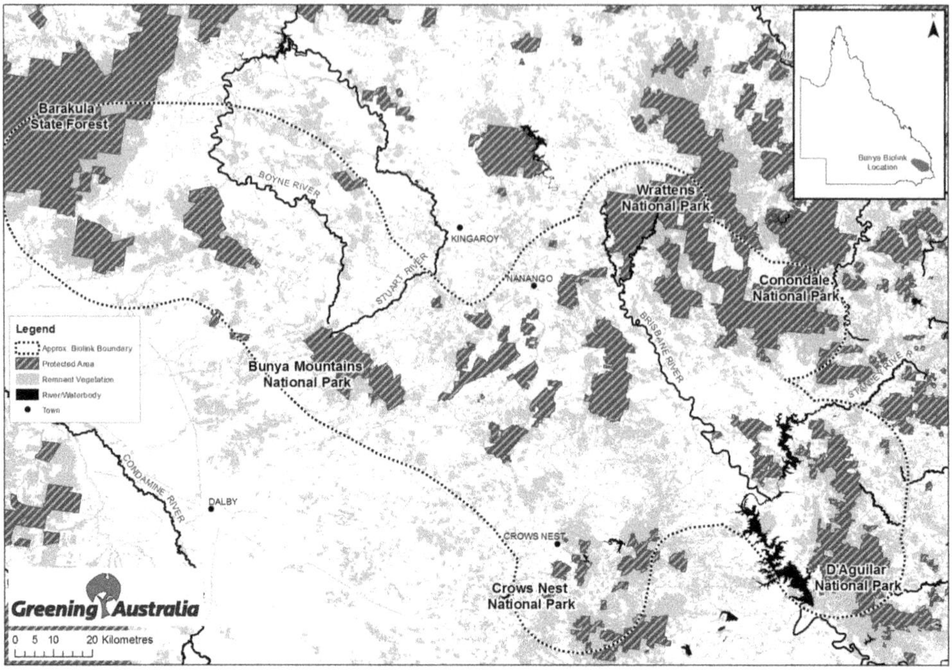

Figure 16.1 The broad outline of the Bunya Biolink that encompasses headwaters of the Brisbane, Condamine and Boyne river systems.

is consistent with the organisation's mission: 'To engage the community in vegetation management to protect and restore the health, diversity and productivity of our unique Australian landscapes' (Greening Australia 2012). The purpose of this strategy is to provide commitment, cohesion and greater effectiveness for Greening Australia's programs and activities. This strategy aims to focus restoration activities and investments to achieve tangible improvements in biodiversity, river health and sustainable livelihoods across highly valued landscapes at the scale of hundreds of square kilometres. Greening Australia acknowledges that this strategy is ambitious and requires the organisation to fully commit, for the long term, to a select number of landscapes and its people in order to achieve quantified criteria of success based on protecting, enhancing and restoring clearly identified conservation assets. The Bunya Biolink is one of 11 Transforming Our Landscapes projects endorsed by Greening Australia's National Board (Greening Australia 2012).

The Transforming Our Landscapes strategy has the following eight enabling strategies that require simultaneous implementation to achieve conservation objectives for a landscape and its people (Figure 16.2).

1 **On-ground restoration** – at the scale of threatening processes;
2 **Science guided** – adaptive landscape planning, implementation and review based on the best available science and research partnerships;
3 **Partnership focused** – transformative outcomes require diverse and enduring partnerships that bring complementary resources and skills;
4 **Practical visionaries** – staff, organisational members, volunteers and partners who have the skills and motivation to achieve this strategy's objectives;
5 **Enthused, educated and empowered communities** – active community participation in implementing practical solutions to environmental and economic sustainability challenges;

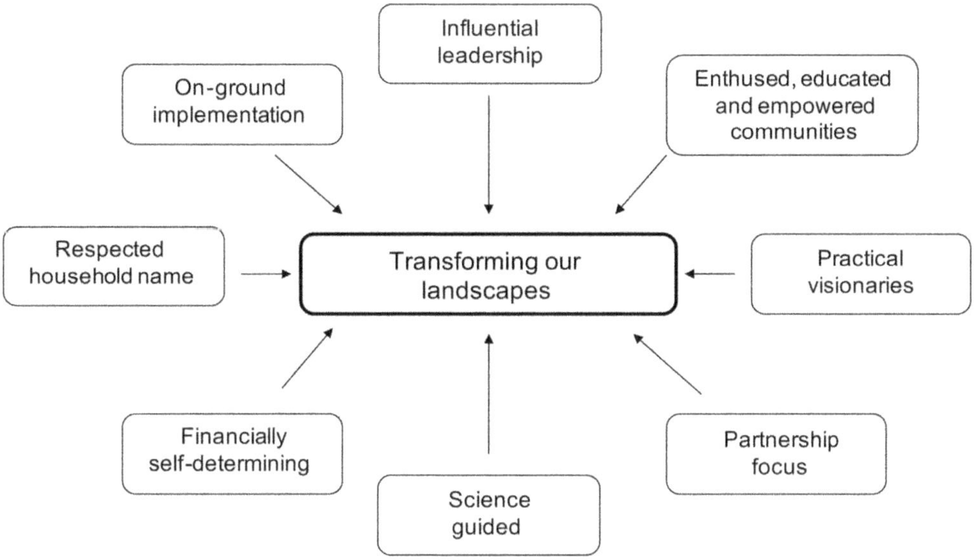

Figure 16.2 How Greening Australia's enabling strategies contribute to a transformative outcome.

6 **Influential leadership** – the capacity to shape and provide direction to environmental policies, programs and practices throughout Australia;
7 **Respected household name** – Greening Australia is widely recognised as a leading on-ground environmental organisation that attracts a diversity of supporters to its transformative landscape projects; and
8 **Financially self-determining** – the organisation is able to attract funds from a range of sources including commercial, corporate social responsibility, government and philanthropic.

This Transforming Our Landscapes strategic intent was developed in part based on Greening Australia's experience gained from being involved in the Gondwana Link initiative (Chapter 3) since its inception.

History and structure of the Bunya Biolink

The Bunya Biolink is one of the first 'green-field' initiatives to be developed after the Transforming Our Landscapes strategy was formally adopted by the Greening Australia Board of Directors. The development of this project had to meet an internal review and approval process based on two categories of criteria: *ecological* understanding based on the best available science, and *feasibility* of implementation.

The four *ecological* criteria for Transforming Our Landscapes projects and how Bunya Biolink met these are summarised below:

1. Presence of functional landscape assets (e.g. large and diverse remnants) in the nominated landscape to act as 'building blocks'

The Bunya Biolink met this criteria as it aims to functionally reconnect large tracts of relatively intact and diverse areas of native vegetation including the Bunya Mountains National

Park, Crows Nest National Park, the vast Barakula State Forest and vegetated slopes of the Conondale Ranges (Figure 16.1).

2. Inclusion of diverse environmental gradients (e.g. altitude, temperature and rainfall)

The Bunya Biolink includes steep environmental gradients particularly altitude (300–1100 m) and rainfall (600–1200 mm).

3. The connectivity potential is understood and mapped

The Bunya Biolink has identified areas of high potential connectivity incorporating existing remnant vegetation between the large state forests and conservation reserves within the project area (Figure 16.1). In addition, this corridor has been highlighted by the Queensland Government as part of their biodiversity strategy (DERM 2011).

4. Identification of key threats and development of abatement strategies

Like most of Greening Australia's transformative landscape initiatives, project planning was supported by using the Conservation Action Planning process and tools (TNC 2012). Conservation Action Planning is a participatory process supported by a planning 'coach' who has training and experience in similar projects elsewhere. This planning identified the following major threatening processes in the Bunya Biolink region: fire suppression, climate change, habitat fragmentation, invasive species, historical vegetation clearing and overgrazing. The Conservation Action Plan (CAP) for the Bunya Biolink has developed broad strategies to reduce these threats.

Bunya Biolink has met Greening Australia's three broad *feasibility* criteria in the following ways:

1. Presence of low cost and high value opportunities to improve biodiversity, river health and livelihoods over relatively small areas

The Bunya Biolink CAP has identified relatively low cost strategies that include management of natural regrowth, grazing management and riparian restoration. The latter is expensive at the site scale, but improves water quality at the catchment scale.

2. Presence of partner organisations and communities that have the willingness, skills and motivation to act at the scale needed to effectively reduce pervasive threatening processes

The Bunya Biolink initiative, through Greening Australia, is actively involved in building partnerships with organisations including the Queensland Department of National Parks, Recreation, Sport and Racing, SEQ Catchments, Condamine Alliance, BirdLife Australia, Somerset Regional Council and the Toowoomba Regional Council.

3. High likelihood opportunities to attract the necessary large investment of funds needed to achieve landscape-scale objectives

The CAP for the Bunya Biolink has identified that at least $12 M is required to achieve the project's objectives over a 5-year time frame. Identified sources of funding include: carbon emissions offsets (biosequestration), environmental offsets for land developers, in-kind support from communities, Queensland and Australian Government environmental programs, corporate social responsibility investments and philanthropic donations.

As at the end of 2012, there is no formal external governance structure for the Bunya Biolink. It is a visionary idea, rather than an organisational entity. Governance is at the level of on-ground projects managed by organisations supporting the Biolink. Individual project governance is based on contractual arrangements with the funder(s), partners and participants. Bunya Biolink is 'open source'. It is not owned by any one organisation, it is in the public domain, though Greening Australia, to date, has been the major champion driving the development of the CAP and the establishment of the pilot projects described below.

Major successes

The Bunya Biolink initiative is proving to be a compelling idea that attracts funders and helps build partnerships, project by project. To date, Bunya Biolink has attracted two major grants from the Australian Government's Caring for our Country Program.

The first project was the Cooyar Creek Riparian Zone Rehabilitation Project that joined with Queensland government agencies, landholders and community groups to restore and protect this creek that is one of the sources of Brisbane's water supply and important habitat for the Platypus (*Ornithorhynchus anatinus*) and native Water-rat (*Hydromys chrysogaster*). The project was highlighted at the time by a joint press release from both the Australian Minister for Environment and Minister for Agriculture, Forests and Fisheries (Burke 2008).

In 2011, Bunya Biolink, through Greening Australia, received a second large ($500 000) grant from the Caring for our Country program. This project is targeting over 100 landholders across 27 700 ha to create a green corridor that links complementary land uses and vegetation management practices to protect and restore Semi-evergreen Vine Thicket, Brigalow regrowth and remnant vegetation.

These pilot projects also include support from corporate social responsibility programs including funding from Perpetual Trustees Ltd for native seed collection and support from Powerlink Queensland for on-ground works in partnership with SEQ Catchments. Conservation Action Planning has been supported with a grant from The Nature Conservancy's David Thomas Challenge.

Greening Australia continues to seek further corporate support through investment in biodiverse carbon sequestration projects to restore priority vegetation communities in the Biolink area. To date, even with the successful introduction of the Carbon Farming and the Clean Energy Futures legislation by the Australian Parliament, investment in carbon sequestration based on locally native vegetation has been slow to develop.

Major lessons

Large-scale land rehabilitation and conservation initiatives such as the Bunya Biolink are relatively easy to invent. The urgent need for large-scale land repair in this region and elsewhere is easy to see from a low elevation fly-over in a plane or via Google Earth. Participatory planning tools like the CAP facilitate buy-in from other organisations working within the region. The challenge is to then 'sell' the Bunya plan to gain long-term institutional support, both within a championing organisation like Greening Australia, as well as support from investors including government agencies, commercial enterprises and individuals. It takes a long time and a lot of dedicated persistence to play this vital role of 'marketing'.

Large-scale conservation is a long-term and multi-dimensional endeavour. We have identified key phases and stages that need to be implemented in a considered and sustained fashion (Figure 16.3). All projects naturally need to start with iterative rounds of participatory planning. The Bunya Biolink planning started with a visioning exercise with a small group within Greening Australia. This included a flying tour at 1000 m altitude over the Bunya Mountain region that provided an overview of the vast restoration challenge. It also helped us with visioning a future alternative for so many hundreds of square kilometres of degraded and fragmented land.

The participatory application of CAP helped us gain a deeper understanding of the conservation challenge, identified potential partners, and helped us to collectively articulate measureable objectives which are needed to define what success looks like. The CAP process was also important for ongoing 'marketing' and it also gives the Bunya Biolink credibility. The Plan captures local and regional scientific and practitioner expertise through the participatory planning process. The CAP is the background document from which Greening Australia has sourced material for a diversity of marketing materials and presentations to many different potential investors. A compelling vision, clear objectives and practical strategies are fundamental to presenting a credible case for investment.

The Bunya Biolink is now in the piloting phase of implementation (Figure 16.3). As noted above, Greening Australia and its partners have achieved the critical seed funding needed to demonstrate what is possible on the ground with keenly interested landholders. We expect this piloting phase to continue for at least another few years, but in the meantime these initial projects help attract increasing support and interest in the Bunya Biolink vision.

Large-scale implementation is still a few years off. The impact of a price on carbon as an incentive for polluters to heavily invest in biodiverse carbon sequestration remains unknown. The carbon pricing legislation did not come into effect until July 2012, and during the first 3 years the price on carbon is fixed as a tax. The price post-tax, and even the future of a price on carbon, is highly uncertain due to potential changes in Australian

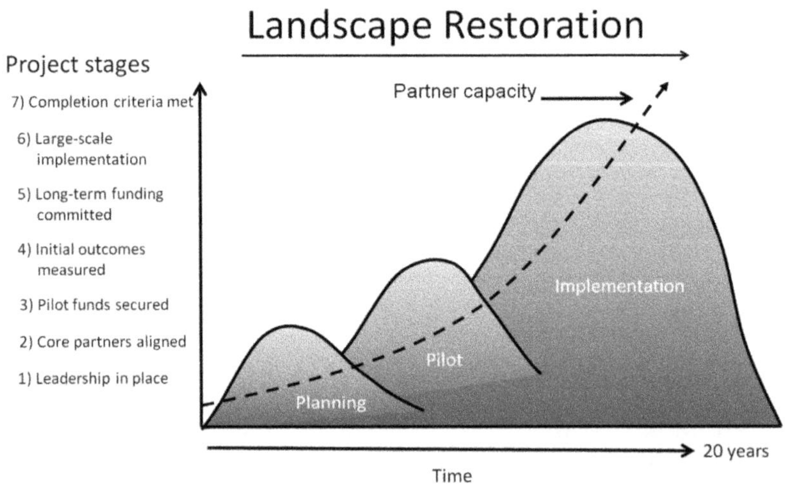

Figure 16.3 The phases required for achieving conservation outcomes at large scales (adapted from P. Walsh and D. Butcher, Greening Australia). Note regular reviews (feedback loops), which are essential, are not included in this diagram in order to maintain clarity.

legislation, risk of global recession, and the failure, to date, for an internationally binding agreement post the Kyoto Protocol that expired in 2012.

The Australian Government's Biodiversity Fund (nearly $1 B over 6 years) provides an opportunity for further support for pilot projects, and could help leverage corporate investment in biodiverse carbon sequestration. However, the Biodiversity Fund, on its own, is unlikely to pay for the cost of achieving landscape and catchment-scale outcomes over the next 20 years.

Applying the lessons

Greening Australia's implementation plan for the Bunya Biolink includes regular review processes needed to help continuously improve strategies and on-ground actions. Review processes have started at the early planning stages. Conservation Action Planning is an iterative process that uses regular reviews to feed into the next iteration. To date Greening Australia has completed a second significant version of the Bunya CAP. The latest Plan iteration was instigated as new Greening Australia staff became involved. The review and update of the Plan helped create ownership by these new staff. An effective CAP should remain a work in progress that is regularly updated with new understanding and greater detail around strategies and actions as projects get funded and implemented.

A detailed on-ground monitoring and reporting plan for the Bunya Biolink is very much a work in progress. The CAP has specific and measureable objectives (e.g. a 30% increase in Semi-evergreen Vine Thicket by 2025, requiring 10 480 ha of restoration works). Monitoring progress to achieving this kind of objectives is essentially an exercise in accounting. More challenging is monitoring the effectiveness of social strategies such as the marketing needed to achieve such demanding objectives.

Demonstrating clear biodiversity or water quality outcomes from specific on-ground restoration efforts is also complex and a long-term undertaking. It is difficult to demonstrate the direct outcomes of 10 km of riparian fencing and revegetation along the Bunya Biolink's Cooyar Creek project. Ideally 10 years of pre-restoration data would be available to quantify the past inherent variation in water quality and wildlife abundance. Then another 10 years of post-restoration data might be needed to detect a significant improvement from the background of natural variation. Greening Australia is seeking collaborations with researchers to help quantify pre- and post-conditions, but funding for research and funding for on-ground activities is rarely linked; rather such funding is often out of synch by a few years at best.

Measuring the biodiversity outcomes of specific restoration strategies in specific places is problematic (Lindenmayer and Gibbons 2012), but measuring the outputs of on-ground actions is far simpler (Freudenberger 2012). Monitoring the immediate consequences (e.g. dead or alive plants) of inputs such as fencing, weed control, revegetation or prescribed fire is far simpler than measuring the outcomes (e.g. response of insect diversity) of such management activities. It should be the responsibility of project proponents such as Greening Australia to monitor the immediate results of their weed control activities, and then improve practices as necessary to ensure effective control. Similarly, restoration plantings should be monitored to ensure they survive and grow well (e.g. Jonson 2010). If not, revegetation practices need to be improved.

There is no sense in monitoring the long-term water quality outcomes of fencing riparian corridors if the fencing fails due to flooding, lack of maintenance or the deliberate re-introduction of livestock – a particular risk during drought. There is much to be learned from every litre of herbicide, every kilometre of fencing, and every planted seedling. Hence the value of pilot projects like the Cooyar Creek project. The perpetual challenge is to provide sufficient incentives and funds for project staff and landholders to revisit sites and record simple data like 'alive' or 'dead'.

One solution to funding long-term site management (which requires long-term monitoring), is the establishment of trust funds. In Queensland, Greening Australia has pioneered the establishment of a vegetation management trust fund, which takes on the responsibility of managing vegetation for up to 20 years to compensate for the loss of vegetation during large infrastructure projects such as oil and gas pipelines, mines and residential and commercial developments. These sorts of trust funds could be a model for projects within the Bunya Biolink.

Conclusions

Large-scale and long-term conservation initiatives need institutional commitment, but not necessarily complex governance arrangements. The Bunya Biolink has successfully moved from early visioning to detailed planning and pilot implementation, all with no formal governance around the Bunya Biolink Plan itself. The Bunya Biolink is proving to be a compelling vision that allows a diversity of individuals and organisations to 'buy-in' and provide support, one project and activity at a time. The governance needed for individual projects within the Bunya Biolink is provided by existing organisations such as Greening Australia or SEQ Catchments that routinely manage contractual arrangements with a diversity of funders including government, corporate and philanthropic entities. No one needs permission from some governance structure to contribute to the vision and plan. Bunya Biolink is 'open-source'.

The challenge is to keep it that way. There are two main risks. There is a tendency for organisations to want exclusivity for claiming successes and mechanisms to avoid blame for failures. This is understandable, but can be managed at the individual project level. The other risk is 'brand abuse'. That is, individuals or organisations claiming to be supporting the Bunya Biolink, but in fact are doing nothing of the kind. This risk can be managed by judging actions against the Bunya CAP. If an action is consistent with the Plan, then it's 'in'; this activity and proponents are part of the Plan. If not consistent with the Plan, then 'name and shame' could be the risk management strategy. Abuse of the Bunya Biolink 'brand' hasn't happened yet, and if it does, we are confident that the loose consortium of Bunya Biolink supporters will successfully deal with such a risk. There is much to learn by doing.

Acknowledgements

We would like to thank the many individuals and organisations that have contributed to developing the Bunya Biolink initiative to date, and particularly thank those who have contributed to the Plan. There are too many people to name individually, but we warmly acknowledge your input and support.

Biographies

David Freudenberger (PhD) was Greening Australia's Chief Scientist. He has 27 years of ecological research experience, including 15 years with CSIRO. He has had research leadership roles in CSIRO and is a graduate of the Australian Rural Leadership Program. David's role in Greening Australia was to ensure the organisation's transformative landscape-scale initiatives are rigorously planned, implemented and evaluated based on the best available science. David was involved with the co-authors in developing the Bunya Biolink initiative from its inception. David, as a qualified CAP 'coach' facilitated the development of the first iteration of the Plan. He is now a senior lecturer at the Australian National University.

Lainie Grigg was the program manager who developed the successful Caring for our Country grant for a second major project under the Bunya Biolink banner. Lainie has extensive experience in the fields of science, landscape ecology, marketing and communications. Lainie has also managed a native plant nursery specialising in endemic South East Queensland species and the supply of information on growing native plants. She left Greening Australia in 2012 to pursue other interests.

Ralf Regeer (MSc, GDip EnvMan) was involved in the development of the Bunya Biolink initiative from its inception and through two iterations of the CAP. Ralf became the general manager of consulting and training services in Greening Australia (Queensland), but left Greening Australia in 2011 to establish his own environmental consulting and ecological restoration firm (Ausecology Pty Ltd).

References

Burke T (2008). Queensland environment and landholders benefit from new funding. Media release; The Hon. Tony Burke MP, Minister for Agriculture, Fisheries and Forestry. DAFF08/156B. 20 November 2008. Available: http://www.maff.gov.au/media_office/media_releases/Burke-media-releases/2008/november_2008/queensland_environment_and_landholders_benefit_from_new_funding [Accessed 20 June 2012].

DERM (2011). *Building Nature's Resilience: A Biodiversity Strategy for Queensland*. Department of Environment and Resource Management, Brisbane.

Freudenberger D (2012). Practical challenges in monitoring and adapting restoration strategies and actions. In: *Biodiversity Monitoring in Australia* (Eds DB Lindenmayer and P Gibbons) pp. 121–126. CSIRO Publishing, Collingwood.

Greening Australia (2012). Visionary landscape-scale projects. Greening Australia, Canberra. Available: http://www.greeningaustralia.org.au/visionary-projects/overview [Accessed 20 June 2012].

Jonson J (2010). Ecological restoration of cleared agricultural land in Gondwana Link: lifting the bar at 'Peniup'. *Ecological Management & Restoration* **11**, 16–26.

Lindenmayer DB and Gibbons P (Eds) (2012). *Biodiversity Monitoring in Australia*. CSIRO Publishing, Collingwood.

Mackey B, Watson J, and Worboys GL (2010). *Connectivity Conservation and the Great Eastern Ranges Corridor*. NSW Department of Environment, Climate Change and Water, Sydney.

Regeer R (2009). Greening Australia grows an ecological partnership. *Nature Refuge News* **12**, 8. Available: http://www.derm.qld.gov.au/register/p02721aa.pdf [Accessed 20 June 2012].

TNC (2012). Conservation Action Planning. The Nature Conservancy, Arlington, USA. Available: http://conserveonline.org/workspaces/cbdgateway/cap/resources/index_html [Accessed 20 June 2012].

SECTION 3. POLICIES AND FRAMEWORKS

17

Scaling up: the policy case for connectivity conservation and the development of Australia's National Wildlife Corridors Plan

Charlie Zammit

What is the policy problem?

Biodiversity conservation has its roots in three sorts of public policy challenges. First is the ongoing protection of sufficient samples of natural ecosystems from conversion to other land uses. Second is the protection of species and habitats that have been reduced to a small proportion of their historical distribution and abundance and are therefore at risk. And third is managing the variety of threats to species, habitats and ecosystems.

Notwithstanding effort over more than 40 years through legislation, the formulation of policies and programs and the attendant efforts of applied research to build knowledge and capacity, and significant public investment, Australia's biodiversity continues to decline (OECD 2007; State of the Environment 2011 Committee 2011). There are inevitably a variety of reasons for this; however, central among them is the mismatch between the scale of the critical threats to biodiversity and those of the institutional policy responses. While the critical threats to biodiversity are most often large-scale and span medium-to-long-term time horizons, many institutional responses to address them are typically local scale and short term. This is not to say that local-scale and short-term interventions are without merit: they demonstrably are. They are a necessary but ultimately insufficient basis for the long-term conservation of Australia's biodiversity.

Coupled with the scale-of-threat:scale-of-response mismatch is the observation that the history of policy interventions has focused more on establishing a system of protected areas and on managing threatened species and less on restoring and maintaining ecological functions across different land tenures. For example, all governments have some form of endangered species legislation and under Australia's *Environment Protection and Biodiversity Conservation Act (1999)* there are over 1600 species listed as nationally threatened, and more than 40 ecological communities similarly designated. Australia's species richness stands conservatively at around 148 000 recorded species, with over 500 000 possible (Chapman 2009). In contrast, most legislative or policy interventions that aim to manage habitat condition or retain ecological function are usually within the bounds of environmental impact assessments for specific development proposals. These are by design focused on local and direct impacts rather than system-wide, longer-term and indirect risks to ecological function or species persistence over their natural range.

Ecological theory and empirical evidence is clear about the complex relationships between interacting populations of species, from soil-dwelling bacteria and Jarrah (*Eucalyptus marginata*) trees to Red Kangaroos (*Macropus rufus*) and Wedge-tailed Eagles (*Aquila audax*), and their assembly into habitats and ecosystems, that over time evolve and adapt under pressure for survival (Peterson and Parker 1998; Loreau 2010; Scheiner and Willig 2011). Within this evolutionary and ecological theatre, space and time are critical, but the nature of the dependence of ecological systems on space and time is driven by the particular life-history attributes of species, and how species interact and assemble into habitats and to form functioning ecosystems. The functioning of ecosystems depends critically on its connections as well as its components.

Accordingly, it is reasonable to conclude that any coherent policy response to threats to biodiversity needs to recognise the scientific evidence for sufficient space and time for species to persist and evolve as interacting ecological assemblages, and for those ecosystems to remain functional over time. The question then arises of what kinds of policy responses can provide purposeful direction to that need at a scale that is effective and within the broader competing pressures on the environment to meet society's needs and expectations.

An emerging focus on connectivity conservation

Since 2005, the three major national conservation policies in Australia, namely *Australia's Biodiversity Conservation Strategy 2010–2030*; *Australia's Strategy for the National Reserve System 2009–2030*; and the *Native Vegetation Framework*, have been reviewed and revised. These revised policies anticipate a greater focus on landscape-scale approaches (DSEWPC in press), they provide agreed national targets for conservation, introduce concepts like resilience and ecosystem services as ways of bringing systems thinking and valuation into conservation policy. They also recognise that successful conservation policy is critically dependent on like-minded human values, attitudes and behaviours, the role of markets and other innovations, and the need for reliable information and systems for monitoring and reporting on the environment. In sum, these revisions recognise the need for conservation efforts to purposefully manage the entirety of Australia's biodiversity as a system – from genes to ecosystems – across all tenures and alongside competing land-use preferences, at a scale that retains ecological functions, allows for evolutionary expression and addresses long-term risks. The key feature of any system is the interaction – the connectedness – of the components of the system. The emerging field of connectivity conservation (Crooks and Sanjayan 2006; Hilty *et al.* 2006; Worboys *et al.* 2010) is based on this recognition.

Over a similar time period, the national natural resource management (NRM) and environment program, Caring for our Country, was created from the legacy of the Natural

Heritage Trust. It established clear national investment outcomes and progressively revised priorities to meet specific targets and to support larger-scale and better-linked conservation projects. Caring for our Country also works to cultivate longer-term partnerships among land managers and community organisations, to create opportunities for multi-year funding and to continue to build the necessary skills and knowledge for effective conservation efforts. While the proritisation and distribution of limited funds for conservation will always be a topic for legitimate and lively public debate, a recent review suggests reasonable progress has been made in meeting the program's objectives (DSEWPC 2012a).

Alongside these conservation programs, a global and domestic policy agenda around climate change was revitalised, with its roots in the IPCC assessment reports, the Stern Review (Stern 2007) and the Garnaut Review (Garnaut 2008). These global and domestic reviews of the climate change problem and the scope for coherent policy interventions injected a fresh, but not altogether novel, perspective into conservation policy thinking by highlighting:

- the need to think seriously about impacts and responses, at scales sufficient to address the risks over decades and centuries;
- the value of a robust evidence-base for both problem calibration and policy responses, including policy delivery on the ground;
- the implicit systems thinking and systems-based analysis linking the biophysical and the socio-economic;
- the economics of doing nothing; and
- the role of valuation and markets.

In mid-2011, the Australian Government released its Clean Energy Futures policy, part of which incorporated significant consideration of the role of, and impacts on, the land sector (Australian Government 2011). In particular, this policy and program architecture foreshadowed new interventions to secure existing stores of natural carbon and to undertake new environmental plantings that would deliver carbon and biodiversity co-benefits. Also implicit in this architecture is recognition of the need to repair and restore, at scale, ecosystems and habitats, and in these efforts to ensure some degree of spatial and functional connectivity that supports the adaptation of species to climate change. An ongoing commitment to funding has been made to drive this policy agenda, and almost 1 billion dollars over the first 6 years has been provided through the Australian Government's Biodiversity Fund.

The other critical policy challenge aligned with connectivity conservation is that of environmental water management, and in particular of river flows in the Murray–Darling Basin, where land-use intensification has led to water diversions for irrigated agriculture. In times of long and severe drought, this infrastructure, and overallocation of water has led to serious disruptions to the health and connectivity of the Basin's ecosystems, as well as placing additional economic pressure on the agricultural sector. The full suite of policy responses to this complex issue is incomplete, but mechanisms are being established to ensure sufficient water is available to meet the environmental flow requirements to re-establish and maintain connected river networks and riparian ecosystems.

The insufficiency of historical approaches to biodiversity conservation, coupled with the emerging threats to ecosystems and species persistence from climate change and water management has created an opportunity to formulate a policy case for connectivity conservation that focuses on maintaining or restoring system connections in the landscape. These connections may be functional (such as movements of water or nutrients) or structural (such as linked mosaics of different vegetation communities). The form of connection that has received most policy attention is connection through vegetation corridors linking different habitats or communities in the landscape.

Developing a *National Wildlife Corridors Plan*

In early 2011, the Australian Government announced an independent Advisory Group to prepare a draft *National Wildlife Corridors Plan* to put into effect their 2010 election commitment (ALP 2010). This work was complemented by an additional election commitment to advance landscape-scale approaches to conservation. In March 2012, the Group's draft Plan was released for public comment (NWCPAG 2012a), and in November 2012 the final Plan was announced by the Australian Government (DSEWPC 2012b).

It is worth noting at the outset that, although called a 'wildlife corridors plan', this national plan is at its core different from the history of wildlife corridor development in the northern hemisphere. In those initiatives the policy problem has been the management of mammals with large home ranges, including those with seasonal migrations over large areas. More formal arrangements have had to be made for these movements across a wide mix of public and private lands (see, for example the influence of the Grey Wolf (*Canis lupis*) and Grizzly Bears (*Ursus arctos*) on the development of the Yellowstone to Yukon Conservation Initiative; Chapter 2). In Australia, there are no equivalent seasonally migrating mammals at this scale, however, the term 'wildlife' has been used here because of its widespread usage as a synonym for 'nature'.

The Plan brings together in a single policy framework, the rationale for connectivity conservation and the role of a national network of wildlife corridors to give effect to that connectivity. In constructing the rationale, the Plan recognises the progressive development of approaches to conservation; it highlights the policy challenges evident in tackling the impacts of climate change and identifies several critical foundation stones that underscore its directions. In particular, the foundation stones comprise two groupings: a spatial set made up of the protected areas estate, other iconic areas including World Heritage properties and Ramsar wetlands, and the almost two-thirds of the continent held in private lands where significant biodiversity persists. A second, process grouping comprises the many formal and informal community and organisationally driven arrangements and partnerships that bring together, often on a voluntary basis and often driven by local motivations, networks of people supported by their various arrangements that provide leadership, social momentum and institutional coherence. The Plan recognises that the ambitions for a national network of wildlife corridors would not be possible without these two foundational elements.

The Plan is constructed in two Parts: Part 1 outlines the policy rationale and provides guiding principles and objectives (see Table 17.1) to inform and enable implementation; Part 2 proposes a five-point Plan of Action for implementation.

The five-point Plan of Action anticipates the delivery of the policy framework through existing or new Australian Government environment program investments. To that extent, the Plan is not specific about funding sources but acknowledges the priorities for landscape-scale approaches and the additional focus on connectivity conservation present in the current funding guidelines.

The Plan of Action proposes that the network of corridors will be made up of initiatives from three distinct scales of corridor development: local, regional and national. It also

Table 17.1. Guiding principles and objectives of the *National Wildlife Corridors Plan*

Guiding principles	***Objectives***
• Building wildlife corridors across Australian landscapes is a cooperative endeavour • Corridors should be designed and implemented in ways that benefit local communities • Healthy functioning landscapes require connectivity at a variety of scales • Effective corridors connect the landscape across the mosaic of land tenures and land uses without affecting property rights • The design and location of corridors should be based on the best available information derived from scientific research, traditional Indigenous knowledge and practitioner experience • Corridors should be designed to assist native species' adaptation to the impacts of climate change • Corridor design recognises and manages for potential risks such as those posed by invasive species and fire	• Protect, maintain and restore native habitats and ecosystems, and their critical processes and functions • Protect natural stores of carbon in native ecosystems to minimise greenhouse gas emissions • Enhance the resilience of Australia's biodiversity and their adaptability to climate change • Support the global and national movement of animals • Assist in managing and protecting Australia's iconic landscapes, and Indigenous and non-Indigenous cultures and heritage • Increase community participation in wildlife corridors and connectivity conservation

concludes that enduring institutional arrangements will be needed to give effect to this network, and in particular to how national-scale corridors might be developed. Specifically, the Plan establishes a mechanism through which voluntary, community-driven proposals for new national wildlife corridors might be nominated, assessed and then declared. This mechanism would include some form of independent advisory council to government, which could be created administratively or under legislation.

Challenges in developing and implementing the policy framework

Developing a policy framework that explains the need for connectivity conservation and a network of wildlife corridors has not been especially contentious. This is the 'why' question and, as noted above, there has been a natural policy creep in this direction for some time. Moreover, there is already a significant body of work around the science associated with connectivity conservation to give standing to this policy framework (Crooks and Sanjayan 2006; Hilty *et al.* 2006; Hobbs and Suding 2009; Worboys *et al.* 2010; Cumming 2011). However, this is not to say that the science community universally endorses its need or its merits (see for example, Gilbert-Norton *et al.* 2010; Pearce 2012). Given the contestable nature of progress in science, this is both expected and welcome, and it will surely stimulate further critical thinking about the kinds of approaches needed (see for example Hobbs *et al.* 2011), likely opportunities and associated risks (Soulé *et al.* 2004; Wyborn 2011).

A complementary case for why formalising the policy case for connectivity conservation and corridors is worthwhile arises from the examples of successful community-driven initiatives at different scales that have been underway in Australia (Whitten *et al.* 2011) and in other countries over the last decade, many of which are illustrated in this publication.

However, in developing the Plan the real challenges have come from the 'who', 'how' and 'where' questions that arise when implementation is the next step to be considered. One significant challenge in developing the Plan has been in describing the voluntary nature of its implementation. At its core, this means that the Australian Government is unlikely to unilaterally, 'from the top down', direct where, and at what scale the network of corridors will be created and managed. There are sound reasons for this, including that under Australia's Constitution, the management of land is the responsibility of state and territory governments. The Australian Government's role, foreshadowed in Part 2 of the Plan, would therefore be two-fold: to provide the necessary leadership and support to strengthen regional community capacity, foster effective partnerships and build enduring institutional arrangements for corridor development and management; and to prioritise funding from existing programs for investments in corridor implementation and ongoing management.

Because connectivity conservation must span the variety of land uses across public and private tenures that make up the working landscapes of Australia, the creation of wildlife corridors will likely require some form of 'social licence' from local communities. Social licence is the level of acceptance or approval continually granted to an organisation's operations or project by the local community and other stakeholders (see http://socialicense.com/index.html). Particular communities, based around their shared interests and aspirations will largely determine how this plays out. Examples of successful bottom-up corridor initiatives reveal this social driver to be critical, and this leads to a speculation that development of large-scale corridors will be unlikely unless there is a community that sees merit in such an initiative. One of the real challenges is therefore in defining this 'community of interest' in connectivity conservation initiatives, and for it to build sufficient trust, credibility and practical expertise to establish an enduring identity that is supported by effective governance. The Plan recognises these challenges, is encouraged by the examples of success evidenced across Australia over the past decade and identifies the crucial ongoing role for community and non-government organisations, private land managers, industry sectors and governments to work in collaborative partnerships. In addressing this challenge, the Plan sees a need for ongoing community development support for governance arrangements over sufficient time for a national network to be progressively established.

Two more technical challenges arise. The first is in determining what kinds of biodiversity (genes, species, ecosystems) connectivity conservation corridors can and should address to enhance existing approaches to conservation. Equally importantly, is the consequential question of the 'design specifications' for wildlife corridors that lead to improved connectivity for different kinds of biodiversity (Rouget *et al.* 2006; Ferrier and Drielsma 2010). Thereafter, there is the matter of identifying where and how different kinds of corridors might be spatially configured so as to better protect and manage biodiversity in working landscapes (Polasky *et al.* 2008). Fortunately, for all of these complex scientific questions, there is an established and growing international research focus and an emerging debate about feasibility and effectiveness (Gilbert-Norton *et al.* 2010; Chapter 23).

Finally, there is the observation that scientific approaches are only one prism through which connectivity conservation and wildlife corridors are designed and managed. The Plan recognises cultural connections with country, both Indigenous and western, as a legitimate foundation stone for a network of corridors. The dreamtime stories and songlines that criss-cross Australia, and the network of travelling stock routes used by early settlers provides another reason for creating corridors that also reconnect people and their country. The challenge is then how best to bring scientific and other kinds of knowledge together to create an enduring community of interest to agree on what a corridor initiative might deliver to a region, and in so doing secure a social licence.

Directions for implementation

Implementation of the *National Wildlife Corridors Plan* will be through existing or new Australian Government NRM and environment programs, and as noted earlier in this chapter, some initial alignments of investment priorities are already evident, as illustrated by recent investments under the Biodiversity Fund (see DSEWPC 2012c).

Implementation through existing programs also opens up opportunities for new kinds of investment partnerships. Two particular opportunities are possible, the first with those philanthropic organisations who traditionally support biodiversity conservation. The second is with industry sectors and Indigenous peoples with large private land holdings and an interest in sustainable land management and biodiversity conservation.

In addition to funding opportunities for implementing programs for connectivity conservation and a national network of corridors, there are two other directions for implementation. First are the ongoing efforts to align connectivity conservation science with the policy directions outlined here. This will build on existing momentum through the National Environment Research Program, the Terrestrial Ecosystems Research Network and the National Climate Adaptation Research Facility to strengthen the capacity of policy makers and program managers to use the best available science to support implementation. Second are the ongoing efforts to strengthen governance arrangements among the NGO and community sectors in order to build their capacity to play an active and long-term role in the co-management of our natural environments.

Conclusions

Here I have argued that traditional approaches to biodiversity conservation, focused on components of the system and often operating at specific habitat or local scale, are a necessary but not sufficient foundation for the long-term protection and management of Australia's biodiversity. A case is made that the physical and institutional scale of response of these approaches is out of step with the scale of threat.

A policy framework for connectivity conservation and the creation of a national network of wildlife corridors is described as the Australian Government's response to this ongoing policy problem. The framework brings together the refocusing of conservation efforts over past years to scale up to landscapes (DSEWPC in press), to have longer implementation and investment horizons, to be more cognisant of links between ecological and socio-economic systems and of the need to better manage these systems in their entirety. The Plan also recognises that the threat of climate change on biodiversity (e.g. Steffen *et al.* 2009) will require both space to move and time to adapt, and that building connections through enduring corridors is one way of providing the necessary ecological resilience.

Acknowledgements

This paper benefited from many conversations with members of the National Wildlife Corridors Plan Advisory Group, with policy and program colleagues across government and from debates with academic colleagues.

Biography

Charlie Zammit holds a PhD in ecology and has spread his professional career between academia and government positions. From 2005 he was Assistant Secretary, Biodiversity Conservation Branch in the Department of Sustainability, Environment, Water, Population and Communities where he was responsible for national biodiversity, vegetation and forest policy issues and for developing market-based approaches to biodiversity conservation. From 1999–2005 he was Professor of Land Use Studies and Director of the Land Use Research Centre at the University of Southern Queensland in Toowoomba. From 1990 to 1999 he held a variety of senior policy positions in the Department of the Prime Minister & Cabinet, the Department of the Environment and the Department of Foreign Affairs & Trade. Prior to then he held academic positions at the Australian National University, the University of San Diego and Macquarie University. He has recently retired.

References

ALP (2010). Green Corridors Plan. Australian Labor Party, Canberra. Available: http://www.alp.org.au/agenda/environment/green-corridors-plan/

Australian Government (2011). *Securing a Clean Energy Future – The Australian Government's Climate Change Plan.* Commonwealth of Australia, Canberra.

Chapman A (2009). *Numbers of Living Species in Australia and the World.* Department of the Environment, Water, Heritage and the Arts, Canberra.

Crooks KR and Sanjayan M (Eds) (2006). *Connectivity Conservation.* Cambridge University Press, Cambridge, UK.

Cumming GS (2011). *Spatial Resilience in Social-Ecological Systems.* Springer Verlag, Berlin.

DSEWPC (2012a). *Caring for our Country Review Report.* Department of Sustainability, Environment, Water, Population and Communities, Canberra. Available: http://www.caringforourcountryreview.com.au/document/index/1

DSEWPC (2012b). *National Wildlife Corridors Plan: A Framework for Landscape-scale Conservation.* Department of Sustainability, Environment, Water, Population and Communities, Canberra.

DSEWPC (2012c). Biodiversity Fund Successful projects – Round One 2011–12. Department of Sustainability, Environment, Water, Population and Communities, Canberra. Available: http://www.environment.gov.au/cleanenergyfuture/biodiversity-fund/projects/round1.html

DSEWPC (in press) *Managing Biodiversity at Landscape Scale.* Occasional Paper No. 2. Department of Sustainability, Environment, Water, Population and Communities, Canberra.

Ferrier S and Drielsma M (2010). Synthesis of pattern and process in biodiversity conservation assessment: a flexible whole-landscape modelling framework. *Diversity & Distributions* **15**, 386–402.

Garnaut R (2008). *The Garnaut Climate Change Review.* Cambridge University Press, Melbourne.

Gilbert-Norton L, Wilson R, Stevens JR, and Beard KH (2010). A meta-analytic review of corridor effectiveness. *Conservation Biology* **24**, 660–668.

Hilty JA, Lidicker WZ, and Merenlender AM (2006). *Corridor Ecology: The Science and Practice of Linking Landscapes for Biodiversity Conservation.* Island Press, Washington, DC.

Hobbs RJ and Suding KN (eds) (2009). *New Models for Ecosystem Dynamics and Restoration.* The Society for Ecological Restoration International, Island Press, Washington, DC.

Hobbs RJ, Hallett LM, Ehrlich PR, and Mooney HA (2011). Intervention ecology: applying ecological science in the twenty-first century. *Bioscience* **61**, 442–450.

Loreau M (2010). Linking biodiversity and ecosystems: towards a unifying ecological theory. *Philosophical transactions of the Royal Society B* **365**, 49–60.

NWCPAG [National Wildlife Corridors Plan Advisory Group] (2012). *Draft National Wildlife Corridors Plan.* National Wildlife Corridors Plan Advisory Group, Canberra.

OECD (2007). *OECD Environmental Performance Reviews: Australia.* OECD, Paris.

Pearce F (2012). Wildlife Corridors could be a waste of space. *New Scientist* **2850**, 26–27.

Peterson D and Parker V (1998). *Ecological Scale: Theory and Applications.* Columbia University Press, New York.

Polasky S, Nelson E, Camm J, Csuti B, Fackler P, Lonsdorf E, Montgomery C, White D, Arthur J, Garber-Yonts B, Haight R, Kagan J, Starfield A, and Tobalske C (2008). Where to put things? Spatial land management to sustain biodiversity and economic returns. *Biological Conservation* **141**, 1505–1524.

Rouget M, Cowling RM, Lombard AT, Knight AT, and Kerley GI (2006). Designing large-scale conservation corridors for pattern and process *Conservation Biology* **20**, 549–561.

Scheiner S and Willig M (2011). *The Theory of Ecology.* University of Chicago Press, Chicago, Ill.

Soulé M, Mackey B, Recher H, Williams J, Woinarski J, Driscoll D, Dennison W, and Jones M (2004). The role of connectivity in Australian conservation. *Pacific Conservation Biology* **10**, 266–279.

State of the Environment 2011 Committee (2011). *Australia: State of the Environment 2011.* Independent Report to the Australian Government Minister for Sustainability, Environment, Water, Population and Communities. Department of Sustainability, Environment, Water, Population and Communities, Canberra.

Steffen W, Burbidge A, Hughes L, Kitching R, Lindenmayer D, Musgrave W, Stafford Smith M, and Werner P (2009). *Australia's Biodiversity and Climate Change.* Commonwealth of Australia, Canberra.

Stern N (2007). *The Stern Review on the Economics of Climate Change.* HM Treasury, London.

Whitten S, Freudenberger D, Wyborn C, Doerr V, Doerr E, and Langston A (2011). 'A compendium of existing and planned Australian wildlife corridor projects and initiatives, and case study analysis of operational experience'. A report for the Australian Government Department of Sustainability, Environment, Water, Population and Communities. CSIRO, Canberra.

Worboys G, Francis W, and Lockwood M (Eds) (2010). *Connectivity Conservation Management: A Global Guide.* Earthscan, London.

Wyborn C (2011). Landscape scale ecological connectivity: Australian survey and rehearsals. *Pacific Conservation Biology* **17**, 121–131.

18

Beyond the Boundaries: Bush Heritage Australia's approach to multi-tenure conservation

Philippa Walsh, Stuart Cowell, Sarah Eccles and Matt Appleby

Introduction

Bush Heritage Australia is a non-profit conservation organisation dedicated to protecting Australia's unique animals, plants and their habitats. In 1991, the organisation was founded with a simple, practical formula for protecting the bush – to buy land of outstanding conservation value, and then manage the land to maintain those values.

In 2005, Bush Heritage Australia (hereafter 'Bush Heritage') initiated a major landscape-scale conservation program called 'Beyond the Boundaries'. This successful program greatly increased the impact of Bush Heritage outside its reserves and was instrumental in allowing the organisation to develop relationships and test tools to underpin its future approach to partnerships.

This chapter describes the origins of the program, program achievements, lessons learnt and future partnership directions.

Description

Beyond the Boundaries was a key program for Bush Heritage over the 5 years from 2005 to 2010. The program evolved as the organisation looked to increase the effectiveness and impact of its conservation work and achieve its ambitious goal to protect or support the conservation management of 1% of Australia (7 million hectares) by 2025. Although the specific program discontinued under this name in 2010, its work and projects continue to shape and drive the work of Bush Heritage.

Bush Heritage forecast that by 2025 over half of the area needed to achieve a goal of 7 million ha of conservation priority land would need to be delivered outside of land it owned. Beyond the Boundaries' mission was therefore to 'deliver large scale, practical and tangible conservation outcomes by inspiring and supporting other landowners to undertake conservation on their own land and by sharing expertise and resources'. The program was also designed to reflect Bush Heritage values of collaboration and pragmatism.

As a pilot, the Beyond the Boundaries program had two initial objectives: (i) establishing partnerships on 10 properties totalling at least 750 000 ha by 2011; and (ii) all properties acquired or managed in partnership used a systemised adaptive management approach. The program explored partnerships at three scales: (a) local 'reserve-based' partnerships where collaborative arrangements are negotiated directly with neighbours; (b) regional projects built on long-term collaboration with many partners across a variety of tenures (e.g. Gondwana Link in Western Australia, Kosciuszko to Coast in NSW and Midlandscapes in Tasmania); and (c) national projects with organisations that own vast areas (e.g. pastoral companies and the Indigenous Land Corporation). The pastoral industry, Indigenous people and government departments were identified as key partners at all levels.

Bush Heritage forecast that the Beyond the Boundaries program would achieve its goals by using five strategies.

1 Through direct management agreements, where Bush Heritage provided assistance to access funds, and then coordinated conservation work on private or public land;
2 By covenanting land, where Bush Heritage assisted the negotiation of conservation covenants on properties on behalf of their owners;
3 By assisting others to buy conservation lands, where Bush Heritage would undertake initial conservation assessments, assist with the land purchase process and support landowners in land management activities;
4 Through conservation planning, where Bush Heritage facilitated conservation management planning on private or public land; and
5 By providing management support, where, in return for covenants or management agreements on significant sites, Bush Heritage provided planning expertise or resources to protect conservation values on the land.

Why establish a partnership program?

As with all other organisations entering the private landholding conservation space, Bush Heritage's primary effort was originally directed at the scale of individual properties. Resources were concentrated on managing specific issues within the boundaries of existing reserves. In the early 2000s, as the organisation's capacity grew from that of a relatively small landholder into a more substantive and influential land management body, it became clear that its management of ecological values could expand and respond to the scale of the processes that sustained or damaged those values. Effective management therefore required, or would benefit from, working more extensively with neighbours and for some issues it would be beneficial to work on developing sound relationships with a much wider group of stakeholders. The long-term threats to Australia's biodiversity, including climate change, require sustained and coordinated action at a scale that matched the scale of the threat – typically across properties, catchments, regions and landscapes.

This drove adoption of a new strategy called 'Anchors in the Landscape'. By establishing conservation areas in particular landscapes – the anchors referred to – Bush Heritage anticipated using their presence in each landscape to create further links and partnerships.

Property acquisition remained an important tool in establishing these anchors. However, farmers, pastoralists and Indigenous communities between them manage nearly 80% of the Australian continent, making them, and the land they manage, crucial to Australia's environmental future. Some of this land is available for purchase, but most is not. Importantly, Bush Heritage was finding that the most critical areas remaining for conservation action rarely come on the market. In many cases people were already interested in managing land to achieve conservation results. Supporting these management roles in active partnerships was seen as a critical mechanism to produce increased conservation outcomes.

The desire to build on existing partnerships and to expand the scale of Bush Heritage activities led to the establishment of the Beyond the Boundaries program. It aimed to build the use of partnerships as the second mechanism to deliver on the Bush Heritage Anchors in the Landscape plan. Beyond the Boundaries targeted partnerships across multiple tenures to create permeable landscapes that buffered existing protected areas and targeted conservation values outside the reach of formal reserve tools (Cowell 2007).

Achievements to date

During the period 2005–2010, Bush Heritage supported Beyond the Boundaries projects across 23 partnership and/or reserve locations (Figure 18.1). In addition, Bush Heritage secured 11 grants to support critical staff and capacity development for the program. The results of this investment were spread across partnerships and reserves.

In 2012, partnership projects developed through this program are now core elements of the Bush Heritage conservation program. They now support management of over 3.5 million hectares of land for conservation. This is more than four times the 750 000 ha goal set for the Beyond the Boundaries program in 2005. A significant proportion of this area is managed as formal Indigenous Protected Areas by Traditional Owners.

The Beyond the Boundaries program also sought to promote the use of an adaptive management framework. The framework chosen was Conservation Action Planning, an adaptive management process developed and used by The Nature Conservancy around the world to plan its conservation projects (based on the Open Standards for Conservation – http://www.conservationmeasures.org). All Bush Heritage projects are managed using this process, and it is also used by the majority of our partners. This enables stronger collaborations by providing a common framework and language for collaboration.

Work done under the Beyond the Boundaries banner contributed to the establishment and strengthening of several significant regional-scale partnerships including Gondwana Link (Chapter 3), Kosciuszko to Coast (Chapter 13), and the Tasmanian Midlandscapes project (Chapter 9).

Case study 1 – Kosciuszko to Coast

Prior to establishment of Kosciuszko to Coast (K2C), a project initially aimed at supporting landscape-scale conservation in the Murrumbidgee River corridor between Cooma and Canberra, several organisations were undertaking projects within the region. Without coordination there was little understanding of how each project might fit together to form a big picture.

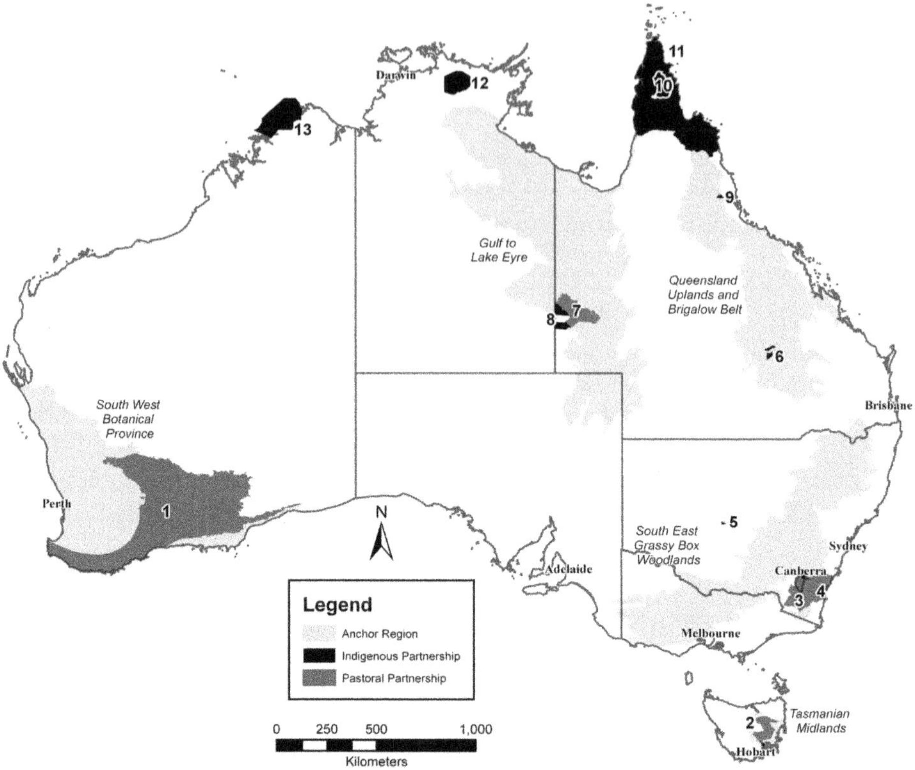

Figure 18.1 Bush Heritage Australia's Anchor Regions and Beyond the Boundaries projects. Key: 1 – Gondwana Link (including Chingarrup); 2 – Tasmanian Midlands; 3 – Scottsdale Reserve; 4 – Kosciuszko to Coast; 5 – Mawonga Station; 6 – Carnarvon Station Reserve & Mt Tabor Station; 7 – Toko Ranges, Mulligan River, Marion Downs & Glenormiston; 8 – Cravens Peak & Ethabuka Reserves; 9 – Yourka Reserve; 10 – Kaanju Ngaachi; 11 – Balkanu/Cape York; 12 – Warddeken; 13 – Wunambal Gaambera.

During 2005–06, there were several meetings between key conservation groups and a government agency (NSW Department of Environment and Conservation) seeking to establish a common vision and work program. An 'anchor' property, Scottsdale, was purchased in December 2006 that has served a valuable role as a venue for events, field days, management trials and volunteer participation. In July 2007, Bush Heritage established a part-time facilitator position (co-funded by the Great Eastern Ranges Initiative – Chapter 13) to develop the project further and encourage collaboration through the Beyond the Boundaries program. Half the funding for this position for the first 3 years was provided by the Great Eastern Ranges Initiative, after which it provided all the funding. A Conservation Action Plan was developed in 2007 with input from all partners and continues to guide the K2C project by establishing objectives and strategic actions.

In addition to direct work on Scottsdale, Bush Heritage also provided conservation assessments for many private landowners and entered into MOUs supported by short-term funding for on-ground works with four private property owners over 817 ha. Two of these landowners have since committed to long-term conservation agreements over more than 500 ha. Other partners such as the Department of Environment, Climate Change and Water and the Nature Conservation Trust also implemented a complementary conservation-covenanting program in the K2C region and Greening Australia offered whole of paddock restoration grants for 3 years from 2007 with Great Eastern Ranges funding. Other conservation groups, such as Upper Murrumbidgee Catchment Coordinating Committee and Friends of Grasslands, were also highly involved. This demonstrates the benefits of common interests merging to deliver a more integrated approach. From 2007, the K2C experience was used to guide the expansion of Great Eastern Ranges Initiative into other areas. Beyond the Boundaries played a vital early role in achieving this.

Case study 2 – pastoral partnerships

The largest area of Bush Heritage reserves (indeed of private conservation lands generally) exists in the pastoral zone, adjacent to several significant pastoral operations.

A substantial effort was made to engage with the pastoral industry at the industry-, company- and property-scale to seek to develop an effective working relationship. In addition to work under the Tasmanian Midlandscapes project, the most successful of these was, and remains, a developing relationship between Bush Heritage and the North Australian Pastoral Co. in far western Queensland, on leases adjoining Bush Heritage's Ethabuka and Cravens Peak reserves. Bush Heritage is working with the company to monitor the impacts of pastoral production on key ecological assets, with each party slowly developing a relationship with the other. It is likely to be some years before the benefits of this work will be known.

There is a critical need to establish effective incentives for the pastoral industry to shift their production systems to be able to incorporate conservation measures while retaining profitability. These opportunities are only just emerging through projects such as the Global Roundtable for Sustainable Beef and the Carbon Farming Initiative. The work of Bush Heritage and others in building relationships on the ground at specific locations will be essential in supporting these higher level policy shifts.

Case study 3 – Indigenous partnerships

Through its implementation of Beyond the Boundaries, Bush Heritage committed to work in partnership with Indigenous people both on Bush Heritage reserves and on Indigenous held lands of high conservation value.

Several significant relationships developed under the Beyond the Boundaries program in Northern Australia with Balkanu (Cape York region), Wunambal Gaambera Aboriginal Corporation (Kimberley region) and Warddeken Land Management (Arnhem Land region) commencing significant partnership projects with Bush Heritage.

The Wunambal Gaambera Healthy Country Partnership covers a significant area, located in the remote north Kimberley. It encompasses some 2.5 million hectares (almost 1 million ha of land and 1.5 million ha of sea) of determined native title country across several tenures. Within this area is the 343 515 ha Uunguu Indigenous Protected Area. For 6 years starting in 2006, Bush Heritage has worked and continues to work in partnership with the Wunambal Gaambera people to collaborate on ensuring that their country remains healthy and protected.

The Beyond the Boundaries program and staff were instrumental in the development of internal Bush Heritage policy to guide work with Indigenous partners and the innovative adaptation of Conservation Action Planning methods. Those involved summed up the benefits of the approach as follows:

> *The Wunambal Gaambera Healthy Country Plan has shown that the success of a collaborative conservation planning process in a cross-cultural context requires support of Traditional Owners' interpretations of 'healthy country' as well as the recognition of cultural, social and economic outcomes. Most significantly, the WGHCP*

> *demonstrates that Indigenous Traditional Owners' aspirations to drive the conservation planning agenda for their ancestral estates can be achieved.* (Moorcroft *et al.* 2012).

The use and demonstrated adaptation of Conservation Action Planning tools by Bush Heritage and Indigenous partners to their local context has provided an empowering 'Healthy Country' planning tool and this process, actively supported and driven by The Nature Conservancy, has led to its use by several other indigenous groups in their Indigenous Protected Area planning projects across Northern Australia (Brown 2011).

Limitations and issues encountered

There are several issues that make implementation of an ambitious program such as Beyond the Boundaries challenging. These include:

1. Opportunism v. prioritisation

Partnerships between a non-government organisation (NGO) and a private landholder are more likely to be driven initially by responding to opportunity, at least until the development of relationships within a geography considered 'strategic' by the NGO. Over time, partnerships are a significant part of a regional strategy, but in their early stages they can appear to sit outside 'strategy' and they can be hard to see within an organisational context.

2. Initial investment of time and resources

Bush Heritage developed a substantial number of relationships through a small specialist team over the term of the program. These staff were responsible for initiating and developing these relationships, in particular with Indigenous interests. As the program grew, this small group was then responsible for numerous relationships in many places, stretching their capacity. Significant time and energy are required to build relationships and trust that required ample resourcing by Bush Heritage to spend enough time at the local level, in pursuing and developing the relationships needed for local partnerships. Conservation values, and therefore many partners, are often located in remote and relatively inaccessible parts of Australia. This meant forming and maintaining partnerships and resourcing relationship-building activities over a long period of time can involve significant cost. Importantly, this is where an 'anchors' approach can be important in supporting partnership activities.

The development of mutual trust together with shared plans and objectives to a point where deeper investment was possible took between 1–3 years for most partnerships. In some cases, projects initiated under Beyond the Boundaries are only now maturing from relatively short-term investments in specific projects to formal consideration of longer-term agreements and memoranda of understanding.

3. Differing partner expectations

The outcomes desired by Indigenous, pastoral and agricultural partners have driven Bush Heritage to acknowledge and incorporate social, economic and cultural outcomes in addition to more traditional 'Western' conservation outcomes. This has challenged us to rethink how we describe and measure what has been achieved with our investment. Using an adaptive management process (Conservation Action Planning), and modifying it for use with specific partners where needed, has provided us with a consistent framework for describing and measuring these outcomes.

4. Maintenance of multi-tenure partnerships in the long term

Bush Heritage reserve management staff have generally been able to develop excellent informal relationships with neighbours and other stakeholders. However, once land management tasks are undertaken on the reserves, they have little spare time left to explore deeper relationships. In the case of Gondwana Link, Midlandscapes and Kosciuszko to Coast, these partnerships were managed by the central (Melbourne-based) Beyond the Boundaries team and significant resources were provided to assist with the coordination of regional partners while local staff focused more strongly on reserve management. The involvement of a central team with partnership expertise and resourcing of shared coordination was vital to establishment of these multi-tenure partnerships.

To manage the load of the central team and to ensure robust relationships at the regional level, earlier planning for transition of these relationships to regional staff and provision of additional time and resources are now viewed as key elements to maintain enduring partnerships and relationships.

Maintaining coordination at the regional level is clearly important for long-term alignment of partner objectives. However, expectations around shared resourcing need to be discussed and managed early to ensure there is clarity on the long-term commitments of each partner.

5. Knowing when to 'end'

Beyond the Boundaries invested significantly in the provision of local partnership 'coordinators' for Gondwana Link, Midlandscapes and Kosciuszko to Coast in the first years of each project. The long-term sustainability of this coordination function also needs to be carefully evaluated and planned for.

Bush Heritage has been faced with questions where we have reduced our commitments based on expectations that long-term resourcing at initial levels would continue indefinitely. Ending or changing partnership activities needs to be considered as clearly as the establishment of those activities. When internal priorities change these need to be clearly communicated to other partners and other stakeholders.

6. Challenges specific to Indigenous partnerships

The capacity of the organisation to work effectively with Indigenous groups has in several places been limited by the fact that native title has not yet been granted or is in dispute. While positive relationships have been established in several of these places, the full partnerships originally envisioned are yet to be achieved. Complexities in the lives and communities of Indigenous partners also mean added logistical challenges when seeking to support effective participation and decision making.

Future directions

Bush Heritage will continue to build partnerships with neighbours, Indigenous people, government agencies and other NGOs. Bush Heritage has now integrated the tools and approaches developed under Beyond the Boundaries into its ongoing strategy.

During the 5 years of the Beyond the Boundaries program, the organisation gained a better understanding of the time and effort required to effectively support partnerships based on its reserves or working with others on their land. Recognition that each relationship has been different and progressed differently due to local circumstances and aspirations of potential partners informs how Bush Heritage now initiates and supports ongoing and potential partnerships.

Of the five strategies initially proposed, the provision of management support through planning expertise or resources to protect conservation values where our partners intend to establish long-term protection through covenants, Indigenous protected areas or other agreements provided the most traction and remains key to our ongoing work. Bush Heritage sees that resourcing its ability to work with partners to apply the Open Standards' adaptive management process will be fundamental to continued development of strong relationships and partnerships.

Bush Heritage will assess potential partnership opportunities with the same rigour we do our property purchases, including using tools such as BioPrEP (Mackey *et al.* 2010) to select target areas within regions. For any new partnership, Bush Heritage evaluates the conservation gains to be made, the financial sustainability, our capacity and that of our partners to manage threats, and the sustainability of the relationship. Partnerships will be secured through a variety of means, including formal agreements and land protection mechanisms.

Within the Great Eastern Ranges Initiative, Scottsdale Reserve remains a key property within the Kosciuszko to Coast project and provides a focal place for people to experiment, volunteer and be inspired. In order to better target our ongoing reserve acquisitions and management strategies, Bush Heritage will continue working with partners to develop shared plans and projects that address threats to conservation at the right scale. Bush Heritage has made long-term formal commitments to the Tasmanian Midlandscapes and Gondwana Link initiatives, both recognised under the *National Wildlife Corridors Plan*. The 10-year partnership agreement with Wunambal Gaambera Aboriginal Corporation in the Kimberley will guide Bush Heritage in developing understanding, tools and processes that work effectively in Northern Australia more broadly and provide approaches that can be adapted and used by others.

In addition to our formal participation in the leadership and governance of large-scale initiatives, we see that there remains significant potential to build on the capacity of our reserves to serve as a focal point for regional conservation activity.

Conclusions

Using the Beyond the Boundaries program to develop a partnership approach has brought many important benefits to Bush Heritage. These include sharing: (i) scarce management resources; (ii) land management knowledge; (iii) differing cultural perspectives; (iv) effort and responsibility; as well as (v) extending conservation action into areas where ownership cannot change; and (vi) creating community ownership of conservation values.

This ambitious program appointed a small team to test and undertake a range of strategies across Australia. While the overall hectare targets have been greatly exceeded, this is largely as a result of a small number of quality partnerships. Lessons learnt through the Beyond the Boundaries program continue to influence communication, position descriptions, budgets and funding strategies.

Acknowledgements

The program was staffed by several creative and committed staff who built the program, refined tools and created many enduring relationships. The commitment and support of former CEO Doug Humann and the Bush Heritage Board were also vital to the program.

Financial and technical support were provided by The Nature Conservancy and the Thomas Foundation, in addition to many dedicated Bush Heritage supporters. Finally we would like to acknowledge the participation and fellowship of people and organisations who worked with Bush Heritage on the Beyond the Boundaries program and through continued partnerships to develop shared ideas on what needed to happen and how it would get done.

Biographies

Philippa Walsh has over 20 years' experience in the not-for-profit sector working for a range of organisations and is currently Senior Manager–Conservation at Bush Heritage Australia. She has worked across all jurisdictions in Australia in the planning and implementation of large-scale conservation programs on private and Indigenous land.

Stuart Cowell has worked in many natural resource management sectors, including production, research, not-for-profit conservation and Indigenous land management over the last 25 years. He was a senior manager with Bush Heritage Australia from 2001–2010 where he led development of the Beyond the Boundaries program. He now works as a consultant supporting the development of landscape management plans and capacity.

Sarah Eccles is a Wathaurong descendent and has 10 years' experience working with Indigenous people throughout Australia in community-based Caring for Country and conservation partnerships. Working in government and non-government organisations including Bush Heritage Australia for the past 7 years, Sarah's passion is partnerships which protect culture and country, while providing livelihoods with social and economic benefits to Indigenous people.

Matt Appleby has a PhD and research background in the field of restoration ecology. Matt has worked on planning and implementation of both the Midlandscapes and Kosciuszko to Coast projects. He has also provided ecological input to management of Bush Heritage reserves across southern Australia. Matt is currently the acting Ecologist Coordinator for Bush Heritage Australia.

References

Brown J (2011). Adapting CAP for Use by Indigenous Groups in Australia and Melanesia and Sharing Lessons Learned. A Coda Global Fellowship supported by the McLean Foundation, 5 June—8 July, 2010. The Nature Conservancy Canada Program, Great Bear Rainforest.

Cowell S (2007). What do you do when the biodiversity you bought gets up and leaves? Challenges facing protected area planning for the private land trust sector due to climate change. In: *Protected Areas: Buffering Nature Against Climate Change. Proceedings of a WWF and IUCN World Commission on Protected Areas Symposium, 18–19 June 2007, Canberra* (Eds M Taylor and P Figgis) pp. 112–116. WWF Australia, Sydney, Australia.

Mackey B, Gilmore S, Pringle H, Foreman P, Van Bommel L, Berry S, and Haseler M (2010). BioPrEP – a regional, process-based approach for assessment of land with high conservation value for Bush Heritage Australia. *Ecological Management & Restoration* **11**, 51–60.

Moorcroft H, Ignjic E, Cowell S, Goonack J, Mangolomara S, Oobagooma J, Karadada R, Williams D, and Waina N (2012). Conservation planning in a cross-cultural context: The Wunambal Gaambera Healthy Country Project in the Kimberley, Western Australia. *Ecological Management & Restoration* **13**, 16–25.

19

Landcare: linking Australia's landscapes by linking its land managers

David Walker

What is Landcare?

Ever since the National Farmers' Federation's Rick Farley and the Australian Conservation Foundation's Phillip Toyne bridged the divide between farmers and conservationists to commence the Decade of Landcare, attitudes and action towards sustainability and conservation on farmland have been transformed. That hitherto unlikely partnership sparked the proliferation of Landcare across rural Australia.

The genesis of Landcare was the realisation that effective action against many land degradation problems impacting upon farm productivity could not be limited to individual farms, rather collaborative action by neighbouring farmers was required. The name 'landcare' was coined in Victoria, and through funding and other resources from the Victorian Government and coordination and technical advice from government agency staff, several 'multidisciplinary, community-based, autonomous groups' became actively involved in land restoration across Victoria (DAFF 2012a).

Launched in 1989, the Decade of Landcare was the nationwide catalyst that stimulated farmers to think about the landscape function that underpinned their farm's productivity and integrity, and to talk with other farmers, scientists and advisers to gain a better understanding of these issues. With government support came resources and the opportunity to work with their neighbours to take coordinated action to address landscape dysfunction and degradation (particularly processes such as soil erosion and salinity). While most of the early work was to staunch the haemorrhage, the ethos of farming has evolved so that land managers acknowledge the important role they must play in nature conservation, and how biodiversity can help support the resilience and productivity of their production systems.

When the National Landcare Program was launched in 1989, it was envisaged there would be 2000 Landcare Groups by the year 2000. Landcare has continued to grow, with some 5000 Landcare and related community groups (such as Bushcare, Coastcare, 'Friends of …') as at 2012, and more than half a million members. 'This army of volunteers has changed the face of Australia's rural and urban landscapes. It has planted millions of trees, shrubs and grasses; repaired riparian zones and restored water quality by reducing erosion and fencing out stock from riverbanks; protected remnants of native vegetation; regenerated areas to provide habitat for native wildlife; improved ground cover, grazing methods and soil management; and rehabilitated coastal dunes and recreational areas' (DAFF 2012b).

While joint action among neighbouring landholders has been highly successful in addressing degradation and loss of productivity of farmland, tree planting and remnant vegetation protection and enhancement has tended to be seen as an outcome in itself, rather than being targeted to enhance biodiversity, and integrated to provide connectivity of habitat across landscapes. According to Lindenmayer (2012), this oversimplification has long been government policy: '[In 1997] I was told by the (Federal) Environment Minister and his minders that "we already knew everything we needed to know about restoring vegetation on farms" and "all we needed to do now was get the trees in the ground"'.

Landcare has realised that effective nature conservation requires much more than that. If you can coordinate activity against degradation, the next step is to coordinate action against habitat loss. This requires integration and joint effort across properties and a much deeper knowledge of conservation ecology than is common among most farmers.

The rate of return on government investment in Landcare has been outstanding. The average ratio of private cash equivalent contribution by Landcare to National Landcare Program funding has been about \$3:1 or \$4:1, with some studies showing an investment multiplier of up to \$12:1 (AFFA 2004).

Much of the recent growth of Landcare, however, has been of groups in urban areas, working on small patches of public land within and near larger population centres. Changes in the focus and funding priorities of the successors to the National Landcare Program, of targeting individuals rather than group effort, and through the implementation of the regional delivery model of natural resource management (NRM) from 2002 and Caring for our Country in 2008, has seen a significant decline in coordinated Landcare activity in rural areas and by farmers (Paton *et al.* 2004; Robins and Kanowski 2011), 'and has, paradoxically, undermined rather than facilitated regionally engaged communities to manage natural resources within their jurisdiction' (Lane *et al.* 2009, p. 70). The unfortunate and unintended result has been a marked decline in the leverage achieved by government funding, with acceptable levels of co-contribution now being \$1:1 or lower.

Why Landcare for connectivity conservation?

The landscape of Australia's agricultural regions shows extreme fragmentation. This includes:

- fragmentation of native vegetation, with most fertile landscapes being highly cleared, and many vegetation communities so reduced as to be classified as 'endangered';
- fragmentation of land use, with the most cleared areas being used for cropping, and less cleared areas for grazing livestock; and
- fragmentation of ownership, with thousands of small to medium 'family farms', occupying most of these regions.

In such landscapes the opportunity for protection of large contiguous areas of land for nature conservation are extremely limited.

Where do we need the nature conservation effort?

According to the Australian Natural Resources Atlas (ANRA 2012), 61.5% of Australia's land area is utilised for grazing or dryland or irrigated cropping, dwarfing the area designated as 'nature conservation', a mere 6.5% of the total.

Not surprisingly, given the pattern of settlement and land utilisation in Australia, most endangered ecological communities are located in the areas that have been most extensively utilised for farming and grazing, and are most strongly held by commercial farmers. It is no accident that this agriculturally productive land is priced accordingly, and increasing the comprehensiveness, adequacy and representativeness of the National Reserve System (NRS) across such landscapes through land acquisition alone is unaffordable and impractical.

The inescapable conclusion is that effective landscape-scale conservation of these highly altered landscapes will rely mostly on the efforts and interest of private landholders. Conservation efforts on private land can augment and complement the NRS, and enhance connectivity between NRS areas.

Who must we rely upon to do it?

There are 134 148 farms in Australia, 90% of which are 'solely dedicated to agricultural production' (NFF 2012). While the average size of all farms is almost 3000 ha (NFF 2012), 63% of farms in Australia are 500 ha or less (OECD 2008). Fewer than 10% are larger than 2500 ha.

Again, the areas of Australia's land that are most in need of significant conservation effort, the areas that have been most altered by development for agriculture and that are home to most of the endangered ecological communities, are the areas also home to these thousands of small to medium farms. Thus achieving effective connectivity conservation across this extensive landscape, highly fragmented in vegetation and tenure, will require the engagement of, and a collaborative effort from, many, many landholders.

Connectivity across farming landscapes

Ecological connectivity has been described, and its importance discussed, elsewhere in this publication (Chapters 2, 23; see also Bennett 2003). The achievement of effective connectivity conservation across landscapes that are fragmented in land use, vegetation and ownership will rely on cooperation and collaboration between neighbouring land managers.

Further, in landscapes that are primarily used for productive agriculture, integration of conservation and production is essential, so farmers can meet their own needs to generate income, as well as the broader population's needs for sustainable food and fibre. Given world population projections, it is unrealistic to expect that large swathes of farming land can be taken out of food production and redirected solely to nature conservation.

What can Landcare achieve?

The taking of Landcare to the next level, from collaboration between farmers in addressing landscape dysfunction, to where action is lifted to high-level landscape conservation, requires integrated planning and support. The transformation in farmers' attitudes and in the landscape itself (Baker 1997; Alexander *et al.* 2000) that has been delivered by Landcare provides an excellent base upon which to build. A number of Landcare organisations have taken it upon themselves to develop the higher level of sophistication and partnerships required. These are outlined below.

Case study 1 – Big Scrub Landcare

The 'Big Scrub' Rainforest is located on the New South Wales (NSW) North Coast (Figure 19.1). At the time of European settlement it was the largest area of lowland subtropical rainforest in eastern Australia, comprising 750 km^2.

Extensive clearing began soon after, initially to harvest the valuable Australian Red Cedar (*Toona ciliata*); thereafter clearing was indiscriminate, with the NSW Government giving allotments of land to farmers provided they cleared the rainforest. Only small scattered remnants of the rainforest survived, with most less than 5 ha in area, and totalling around 750 ha, just 1% of the original area (BSRLG 2012).

Concerned locals formed Big Scrub Rainforest Landcare Group (BSRLG) in 1992 in response to the fragile nature and high conservation value of the remaining Big Scrub remnants. BSRLG have identified more than 60 significant Big Scrub remnants, ranging from 0.4 ha to 171 ha. The major work of the group has been extending and enhancing these remnants, with 39 of the remnants being on private land and 34 on properties engaged in commercial primary production.

It is estimated that 1.5 million trees have been planted over the last 20 years, providing 250 ha of revegetation (T. Parkes, BSRLG, pers. comm. 2012). This exceptional effort has increased the area of significant remnants by 33% over 20 years.

Partnerships are critical to the success of landscape-scale cross-property conservation, and BSRLG works with a consortium of 10 regional partners (see www.bigscrubrainforest.org.au).

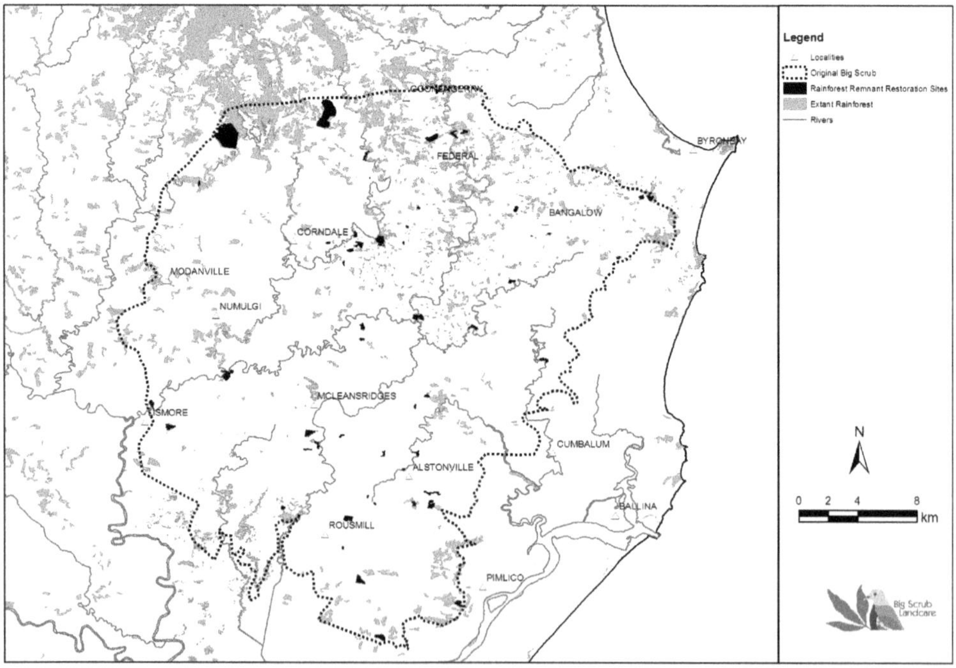

Figure 19.1 The Big Scrub Rainforest – before 1750 and present (Copyright Big Scrub Landcare 2012).

Case study 2 – Communities-in-Landscapes

The Communities-in-Landscapes project targets the nationally endangered Box-Gum Grassy Woodland landscapes west of the Great Dividing Range in NSW in three NRM regions; Central West, Lachlan and Murrumbidgee. Landcare NSW and eight partners (see cil.landcarensw.org.au) have collaborated to increase landscape-scale conservation across multiple land tenures. Underpinning the project has been the provision of quality, practical and relevant information on conservation ecology (C-i-L 2012).

The project used existing Landcare networks to recruit landholder groups or developed new groups. The procedure has been an initial meeting to canvass interest, followed by training activities to improve farmers' knowledge of sustainable grazing practices and of conservation ecology, a 'connectivity workshop' to see how remnant vegetation on individual farms can be extended and enhanced to provide linkages across their local landscape, and flora and fauna surveys. 'Seed funding' was provided to assist farmers with the cost of their on-ground works.

Each Cross Property Conservation Planning (CPCP) group developed their revegetation proposals as part of individual whole farm plans, had them critiqued by their peers, and then integrated them into a joint landscape CPCP (see also Chapter 12).

Seven CPCP groups were supported across the three NRM regions (Figure 19.2), integrating conservation activities across 67 landholders' properties, extending across

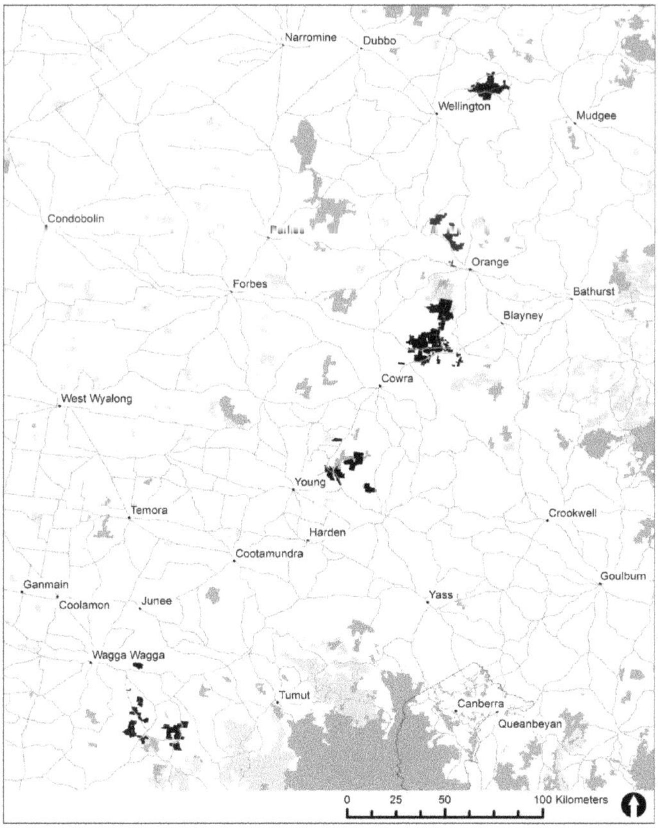

Figure 19.2 Location and extent of land involved in Communities-in-Landscapes (CiL) cross-property conservation. CiL properties displayed in black, protected areas in dark grey and state forest in light grey.

65 624 ha. Formation and effective functioning of CPCP groups involves building communal ownership of, and a joint vision for, the landscape. In areas where there was an existing Landcare network, the time to establish a CPCP group was halved.

This project has provided multiple 'wins'; to the environment, to the productivity and resilience of farms, and to the social fabric of rural communities. Farmers expressed particular satisfaction with the multi-agency approach and the project being driven by the landholders rather than the usual 'top down' approach.

Case study 3 – Liverpool Plains Koalas

Liverpool Plains Land Management Inc. (LPLM) was incorporated by local landholders in 1992 to coordinate action against land degradation and to foster Landcare across the Liverpool Plains (see http://www.lplm.org.au/). A major project has involved local Landcare Groups in a series of revegetation programs in the 1990s to address salinity (LPLM 2012), and to also provide enhanced habitat for the local Koala (*Phascolarctos cinereus*) population.

The tree plantings have been successful in mitigating the impacts of salinity and in supporting an expanding and relatively healthy Koala population, as well as helping to position Gunnedah to sustain its Koala population in the face of loss of habitat from coal mining (Lunney *et al.* 2009, 2012a; Avolio 2012).

LPLM implemented a research project from 2008 to 2011, with NSW Government and Sydney University zoologists to investigate the extent to which Koalas utilised the 1990s plantings, by fitting koalas with GPS radio-tracking collars. It was found that revegetation areas as young as 8 years old provide valuable koala habitat (Figure 19.3), but that there was significant variation in the 'attractiveness' of particular trees to Koalas and the extent to which they are utilised (LPLM 2012).

A new Environmental Trust-funded research project is due to commence in 2013 by LPLM and the same team of zoologists to identify the best patterns for planting and the factors that produce trees attractive and nutritious for Koalas, so that future plantings ensure the best outcomes for koala conservation. The project will inform a concurrent LPLM/

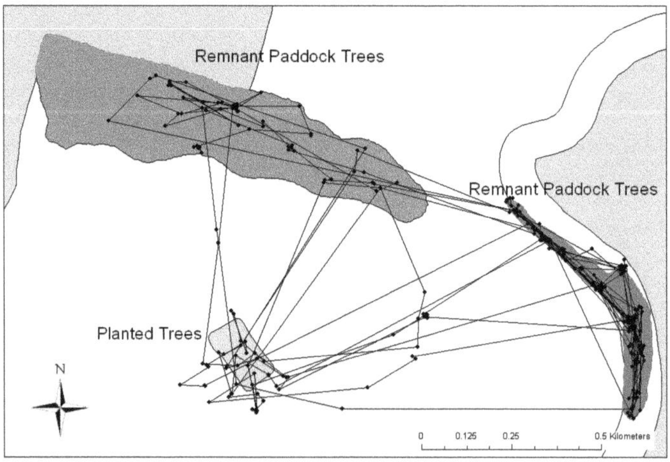

Figure 19.3 The movement of a Koala plotted from the data from the GPS units in the collars. (Source: Lunney *et al.* 2012b).

Australian Government Biodiversity Fund project working with Landcare groups to plan and coordinate cross-property revegetation, with species that will enhance biodiversity generally, and koala tree species in particular.

There are numerous Landcare groups and networks across Australia that have moved past focusing purely on land degradation to seeing the need for coordinated conservation effort across multiple and varied land tenures. In most cases the limiting factor is not lack of vision, but lack of capacity and resources. The evolution in thinking and activity of Landcare groups towards conservation connectivity is typified by several of these groups, particularly in Victoria, evolving into Conservation Management Networks (see Chapter 21).

What works, and what went wrong?

Landcare has been successful in transforming the way landholders 'look at their landscape' (Alexander *et al.* 2000). By the end of the 1990s, some 47% of landholders were active members of Landcare groups (Curtis and De Lacy 1997), with a further 35% influenced by the activities of Landcare groups in their area and the changed practices that they were adopting (AFFA 2004; DAFF 2012a). In the 2012 survey of the health of the Landcare movement, 73% of farmers consider themselves part of Landcare (de Hayr 2012). Landcare is universally seen as 'the trusted volunteer organisation' (Gillard 2010, p. 2), and is seen by landholders to be 'theirs', rather than some policy imposed from above.

Unfortunately, the factors that enhance participation and adoption among farmers are the elements of Landcare support that have been most impacted by changes to NRM policy implementation across most levels of government in Australia.

The characteristics of Landcare that have underpinned its success in community engagement, and ensure that land management practices change becomes embedded are outlined below:

1. Local focus within local communities

Landcare learning and activities are driven by the local landholder community, so they are relevant to the group and address their identified priorities. According to Colliver (2011), private landholders need to be 'in the mix' as initiators of action, not just as implementers of government priorities. It is this characteristic that engenders 'community ownership', changes social norms and builds social capital.

The focus by government on 'national' or 'statewide' priorities and targets has disempowered communities: their local priorities are seen to be unimportant.

2. A necessary level of support and facilitation

The spread and growth in capacity of Landcare was not spontaneous. The deployment of Landcare coordinators to canvass 'communities of interest', arrange initial meetings, facilitate the identification of the local NRM issues, coordinate property planning and assist groups identify and source the information and expertise they needed was vital. Many farmers are time-poor, under-staffed and feel under pressure to focus on business survival. Without the support and encouragement provided by coordinators, groups struggle to maintain enthusiasm and direction.

Coordinators are also vital to the formation of the partnerships that have been integral to the success of the three case studies described above. According to Colliver (2011, p. 171), 'community-based coordinators are essential to the formation of partnerships' and 'skilled and enthusiastic coordinators drive those partnerships, maintaining commitments from government agencies even as programs and roles changed', while Compton *et al.* (2009, p. 12) add 'a facilitator contributes to many aspects of community functionality'.

Landcare coordinators are often young, enthusiastic and with tertiary qualifications, adding significantly to the 'social capital' of local communities. Government decisions, at both state and national level, to withdraw financial support for Landcare coordinators has had a significant detrimental impact on participation in and the level of activity of Landcare.

3. Funding

In the early days of Landcare, seed funding for the commencement of on-ground works provided groups with the catalyst to trial and implement the changed land management practices they had proposed in their property plans. However, provision of funding for on-ground works is not the 'make or break' that it is often considered to be.

According to Nabben (1999, p.7), it is the group dynamics that are 'instrumental in changing landholder attitudes, understanding and motivation. The financial incentives just assisted in the accelerated adoption of new farming systems.' Once farmers see the advantages for themselves and their businesses in sustainable production systems and a healthy biodiverse landscape, they tend to find the money from within their own cash flow (Alexander *et al.* 2000; Ambrosio *et al.* 2009).

The 2012 survey of Landcare groups and 'Landcarers' found funding was not a barrier to adoption and maintenance, but that activity increased significantly when groups had access to staff and projects funding. A majority of groups indicated they required some funding for both support staff and on-ground works (de Hayr 2012).

4. Shared oversight

Landcare groups have a shared interest in the success of changed practice adoption by their members and tend to jointly co-mentor members through the change process and hold each other accountable for the effective use of funds.

The recent practice of many regional bodies preferring to deal 'one-on-one' with individual landholders, rather than with groups of their peers, loses that shared vision of landscape and accountability.

The withdrawal of funding, support and recognition for Landcare that has taken place since 2000 has resulted in a general decline of Landcare activity in the farming community. Certainly the continued efforts of passionate individuals have kept some groups such as BSRLG active. Such groups, however, tend to have a more urban-based demographic: less time-poor and more 'activist'. Location in the more populous (and wealthy) local government areas has also helped, as city-based councils in particular have recognised the contribution that Landcare can make to their environmental targets, and have provided staffing or resources to help.

By contrast, many Landcare groups in the more agricultural areas have become less active and have ceased to meet regularly. Lack of external support, volunteer burnout and concerns with economic survival through recent drought and flood events has seen many Landcare members losing focus on innovation towards sustainability and conservation.

Where to from here?

Landcare needs to be a partnership between all levels of government and the community. The progress and successes, and the trials and tribulations of Landcare over 25 years show us what we can do, and what we need to do, to encourage farmers to 'take a landscape-scale view' of their remnant vegetation, and to work with their neighbours to improve connectivity and habitat quality across larger and larger areas. The potential for large-scale connectivity conservation across farmland was demonstrated by one of the Communities-in-Landscapes landholders from Limestone Creek near Mandurama, NSW, who said:

> *'Wouldn't it be terrific if we could build this project to have connectivity from Mount Canobolas to Lake Wyangala' (a distance of 60 km).*

This will not happen without a vibrant, forward-looking and engaged Landcare movement.

Hopefully the nadir of Landcare support and recognition is past. In particular, two changes to Australian Government policy since 2010 indicate a recognition of the disempowerment of communities and Landcare that had inadvertently resulted from the implementation of the regional model for NRM and Caring for our Country, and a willingness to remedy.

The redeployment of some Landcare support personnel (embedded in the community rather than in government agencies), through the Caring for our Country Regional Landcare Facilitator Initiative (in each of the 56 regions), and the re-introduction of a small grants pool (i.e. Community Action Grants), tailored to the needs of smaller community and Landcare groups, have thrown a lifeline to Landcare groups and their activities.

The Australian Government's 2012 review of the Caring for our Country Program provides an opportunity to further redress the shortcomings of recent policy, and includes an undertaking to focus more on 'capacity building and developing leadership in our communities' and also states 'regional NRM will deliver community priorities from the regional NRM plans' (DAFF 2012c).

State governments in Victoria, particularly, but also New South Wales and Tasmania, have moved to address the lack of support for Landcare in those states. There seems to be greater recognition universally of the role that Landcare can play in meeting numerous state and territory government NRM targets.

Regional bodies also are beginning to understand that they need an aware, engaged and active community working in equal partnership with them in order to meet their responsibilities effectively. The signing of the Statement of Common Purpose between the National Landcare Network and the National NRM Regions Working Group in April 2012 acknowledges the individual focus as well as the complementary strengths of Landcare and NRM regional bodies and hopefully will herald a new era of collaboration (see http://www.nln.org.au/files/StatementCommonPurpose20April2012.pdf).

Achieving landscape-scale connectivity conservation across farming landscapes, when landholders are under increasing pressures to survive economically, adapt to climate change and also to increase food production will not be easy, or cheap. But neither are any other schemes for conservation, and the predominance of farming across most of the areas of greatest habitat loss means that farmers must play a key role.

Conclusions

Most of Australia's land mass is 'solely dedicated to agricultural production', and managed by commercial agricultural businesses. The landscapes most suited to this commercial land use are the most highly altered, and consequently the most in need of renewed focus on nature conservation. These landscapes are also made up mostly of small to medium-sized farms.

It is inescapable that landscape-scale conservation in agricultural landscapes must engage multiple landholders, and must see coordination and education of conservation planning across numerous neighbouring farms. The model for delivering such integrated and planned conservation effort is Landcare. Investment in building the capacity of land managers through the Landcare model has leveraged far greater co-contribution from those landholders than any other NRM program.

Landcare has transformed the way that farmers relate to their land and its sustainability. It has produced thousands of environmentally aware, conservation-focused farmers, who understand that addressing land management issues is best achieved by collaboration. The achievement of integrated, coordinated planning and activities for landscape-scale connectivity conservation across farmland requires investment in the people who manage that land.

The prerequisite is to enable farmers to develop a shared vision of their landscape with their neighbours. Then the farmers need conservation ecology information that is relevant, practical and plausible with support and assistance to plan, for their own properties and with their neighbours, the changed management and remnant protection or revegetation that is necessary. They need the provision of seed funding to implement the changed management, revegetation and remnant protection plans so developed.

The catalyst for these changes, and the embedding of the changes as their social norms, is continually shown to be the provision of effective Landcare support personnel. They are the ones who can facilitate the awakening, the attainment of understanding, the coordinated planning, the implementation of changed management and the ongoing support necessary to entrench the changed attitudes that indicate 'ownership' rather than obedience.

Acknowledgements

Sincere thanks are due to Tony Parkes and Paul O'Connor of Big Scrub Rainforest Landcare Group, and Tony Cox, NSW Department of Primary Industries Communities-in-Landscapes Project Officer for providing the information and maps for the Big Scrub and C-i-L case studies. Most of all, though, the thousands of Landcare volunteers and staff who have worked together to build the social capital of their communities, and improve the sustainability, productivity, integrity and resilience of their local landscapes, deserve the commendation, recognition and thanks of the Australian nation.

Biography

Twenty or more years working and observing as a practising farmer saw **David Walker** becoming more interested in broader natural resource management and the importance of community self-determination. He is currently employed as Executive Officer of Liverpool Plains Land Management Inc. and of the Gould League, Australia's oldest environmental education organisation. He has a BRurSc (Hons), Dip Ag Econ. He is currently studying for a MEnvSc & Mgt. He has a strong belief in the value of an empowered landholder community, 'aware, engaged and active', as the best catalyst for the adoption of sustainable land management practices. He has been foundation Chairman of Landcare NSW Inc., and is currently Chairman of the National Landcare Network, the federation of Landcare Peak Bodies across Australia.

References

AFFA (2004). *Review of the National Landcare Program*. Department of Agriculture, Fisheries and Forestry Australia, Canberra.

Alexander F, Brittle S, Ha A, Gleeson T, and Riley C (2000). *Landcare and Farm Forestry: Providing a Basis for Better Resource Management on Australian Farms*. ABARE Report to the Natural Heritage Trust, Canberra.

Ambrosio C, Linehan C, and McAllister A (2009). 'Privately funded contributions to NRM activities in the Goulburn Broken Catchment: testing the "x2 Assumption"'. Department of Primary Industries, Tatura.

ANRA (2012). Australian Natural Resources Atlas. Land Use – Australia. Available: http://www.anra.gov.au/topics/land/landuse/index.html [Accessed 21 May 2012].

Avolio C (2012). Gunnedah may hold key to reversing koala decline. *Australian Geographic* **June**. Available: http://www.australiangeographic.com.au/journal/gunnedah-may-hold-key-to-reversing-koala-decline.htm [Accessed 23 June 2012].

Baker R (1997). Landcare policy, practices and partnerships. *Australian Geographical Studies* **35**, 61–73.

Bennett AF (2003). *Linkages in the Landscape: The Role of Corridors and Connectivity in Wildlife Conservation*. IUCN, Gland, Switzerland and Cambridge, UK.

BSRLG (2012). Big Scrub Rainforest Landcare Group. Available: www.bigscrubrainforest.org.au [Accessed 21 May 2012].

C-i-L (2012). Communities in Landscapes. Available: http://www.cil.landcarensw.org.au [Accessed 21 May 2012].

Colliver R (2011). Community-based governance in social-ecological systems: an inquiry into the marginalisation of Landcare in Victoria, Australia. PhD thesis, Murdoch University, Perth.

Compton E, Prager K, and Beeton B (2009). Landcare bowling alone: finding a future in the 'fourth' phase. In: *Contested Country: Local and Regional Natural Resources Management in Australia*. (Eds M Lane, C Robinson and B Taylor) pp. 147–160. CSIRO Publishing, Melbourne.

Curtis A and De Lacy T (1997). Examining the assumptions underlying Landcare. In: *Critical Landcare*. (Eds S Lockie and F Vanclay) pp. 185–199. Centre for Rural Social Research Key Papers No. 5. Centre for Rural Social Research, Charles Sturt University, Wagga Wagga.

DAFF (2012a). What is Landcare. Department of Agriculture, Fisheries and Forestry, Canberra. Available: http://www.daff.gov.au/natural-resources/landcare/publications/making_a_difference_a_celebration_of_landcare/section_1_-_where_it_all_began [Accessed 21 May 2012].

DAFF (2012b). Australian Framework for Landcare. Department of Agriculture, Fisheries and Forestry, Canberra. Available: http://www.daff.gov.au/__data/assets/pdf_file/0004/2166745/framework-for-landcare.pdf [Accessed 21 May 2012].

DAFF (2012c). Caring for our Country Q&A. Department of Agriculture, Fisheries and Forestry, Canberra. Available: http://www.daff.gov.au/about/budget/2012-13/cfoc-q-and-a [Accessed 23 June 2012].

de Hayr B (2012). *Health of the Landcare Movement Survey Results*. National Landcare Facilitator Program. Australian Government, Canberra. Available: http://www.landcareonline.com.au/wp-content/uploads/2012/09/Health-of-Landcare-Movement-Survey-Report-Summary-August-2012.pdf [Accessed 13 September 2012].

Gillard J (2010). Carbon Farming Initiative. 2010 Labor election policy announcement. Australian Labor Party, Canberra.

Lane M, Haygreen A, Morrison TH, and Woodlands J (2009). Will regionalisation achieve integrated natural resource management? Insights form recent South Australian experience. In: *Contested Country: Local and Regional Natural Resources Management in Australia*. (Eds M Lane, C Robinson and B Taylor) pp. 59–74. CSIRO Publishing, Melbourne.

Lindenmayer D (2012). There's more to successful revegetation than 'getting trees in the ground'. *The Conversation*. Available: http://theconversation.edu.au/theres-more-to-successful-revegetation-than-getting-trees-in-the-ground-6844 [Accessed 20 April 2012].

LPLM (2012). Restoring Koala habitat around Gunnedah: building on a 1990s success. Liverpool Plains Land Management Inc. NSW Environmental Trust (Restoration) Report. Available: http://www.lplm.org.au/koalareport.html [Accessed 21 May 2012].

Lunney D, Crowther MS, Shannon I, and Bryant JV (2009). Combining a map-based public survey with an estimation of site occupancy to determine the recent and changing distribution of the koala in New South Wales. *Wildlife Research* **36**, 262–273.

Lunney D, Lemon J, Crowther MS, Stalenberg E, Ross K, and Wheeler R (2012a). An ecological approach to Koala conservation in a mined landscape. In: *Life-of-Mine Conference Proceedings, Brisbane, 10–12 July 2012*. pp. 345–354. Australasian Institute of Mining and Metallurgy, Carlton.

Lunney D, Crowther MS, Wallis I, Foley WJ, Lemon J, Wheeler R, Madani G, Orscheg C, Griffith JE, Krockenberger M, Retamales M, and Stalenberg E (2012b). Koalas and climate change: a case study on the Liverpool Plains, north-west NSW. In: *Wildlife and Climate Change: Towards Robust Conservation Strategies for Australian Fauna*. (Eds D Lunney and P Hutchings) pp. 150–168. Royal Zoological Society of NSW, Mosman.

Nabben T (1999). Funding to community Landcare groups in Western Australia: a powerful "New Generation" of soil conservation incentives. In: *Incentives in Soil Conservation: From Theory to Practice*. (Eds D Sanders, P Huszar, S Sombatpanit and T Enters) pp. 197–214. World Association of Soil and Water. Oxford and IBH Publishing, New Delhi and Calcutta.

NFF (2012). National Farmers Federation. NFF Farm Facts: 2012. Available: http://www.nff.org.au/farm-facts.html [Accessed 21 May 2012].

OECD (2008). *OECD Environmental Performance Reviews: Australia 2007*. OECD Publishing, Paris.

Paton S, Curtis A, McDonald G, and Woods M (2004). Regional Natural Resource Management: Is it sustainable? *Australasian Journal of Environmental Management* **11**, 259–267.

Robins L and Kanowski P (2011). 'Crying for our Country': eight ways in which 'Caring for our Country' has undermined Australia's regional model for natural resource management. *Australasian Journal of Environmental Management* **18**, 88–108.

20

Biolinks in south-eastern Australia – changing land use for a changing climate: Victorian policy responses

Ian Mansergh

'... Arrayed with a huge variety of creatures, plants and land forms, in spite of its use and misuse at the hands of settlers and occupiers, it is a unique natural estate of great beauty, spirit and strength.'

Michael Leunig
'Sleep on country', *The Age*
26 January 2012, p. 1

Biolink zones as a framework

Changing global climates are causing complex environmental change at a variety of scales. Projections indicate that these changes will accelerate during this century and include the risk of mass extinctions (Thomas *et al.* 2004; IPCC 2007). Distribution and abundance of biota will change and impacts of habitat and population depletion and fragmentation will be exacerbated in south-eastern Australia. For south-eastern Australia, a system of biolink zones, representing large-scale land-use change, was an early science and policy response to the threats of climate change and to provide better ecological linkages between existing large protected areas (DCE 1992; Brereton *et al.* 1995). A comprehensive statewide biolink zoning achieved formal policy adoption in 2009, making Victoria the first jurisdiction, internationally, to have such an adaptation response to biodiversity and climate change

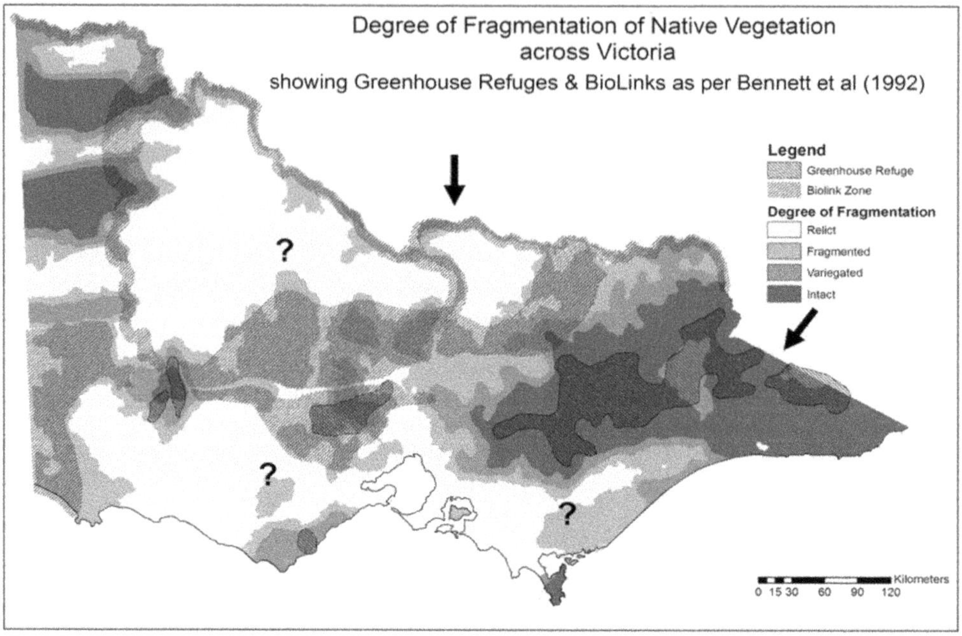

Figure 20.1 Biolink zones and climatic refugia in Victoria overlaying relative degrees of habitat fragmentation. Arrows = climatic migration; Question marks = potential areas for biolinks. Mapped biolink zones are from Bennett *et al.* (1992) and these are essentially the same as those presented in Victorian Government (2009).

(Victorian Government 2009; Figure 20.1), although a change of government since that time has made the position of biolinks in formal policy less certain.

Biolinks are identified broad areas of the landscape at the subcontinental scale in which the functional ecological connectivity for biodiversity is enhanced and/or restored in order to provide space for species to 'self adjust' to changing environmental conditions (Mansergh *et al.* 2008). Derived from projected changes to biota under climate change, the system of linking zones is a purposeful adaptation to ameliorate the 'island' character of the National Reserve System (NRS) and areas of native vegetation by restoration of the potential for evolutionary adaptation along gradients between climatic refugia. Biolinks are areas (10–100s km in width/length) that may encompass corridors, 'stepping stones' and entwined networks of habitat where the spatial density of high quality native vegetation is between 30 to ≥50% embedded within a variety of land uses (Mansergh *et al.* 2008). The texture of the landscape is 're-roughened' across past agricultural landscapes, 'maintained or enhanced' in existing forests, and 'rebuilt' out from areas currently retaining (or with the potential to retain) a diversity of biota and local refugia (e.g. topographical features, ridges, hilltops, riparian and remnant vegetation exhibiting resilient capacity). Biolinks represent a novel land use and create long-term permanent ecological infrastructure that includes bio-sequestration of carbon for changing climates and environmental conditions. At various scales, they provide and reflect a new, emerging social meaning for landscape, including intergenerational equity (Mansergh *et al.* 2006; Mansergh 2010).

Porous bordered zones, biolinks are *not* corridors in the traditional sense, rather a land-use example of what can now broadly lumped under 'connectivity conservation' (Worboys and Pulsford 2011). Traditional wildlife corridors are usually species specific, directly joining two areas, reactive to past problems and tend to reinforce a harsh dichotomy between conservation and 'productive' land (see Hilty *et al.* 2006). The *National Wildlife Corridors Plan* has a more expansive definition of corridors (DSEWPC 2012). Whatever the ultimate national nomenclature, the 20-year journey of biolinks illustrates some basic concepts for inclusion of ecological connectivity into improved national policies. 'Biolinks' is used here but the 'zone' component should be kept front of mind – people live and work in 'biolinks' – which manifest as a multi-purposeful framework for land use change.

Derivation, rationale and feasibility of biolink zones

The biota of south-eastern Australia was recognised early as vulnerable to global warming particularly within the cool-moist Bassian bioclimatic zone of the driest inhabited continent (Figure 20.2). Attempted resolution of this 21st century challenge began when the theoretical species responses to the Greenhouse Effect of Peters and Darling (1985) were contextualised for south-eastern Australia (Mansergh and Bennett 1989). In order to improve biodiversity conservation, they recommended bio-climatic modelling of species responses and ecological linkages between large areas of existing vegetation as successful ecological restoration had recently been shown to work in the Victorian alpine region (Mansergh and Scotts 1989). The importance of genetics and the potential and desirability of large areas of regenerating native vegetation to drawdown atmospheric CO_2 were also highlighted.

The resultant bioclimatic modelling of vertebrate fauna, coined 'biolink zones' to encapsulate a potential new land use response to projected climate changes (Bennett *et al.* 1992;

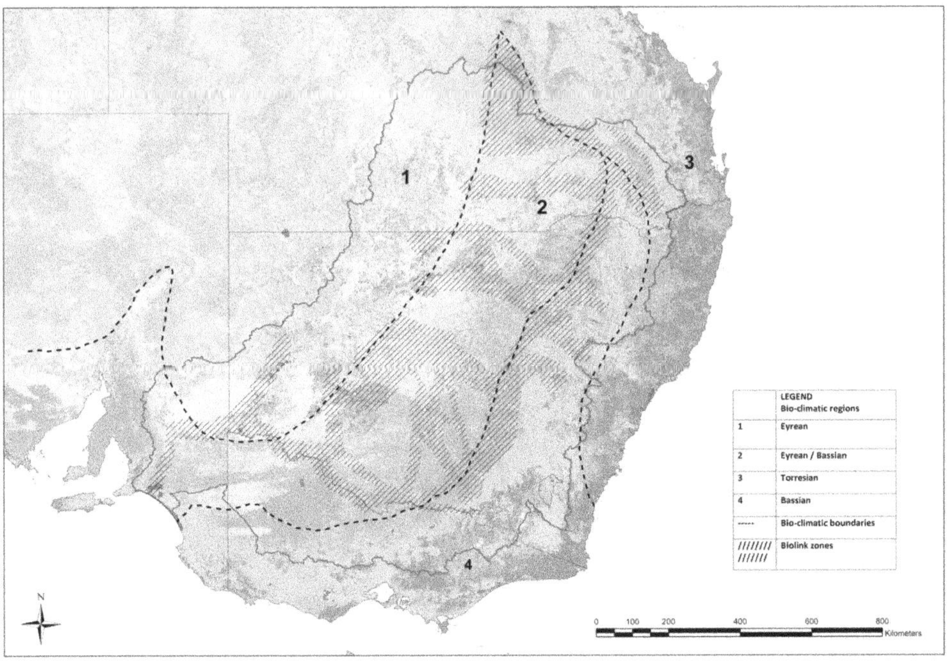

Figure 20.2 Biolinks in Bassian–Eyrean biocimatic transition zone over typography and dynamic land cover of eastern Australia and Murray–Darling Basin (Data: Bennett *et al.* 1992, Geoscience Australia (250 × 250 pixels), CRC for Spatial Information). (See Figure 20.1 for Victoria.)

Brereton *et al.* 1995, Figure 20.1). This was an attempt to link the two themes of the 1992 Earth Summit (i.e. biodiversity conservation and climate change) into an adaptation response for land use change at the subcontinental level. Mapped for Victoria, the analysis also included southern New South Wales (NSW) (Figure 20.2). At inception, biolinks were based on a multi ecosystem services and socio-ecological approach to ameliorate adverse effects of both past and foreseeable future land use.

Biolinks were conceived as providing the *necessary space for the operation of evolutionary processes for biota to self adapt to change and for ecological communities to rebuild.* The purpose, outcome and broad dimension were defined but activities within biolinks were purposely left non-prescriptive. Scope for increased incorporation of Koori (Indigenous) knowledge and management was presumed. Establishing a system of zones rather than narrow corridors recognised the potential for medium- to long-term environment transitions that could allow the spatial availability of habitat types (e.g. open forest to woodlands) to regenerate and evolve across the biolink allowing a multitude of potential gene flow pathways over time. In the additional space provided within the zone, natural processes could self-build out from the existing reservoir of ecological communities and latent resilience. *Being there* is a key aspect of resilient landscapes, incorporating both adaptive and transformative capacity (Mansergh and Cheal 2007) and novel native communities can evolve from available (or transported) genetic material. As envisaged, biolinks neither forced the re-establishment of specific past vegetation communities, nor neglected the importance of reservoir vegetation outside biolinks (which are all local refugia). The urgency to optimise regeneration of trees across the depleted forests and woodlands of south-eastern Australia was emphasised because of their long lead time to maturity and the critical buffering microclimates provided to the above and below ground biota (Vesk and Mac Nally 2006; Mansergh 2010).

Explicit within the biolinks concept was the need for societal recognition (policy) of a new land use, particularly over land initially alienated for agriculture. This new use was in the form of a *broad zone,* rather than a specifically defined area.

Subsequent research in biophysical sciences (e.g. ecology and climate) and policy (e.g. carbon, adaptation) have confirmed the basic premises, definition and magnitude of the problems biolinks sought to address and resolve (IPCC 2007; Steffen *et al.* 2009). More recent research shows that carbon in native vegetation in south-east Australia has been grossly underestimated (Paul *et al.* 2012). Land use change to biolinks merges mitigation, drawdown of CO_2, and adaptation to future environmental change.

Ecological resilience persists, but without active conservation management will diminish rapidly in potential biolinks. In northern Victoria and probably NSW, pastoral areas with as little as 2.5% native vegetation cover retain the ability for natural regeneration to expand by up to 40% if grazing pressure is removed. However, this capacity is likely to decline over the coming decades (Dorrough and Moxham 2005; Davidson *et al.* 2011). In areas of land use change, 'benign neglect' has seen increased resilience in these regenerating communities since the mid-1960s.

In short, research across a range of disciplines has supported: (i) the rationale for biolinks; (ii) the possibilities for regeneration and increasing landscape resilience; and (iii) the urgency to act. Research from other continents has reached similar conclusions for the need for biolinks, e.g. 'bold connectivity zones' for Europe (Opdam and Wascher 2004). A global review of wildlife corridors concluded that continental scale re-connectivity of nature was the only realistic alternative for the global conservation issues of habitat fragmentation under climate change (Hilty *et al.* 2006).

Developing policy

Policy development was influenced within four ongoing, interlinked and self-supporting themes: influencing the broad policy context; development of formal government policies; resonance with the broader community; and manifestations on-ground. Evidence was refined and expanded for biolinks as an appropriate landscape response to climate change, integrating both adaptation and mitigation. A narrow 'for biodiversity only' perspective was avoided – as new supportive evidence came from each of the underpinning ideas to evolve the biolink narrative.

Cultivating the policy context

Policy change needs a broad multi-layered constituency that comprehends the evidence base and underlying values. Over a decade, the author (as an officer in the Department of Sustainability and Environment) and others provided over 100 seminars and papers to various academic, institutional and sector forums (international, all levels of government and regional planning, policy, non-government organisation (NGOs), local government, industrial) (Table 20.1). It became apparent that policy adoption of biolinks was more a socio-ecological question of land use change rather than a technical refinement of bio-physical evidence. A broadened audience base was cultivated in forums and the literature (biodiversity, carbon, soils, water, amenity, production) (Mansergh *et al.* 2006). The trajectories of future land uses transitioning out of past agriculture (e.g. Barr 2008) provided fertile socio-economic 'ground' for biolinks to be heard (also adopted by Steffen *et al.* 2009). As adaptation to climate change became a societal and academic focus, biolinks fitted robust policy for good adaptation incorporating such principles as: *no regrets*; *reversible*; and *built-in safety margins* with the *capacity for strategic synergies* (Hallegatte 2009). The biolink framework incorporates new landscape services and their changing relative value and risks evolving from the societal responses to future climate change (Mansergh *et al.* 2008). As the massive CO_2 emissions from land clearing were quantified, biolinks showed that mitigation and adaptation could and should merge in landscapes, while strategic policy should fuse both (see also Garnaut 2008; Mansergh 2010).

Table 20.1. Chronology of the development of biolink zones of south-eastern Australia in the literature, policy and some projects

Develop and disseminate	1989	Bioclimatic modelling and climate change corridors advanced as a concept.	Mansergh and Bennett (1989)
	1992	Bioclimatic modelling of 42 vertebrate species, latitudinal and altitudinal distribution change and extinctions projected, climatic refugia and biolink zones defined and identified. Biolink zones remained essential the same for succeeding science and policy.	Bennett *et al.* (1992)
	1992	Victorian policy option in *Draft Flora and Fauna Guarantee Strategy* – each biolink mapped and explained.	DCE (1992)
	1995	Refined (shortened) version of Bennett *et al.* (1995) published in *Biological Conservation.*	Brereton *et al.* (1995)

(Continued)

Table 20.1. Chronology of the development of biolink zones of south-eastern Australia in the literature, policy and some projects (*Continued*)

Refine, promote as policy	1997	Victoria's Biodiversity Strategy sees biolinks as future asset under climate change. Policy recognised by both major parties.	NRE (1997)
	1998–2011	Over 100 presentations to national science workshops, state, regional and NGO forums of stakeholders – rationale for biolinks and holistic landscape response to climate change was outlined.	Mansergh *et al.* – available from author
	2002–2006	Biolinks incorporates land use changes in 'traditional' agricultural landscapes and placed in historic context.	e.g. Mansergh *et al.* (2006)
	2004	Ecological connectivity (but not biolinks) recognised in first national Action Plan for biodiversity under climate change.	NRMMC (2004)
	2006	Key development area under climate change adaptation in Victorian Sustainability Action Statement policy.	Victorian Government (2006)
	2007	Genetic component and resilience expanded; existing land use change in context of reserve system, particularly woodlands, and carbon sequestration.	Mansergh and Cheal (2007)
	2007–present	Biolinks enters common usage. Beginning to be used in local government planning schemes, predominantly, but not only in Victoria. Fuller incorporation of carbon sequestration.	Doorman (2007)
	2008	Biolinks taken internationally, and new tools demonstrated (visualisation) for implementation.	Mansergh *et al.* (2008)
	2009	Formally recognised as one of three pillars in Victorian Government White Paper policy on Land and Biodiversity (policy withdrawn from website in 2012).	Victorian Government (2009)
	2009–2011	Biolink projects emerge from large NGOs over substantial areas of Victoria. Regional biolinks emerge to link into statewide system e.g. Central Victorian Biolinks.	Central Victorian Biolinks (2012)
	2010	Biolinks in regional context of historic land use change (trends in eastern Australia), demographic change and indicative social value of ecosystem services over time, land use and greenhouse gas sequestration. Threatening process (weeds) under climate and future land-use change outlined.	Mansergh (2010)
Practical use and community uptake	2012	Publication of *National Wildlife Corridors Plan*. Many projects which gained funding under Commonwealth Biodiversity Fund mentioned biolinks (e.g. Mornington Peninsula and Western Port Biosphere Reserve).	DSEWPC (2012)

Developments within formal policy

Based on Bennett *et al.* (1992), biolinks achieved its first major policy recognition in the statutorily required *Draft Flora and Fauna Guarantee Strategy* in 1992 (DCE 1992), just before a change from Labor Government. This transitioned biolinks from science to the Victorian policy agenda and, significantly, into a broader community awareness. Biolinks achieved bi-partisan support in 1997 when a Coalition Government published *Victoria's Biodiversity Strategy* (NRE 1997). Here it was foreshadowed that: 'Through restoration in "biolinks", Victoria had obtained an international reputation for ameliorating the potentially adverse effects of the enhanced greenhouse effect on biodiversity' (NRE 1997, p. 3). At the time other strategic priorities were deemed more pressing (e.g. net gain in vegetation, environmental metrics, BushTender etc.) as climate change had yet to achieve societal prominence. Other Australian jurisdictions were slow to follow Victoria's lead although the Australian Government's *National Biodiversity and Climate Change Action Plan* highlighted the basic conceptual need (NRMMC 2004, Actions under Objective 5).

Biolinks then resurfaced on the formal policy agenda in the *Victorian Sustainability Action Statement*, significantly under climate change adaptation, not biodiversity conservation (Victorian Government 2006). Internationally, the 4th IPCC Report confirmed that anthropogenically caused climate change was happening and Australia's biodiversity was the most vulnerable sector with low adaptive capacity and small coping range (IPCC 2007). The Victorian Government policy response was a Green and White Paper process, informed by both scientific and community reference committees. In the final White Paper, *Securing our Natural Future*, mapped biolinks were designated as one of three key policy pillars for land management to increase landscape resilience and biodiversity conservation in adapting to future climates (Victorian Government 2009). In November 2010, a new Coalition Government was elected without a formal environmental policy – and no official view on biolinks. However, proposed institutional arrangements (irrelevant to biolinks) in *Securing our Natural Future* were at variance with Coalition policy and the entire publication was withdrawn from the Department of Sustainability and Environment's website in January 2012.

Broader community and on-ground works

Biolinks is now in common use suggesting a positive resonance within diverse communities across Australia – policy makers, planners, ecologists and community groups. Across Victoria, the biolink brand is used by a plethora of projects generated from local communities, NGOs, local planning schemes and regional catchment strategies. For example, the Victorian Naturally Alliance's Central Victorian Biolinks Project provides an over-arching vision that gives local biodiverse plantings a wider context and meaning, while the Mornington Peninsula and Western Port Biosphere Reserve uses biolinks as a key platform of its sustainability agenda and extends the connectivity of the statewide framework.

Major hurdles around implementing biolinks

Dominant land use industries are often reluctant to share space, particularly in Australia, where agriculture and pastoralism were integral mechanisms for populating what was perceived as *terra nullius* following European colonisation. Over the last 50 years, the national economic predominance of agriculture has steadily declined and most of the wealth now comes from a fraction of the land allocated to it (NLWRA 2001; Mansergh 2010). Over-allocation of land and water to agriculture has led to the 'paradox' of agricultural land use where the related legacies – dispossession, de-naturalisation and de-carbonisation are deeply ingrained and difficult to resolve (Hamblin 2009). In a land use policy context, biolinks was perceived to potentially threaten traditional agricultural uses/property rights which remained unburdened by associated social and environmental responsibilities. Changes in resource allocation or social preferences for public good often elicit strong community resistance, currently evident in the debates on water allocations in the Murray–Darling Basin. That said, landholders' perceptions are changing through programs such as Landcare (Chapter 19). Past Victorian Government policies on biolinks from both sides of politics are hopefully precursors of change that will eventually overcome the institutional inertia of land use policy and planning. However, at the time of writing the current Victorian Government's policy position is unclear.

The concept of an outcome focused, non prescriptive 'zone' allowing multiple uses has been difficult for some planners to conceive as most Victorian private land is zoned for agriculture and therefore there is no impetus for change from that sector. In the 1990s, the Victorian agricultural lobby sought both the 'right to farm *and* the right to sub-divide'. This perplexing and seemingly contradictory demand becomes understandable as ~50% of private land Victoria has land use trajectories away from agriculture towards amenity or transitional landscapes (Barr 2008). In this vast area, a permanent policy re-visioning beyond that of agriculture of the past or relentless peri-urban subdivisions, remains elusive. On the other side, some conservation policy makers considered that these zones may lessen the perceived importance and/or protection afforded to reservoir vegetation 'outside' of biolink zones. This contention was neither the intention nor supported in any of the relevant literature on biolinks; however, this tended to dissipate a united front for conservation.

Future directions

Progress towards developing biolinks will be based on actions of local–regional communities, policy makers and scientists (probably in that order). The ongoing debate should encourage embedding a new societal landscape vision across jurisdictions; monitoring on-ground activities and government initiatives; and targeted use of science (especially genetics and soils).

Legitimising biolinks, or similar, as a nationally supported land use is required to redress the historic overallocation of our land and water resources. The *National Wildlife Corridors Plan* attempts this with the virtue of inclusion of Indigenous culture but would benefit from incorporation of the broader philosophical underpinnings outlined above. Victorian biolinks need to be connected to similar land uses in adjacent jurisdictions, especially the Murray–Darling Basin where the absence of linkages is of national concern and dilutes Victoria's efforts (Figure 20.2).

Where skeletal biolinks are already emerging, typically within communities in 'amenity' and 'transitional landscapes', they need to be nurtured and initially concentrated on areas where conservation values and resilience are high. Protecting the biodiversity gains of past 'benign neglect' on disused farmland need to be recognised and included in the broader re-carbonisation/biolinks agenda. At the Victorian level, programs such as BushTender offer opportunities for more targeted investment in biolink zones. Improved understanding of biodiversity benefits to landscapes containing high value agriculture (e.g. vineyards) will assist integration of biolinks. Perceived increased risks of fire could be ameliorated by improved planning, settlement and building design.

Over time, projects under the Australian Government's Biodiversity Fund and the Carbon Farming Initiative offer opportunities to demonstrate a broad range of interrelated benefits. At the landscape level, it is highly probable that more cost-effective carbon sequestration will come from changing land use i.e. restoring biodiversity across the landscape in biolinks, than adjusting agricultural management *per se* (Garnaut 2008). Indeed large-scale agricultural expansion and/or intensification will likely release more atmospheric carbon. Better quantification of ecosystem services and their benefits is vital to reinforce the improved policy directions now emerging.

Conservation of genetic diversity is a fundamental requirement of biodiversity conservation yet it has received little attention in landscape restoration – we must maintain the potential for evolutionary development, particularly as environmental conditions alter in response to changing climates (see Chapter 23). Fitness and survival of species may be enhanced by making depleted populations more genetically robust and resilient rather than relying solely on local diversity (Hoffmann and Sgro 2011; Weeks *et al.* 2011). These insights will now be harnessed within biolink establishment and management. In terms of climate science, our understanding continues to rapidly evolve. However, it may be a false hope to wait for more definitive and precise answers before acting upon the broad trajectories already established.

Conclusions

The ecological impacts of past clearing and fragmentation in south-eastern Australia means native biodiversity requires more 'space' than currently allocated to it to enable it to self-adjust and optimise its evolutionary potential. The historic overallocation of land to agriculture and the socio-economic trajectories of land use change indicate that there is space for the creation of biolinks and that critical resilience has been retained in many areas. Broad biolinks zones offer opportunities to optimise cultural and socio-ecological outcomes as Australia adapts to, and tries to mitigate, climate change.

Our sense of place evolves from our collective past and looks to a future where we face a range of uncertainties, not the least of which are the impacts of climate change. Our responses to these challenges should facilitate the adoption of biolinks or a similar framework. Some Australian 're-dreaming' is required and fortunately this is already emerging in rural communities and national and state policies. This will, hopefully, provide a new conservation management framework for urgent action across this 'wide brown land'.

Acknowledgements

The opinions expressed here do not necessarily reflect the views of La Trobe University. Many people contributed to the ideas in this chapter and I thank the following for encouragement, constructive criticism and comraderie over the years: Wayne Chamley, Fiona Mackenzie, Peter Woodgate, Ary Hoffmann, David Cummings, Serenity Hill, Alex Lau, Jaymie Norris and Kate Stothers.

Biography

Ian Mansergh is a Victorian ecologist, researcher and policy analyst with over 30 years' experience in ecology and conservation and climate change policy development in south-eastern Australia. Initially trained as an historian, Dr Mansergh is an Honorary Research Fellow at La Trobe University and worked for the Department of Sustainability and Environment for over 30 years. He has extensive experience in alpine, forest and woodland ecology, land use and conservation management and he has published over 70 scientific papers. An early researcher on climate change, biodiversity and adaptation, he has evolved this to encompass the socio-ecology of resilient landscapes under climate change. He now chairs the implementation committee for 'Growing Connections for Sustainability' Forum, part of the Mornington Peninsula and Western Port Biosphere Reserve.

References

Barr NF (2008). Social landscapes in Victoria. In: *Landscape Analysis and Visualisation* (Eds C Pettit, W Cartwright, I Bishop, K Lowell, D Pullar and D Duncan) pp. 305–325. Springer, Berlin.

Bennett S, Brereton R, and Mansergh I (1992). Enhanced greenhouse and the wildlife of south-eastern Australia. *Arthur Rylah Institute Technical Report* **127**. Department of Conservation and Environment, Melbourne.

Brereton R, Bennett S, and Mansergh I (1995). Enhanced greenhouse climate change and its potential effect on selected fauna of south-eastern Australia: A trend analysis. *Biological Conservation* **72**, 339–354.

Central Victorian Biolinks (2012). Central Victorian Biolinks: Rebuilding Healthy and Resilient Landscapes. Available: http://www.centralvicbiolinks.org.au [Accessed 1 December 2012].

Davidson I, Sheahan M, and Thackway M (2011). An innovative approach to local landscape restoration planning: lessons from practice. *Ecological Management & Restoration* **12**, 175–188.

DCE (1992). *Draft Flora and Fauna Guarantee Strategy.* Department of Conservation and Environment, Melbourne.

Doorman CF (2007). Promising the future? Global change projections of species distributions. *Basic and Applied Ecology* **8**, 387–397.

Dorrough J and Moxham C (2005). Eucalypt establishment in agricultural landscapes and the implications for landscape scale restoration. *Biological Conservation* **123**, 55–66.

DSEWPC (2012). *National Wildlife Corridors Plan: A Framework for Landscape-scale Conservation.* Department of Sustainability, Environment, Water, Population and Communities, Canberra.

Garnaut R (2008). *The Garnaut Climate Change Review.* Cambridge University Press, Melbourne.

Hallegatte S (2009). Strategies to adapt to an uncertain climate future. *Global Environmental Change* **19**, 240–247.

Hamblin A (2009). Policy directions for agricultural land use in Australia and other post-industrial countries. *Land Use Policy* **26**, 1195–1204.

Hilty JA, Lidicker WZ, and Merenlender AM (Eds) (2006). *Corridor Ecology: The Science and Practice of Linking Landscapes for Biodiversity Conservation.* Island Press, Washington, DC.

Hoffmann A and Sgro CM (2011). Climate change and evolutionary adaptation. *Nature* **470**, 479–485.

IPCC (2007). *IPCC Fourth Assessment Report, Working Group II Report: Impacts, Adaptation and Vulnerability.* Intergovernmental Panel on Climate Change/Cambridge University Press, Geneva and Cambridge.

Mansergh I (2010). North Central Victoria: Climate change and land use: Potential for third century in a timeless land. *Proceedings of the Royal Society of Victoria* **122**, 161–183.

Mansergh I and Bennett S (1989). "Greenhouse" and wildlife management. *Victorian Naturalist* **106**, 248–252.

Mansergh I and Cheal D (2007). Protected area planning and management for eastern Australian temperate forests and woodland ecosystems under climate change – a landscape approach. In: *Protected Areas: Buffering Against Climate Change, Proceedings of a WWF and IUCN World Commission on Protected Areas Symposium: Canberra.* (Eds M Taylor and P Figgis) pp. 58–72. WWF Australia, Sydney.

Mansergh IM and Scotts DJ (1989). Habitat continuity and social organization ot the Mountain Pygmy-possum (*Burramys parvus*) restored by tunnel. *Journal of Wildlife Management* **53**, 701–707.
Mansergh I, Anderson H, and Amos N (2006). Victoria's living natural capital – decline and replenishment: 1880–2050 (Part 2). *Victorian Naturalist* **123**, 288–323.
Mansergh I, Cheal D, and Fitzsimons JA (2008). Future landscapes in south-eastern Australia: the role of protected areas and biolinks in adaptation to climate change. *Biodiversity* **9**, 59–70.
NLWRA (2001). *Australian Agriculture Assessment – 2001.* vol. 2. National Land and Water Resources Audit, Canberra.
NRE (1997). *Victoria's Biodiversity Strategy* (3 vols). Department of Natural Resources and Environment, Melbourne.
NRMMC (2004). *National Biodiversity and Climate Change Action Plan.* Natural Resource Management Ministerial Council, Canberra.
Opdam P and Wascher P (2004). Climate change meets habitat fragmentation: linking landscape and biogeographical scale levels in research and conservation. *Biological Conservation* **117**, 285–297.
Paul K, Roxburgh S, Raison J, Lamour J, England J, Murphy S, Norris J, Ritson P, Brooksbank K, Hobbs T, Neuman C, Lewis T, Read Z, Clifford D, Kmoch L, Rooney M, Freudenberger D, Jonson J, Peck A, Giles R, Bartle J, McArthur G, Wildy D, Linsay A, Preece N, Cunningham S, Powe T, Carter J, Bennett R, Mendham D, Sudmayer R, Rose B, Butler D, Cohen L, Fairman T, Law R, Finn B, Brammer M, Minchin G, Van Oosterzeeand P, and Lothian A (2012). 'Improved estimation of biomass accumulation by environmental plantings and mallee planting using FullCAM'. Report for Department of Climate Change and Energy Efficiency. CSIRO, Canberra.
Peters RL and Darling JS (1985). Potential effects of greenhouse warming on natural communities. In: *Effects of Changes in Stratospheric Ozone and Global Climate v.3. Climate Change.* (Ed. J Titus) pp. 137–150. EPA, Washington, DC.
Steffen W, Burbidge A, Hughes L, Kitching K, Lindenmayer D, Musgrave W, Stafford Smith M, and Werner P (2009). *Australia's Biodiversity and Climate Change.* CSIRO Publishing, Melbourne.
Thomas C, Cameron A, Green R, Bakkenes M, Beaumont L, Collingham Y, Erasmus B, de Siqueira M, Gralnger A, Hannah L, Hughes L, Huntley B, van Jaasvel A, Midgley G, Miles L, Ortega-Huerta M, Peterson A, Phillips O, and Williams S (2004). Extinction risk. *Nature* **427**, 145–148.
Vesk PA and Mac Nally R (2006). The clock is ticking: revegetation and habitat for birds and arboreal mammals in rural landscapes of southern Australia. *Agriculture, Ecosystems & Environment* **112**, 356–366.
Victorian Government (2006). *Our Environment Our Future: Environmental Sustainability Action Statement.* Victorian Government, Melbourne.
Victorian Government (2009). *Securing our Natural Future: A White Paper for Land and Biodiversity at a Time of Climate Change.* Department of Sustainability and Environment, Melbourne.
Weeks A, Heinze D, Kelly T, and Mansergh I (2011). 'Wild translocation of mountain pygmy possums for the genetic rescue of the Mount Buller population'. Report to Victorian Department of Sustainability and Environment.
Worboys GL and Pulsford I (2011). 'Connectivity conservation in Australian landscapes'. Report prepared for the Australian Government Department of Sustainability, Environment, Water, Population and Communities on behalf of the State of the Environment 2011 Committee, Canberra.

21

'Networking the networks': coordinating Conservation Management Networks in Victoria

Jim Crosthwaite, James Fitzsimons, Julia Stanley and Jane Greacen

Description of the policy framework

In this chapter, we outline the history and challenges in coordinating Conservation Management Networks (CMNs) in Victoria. We also touch on how CMNs have networked, largely informally, at a national level.

A CMN is 'both a land-based network of remnant areas of a particular ecological community (usually threatened, and/or under-represented in the reserve system) or a group of related communities and a human network of the land owners (public and private), land managers and other interested individuals and agencies' (Context Pty Ltd 2008). The CMN helps to coordinate the protection and management of fragmented ecological communities across a range of tenures and with a variety of protection mechanisms (Binning and Young 1997; Thiele and Prober 1999, 2000).

CMNs are a recognised way of linking landscapes in Victoria (Mansergh *et al.* 2008). Since the first Victorian CMN was established in 1999, the number of networks in Victoria has grown steadily to 13 (Figure 21.1). With this growth has come increased coordination and communication between CMNs, agencies and facilitators across Victoria. Victoria is currently the only Australian state that has gone to a statewide coordination body for multi-tenure conservation networks.

The Statewide Advisory Group (SWAG) is the peak body for CMNs in Victoria. It is a forum for CMNs to exchange ideas, as well as for interaction between CMNs and program managers in government departments, statutory authorities and other groups associated with CMNs. It provides strategic leadership; shares information about CMN activities and establishes biodiversity priorities; guides the establishment of new CMNs; supports consistency of

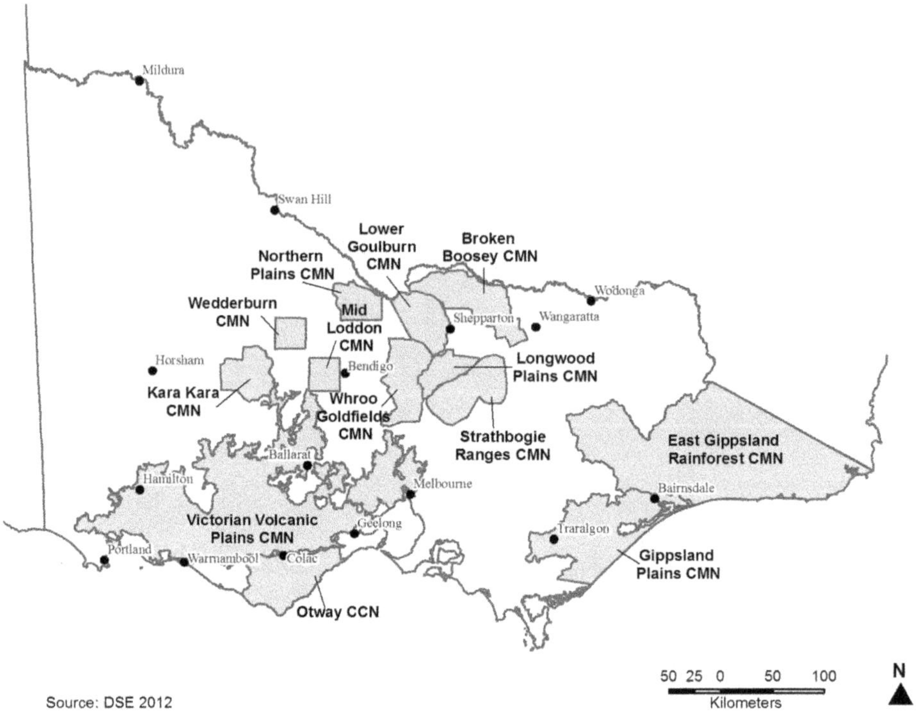

Figure 21.1 Locations of Conservation Management Networks (CMNs) in Victoria (as at early 2012). Note: CMNs map their areas differently. Some focus on all remnant vegetation within a defined geographic boundary, whereas others (such as the East Gippsland Rainforests CMN) focus on a particular vegetation type within a broad geographic area.

CMN planning, monitoring and evaluation; and oversees the implementation of the *Strategic Plan for Conservation Management Networks in Victoria* (Context Pty Ltd 2008). Current membership of the CMN Statewide Advisory Group is listed in Appendix 21.1.

When and why

The concept of CMNs was originally formed for landscapes with both fragmented vegetation remnants and fragmented land tenure (Thiele and Prober 1999). The first CMN was created in New South Wales (NSW) for the extensively cleared and fragmented Grassy Box Woodlands of the inland slopes of the Great Dividing Range (see Prober *et al.* 2001 and Chapter 12). The approach is particularly suited to Victoria where some 79% of the landscape is considered fragmented (VEAC 2010).

Victoria's first CMN was formed in late 1999 on the Gippsland Plains by the Trust for Nature (see Chapter 11 and Fitzsimons and Wescott 2005). Then, in 2001, the Environment Conservation Council (ECC) identified eight areas potentially suitable for the establishment of CMNs in central Victoria, and proposed a trial CMN on the Broken and Boosey Creeks (ECC 2001). The Victorian Government accepted this recommendation (Government of Victoria 2002) and, as part of a multi-million dollar Box-Ironbark program, provided funding for two trial CMNs – Broken–Boosey and Wedderburn–Wychitella (Garbutt 2002; and see Chapter 10).

The origins and drivers of these networks have varied from non-government organisations to government to landholder initiated. Considering the devolved and organic development of these networks, some form of coordination was considered important as early as 2002, to improve communication between networks, exchange information and learn from common experiences, and further develop this still new concept. The evolution of the coordinating framework is described in four stages as follows.

The **first stage** (2002–2004) followed the Victorian Government's commitment to establish two Box-Ironbark CMNs in central Victoria. The Victorian Department of Sustainability and Environment's (DSE) Protected Area Policy group and later Biodiversity Conservation Strategies group was tasked with initiating the CMNs. Informal discussions were held with the Trust for Nature and CMNs in NSW. The two trial CMNs were managed within the larger Box-Ironbark program, and both provided quarterly reports to the program's steering committee.

The **second stage** (2004–2005) saw the formation of a 'Statewide Stakeholder Group' for CMNs. It focused on aligning activities between the CMNs managed by Trust for Nature and those operating under DSE's aegis; fostered stronger integration between the programs of members of the group and biodiversity planning framework; and more generally promoted CMN development. The statewide CMN Stakeholder Group met three times from September 2004 to November 2005. Participants included representatives from organisations directly involved in on-ground CMNs, and broader peak interest groups (listed in Appendix 21.1). By this time, there were four CMNs in Victoria.

The development of a strategic plan for CMNs in Victoria, and active participation by CMNs in governance was the **third stage** (2006–2008) of coordination. Thinking around a Strategic Plan was focused initially on the Box-Ironbark CMNs. At this time, the Victorian Government's Box-Ironbark program was due to end, and a Steering Committee was no longer required. The Box-Ironbark CMNs were one of the few elements of this program that were planned to continue. In 2006, a report was commissioned to consider issues involved in the Box-Ironbark CMNs becoming more self-sustaining (FWR 2006). Soon after this, their two CMN facilitators began to set up two new CMNs (Mid-Loddon and Whroo Goldfields). A Project Control Board was established to govern the Box-Ironbark CMN program. Crucially this new body included the four Box-Ironbark CMNs which gave them input into the governance of the program.

In 2007, a statewide strategic planning process for CMNs commenced. This was initiated and supported by DSE, and included a strategic planning workshop in May 2007, followed by the development of a strategic plan by consultants (Context Pty Ltd 2008), with the active involvement of the CMNs. The broad aim was to help position CMNs in relevant statewide programs to achieve conservation outcomes.

The Strategic Plan was endorsed by the eight CMNs and eight agencies (including four Catchment Management Authorities (CMAs)). It included the establishment of a Statewide Advisory Group to oversee its implementation.

The formation of the Statewide Advisory Group (SWAG) is the **fourth stage** (2008–present) in coordinating CMNs in Victoria. The inaugural meeting of SWAG was held in 2008, and agreed on the roles for SWAG that would help the CMNs to achieve the eight directions set out in the Strategic Plan (see Appendix 21.2).

SWAG's roles were further clarified in the development of its terms of reference, so that 'the SWAG will support information sharing for CMNs, communicate with stakeholders, support and advocate for CMNs, and conduct state-wide planning for CMNs'.

The current SWAG membership represents all Victorian CMNs and key agencies (Appendix 21.1). Unlike the earlier CMN Statewide Stakeholder's Group, it includes CMNs but not the peak bodies with lobbying power (Appendix 21.1). Meetings are held four times a year, alternating between face-to-face in Melbourne, and teleconference. Unlike the previous Statewide Stakeholder Group, SWAG:

- is not run or chaired by DSE, however, DSE does provide executive support;
- is attended by CMN facilitators as well as other representatives who the CMN chooses;
- has invited additional people based on their expertise or connections, rather than who they represent; and
- is attended by fewer divisions of DSE.

The establishment of SWAG along with the Strategic Plan has provided a structural framework that has given confidence to those establishing new CMNs, and to their funders, by providing a consistent set of objectives and broadly defined ways of operating.

National networking

The concept of 'informal protected area networks' was recognised in the *National Objectives and Targets for Biodiversity Conservation 2001–2005*, although no definition was provided (Commonwealth of Australia 2001). An attempt to have a nationally accepted framework for the establishment and definition of CMNs (i.e. Thiele *et al.* 2002) has yet to be accepted by government institutions. Nevertheless, the Australian Government has funded on-ground activities of CMNs through competitive schemes, most recently the Caring for our Country program.

While there has not been formal coordination of CMNs at a national level, there have been several national workshops under the banner of the 'Conservation Management National Network'. In early 2007, there was a joint effort by all CMNs in Victoria and NSW to increase the profile of CMNs at the national level.

Achievements to date

CMNs in Victoria are now well established and recognised, the numbers are continuing to increase (Appendix 21.3). Each new CMN brings new stakeholders into coordinated conservation planning and activity. This is a major achievement for biodiversity conservation.

Statewide coordination of CMNs has played a role in supporting each new CMN, while also contributing to biodiversity policy at a statewide level. The advocacy for CMNs by SWAG is one reason why the contribution of CMNs towards achieving biodiversity policy objectives has been recognised in many reports making recommendations to the Victorian Government, such as:

- reports by the Victorian Environmental Assessment Council (VEAC) on river red gums (VEAC 2008) and remnant vegetation (VEAC 2011);
- the White Paper on Land and Biodiversity (Victorian Government 2009); and
- publicly released documents for the renewal of Victoria's Biodiversity Strategy (DSE 2010).

In 3 years, SWAG has largely achieved what it set out to achieve. The most important outcomes are outlined below.

1. Information sharing and communication with stakeholders

SWAG has provided CMNs with a regular forum to share their learnings and knowledge; to compare progress; to be kept informed of statewide and national developments; and to provide information about funding opportunities. CMNs typically operate at the landscape scale and SWAG provides a forum to learn from networks in other parts of the state. Beneficial relationships with associated statewide bodies have been formed through SWAG because its membership has been open to all relevant agencies as well as the CMNs.

2. Support and advocacy

SWAG does not provide funds to CMNs. However, SWAG has provided CMNs with a statewide and national voice. This has manifested through policy input at state and federal government levels, representation at various biodiversity forums as well as preparing documents to promote CMNs.

At the end of 2008, SWAG compiled information about the activities of each CMN into an important paper (SWAG 2008) that helped streamline the way CMNs report on their work, and assisted inexperienced CMNs by providing language and content for funding submissions. In 2010, SWAG facilitated the next step in supporting the CMNs achieve their strategic plan with the production of a document outlining progress and achievements to date (SWAG 2010).

The definition of a CMN was revised by SWAG in 2010. This agreement was reached after many meetings and discussions and was a significant achievement. It was only possible because SWAG had the capacity to bring the CMNs together despite them having evolved differently and operating to different definitions, aims and objectives.

3. Statewide planning for CMNs

The CMN Strategic Plan generated excellent discussion and enthusiasm for the evolving CMNs, and created an environment of optimism and willingness to coordinate relevant activities at the statewide level. Preparing a model framework for new CMNs has encouraged CMNs to review and, where appropriate, upgrade their governance.

Limitations

Different interpretations of what is a CMN

Although CMNs now have a Strategic Plan as a guide, there are considerable differences between the purpose of some CMNs and the activities they focus on. Arguably, for some CMNs, there is now less focus on the original CMN purpose of identifying and protecting the remaining remnants of ecological communities. Rather, some CMNs have focused their attention on a wider set of activities (e.g. pest plant and animal control), often associated with protecting iconic species, for example Malleefowl (*Leipoa ocellata*) and Bush Stone-curlew (*Burhinus grallarius*). This necessarily involves management of those associated remnants. It also attracts more support and involvement from local communities, and it is a clear differentiation from local Landcare groups. SWAG has not attempted to 'impose' a purpose – rather to accept all CMNs that agree to work broadly to its Terms of Reference and to the Strategic Plan. This shift has the potential to make it harder to communicate a unified perspective on the purpose and achievements of a CMN, and it may have reduced opportunities for CMNs becoming more recognised at state and national levels. Nevertheless, as CMNs that are members of SWAG are working consistently with the Strategic Plan, this diversity could also be considered a success and a validation of the CMN model.

Comparisons with Landcare have led to intense questioning about what CMNs are, and how they are differentiated from Landcare. It has been one of the challenges for developing a strong and recognisable brand for the CMN model. Many Landcare groups now treat biodiversity conservation as a priority, and they also work on both public and private land (Chapter 19). There has been concern by some that there may be competition between CMNs and Landcare for funding. Generally this has not been an issue, and has been coordinated at a local level. At the recent 25th Anniversary Conference of Victorian Landcare, the four CMN representatives on a panel discussion about CMNs and Landcare each had a very strong Landcare background. Their CMNs were the four most recently established ones, testifying to the growing interest by local communities in establishing CMNs (see Appendix 21.3).

The growth and increasing diversity of CMNs in Victoria, and their formation independently of the Victoria's conservation planning bureaucracy, provide a prime example of the tension Fitzsimons (2004) highlighted between encouraging community involvement in nature conservation and ensuring limited resources are used in the highest priority areas, as determined by scientific research. Ultimately both aspirations will be accommodated when these models become well established.

Who is leading CMN policy in Victoria?

CMNs are entirely consistent with Victorian Government policy. For at least 25 years (when Landcare emerged), the Victorian Government fostered and supported community-based initiatives in land management, natural resource management and biodiversity conservation.

However, the role of CMNs has yet to be explicitly recognised in official government policy for biodiversity, apart from setting up two CMNs as part of the Box-Ironbark program (see above). Resourcing for CMNs is very low when compared to other programs. Most do not have the ongoing core funding that has been recommended as a minimum (EWR 2006).

Policy development in relation to CMNs, rather than biodiversity conservation, is being led through the organic relationship of SWAG membership between CMNs, DSE, Trust for Nature, Parks Victoria and other SWAG members. This would be strengthened considerably with explicit government recognition and greater financial support for CMNs, and would increase the formal status of the CMN program within DSE. This would increase the likelihood of program continuity if there are organisational changes in DSE or staffing changes, or when overall funding is constrained.

SWAG member organisations (including CMNs) contribute financially to the SWAG role, with DSE making up the shortfall. SWAG is potentially vulnerable as there is little certainty about its funding from year to year. CMNs themselves rely mostly on winning grants, and cannot guarantee to support SWAG. The funding is now primarily used to pay for an executive officer, for two days a week based in a DSE office (as at June 2012). This issue of funding has caused some tension, with some SWAG members asking why there isn't core government funding for such an important role. For others, it raises 'city *v.* bush' tensions; 'why are CMNs giving $ back to the city when the need for spending their own funds is perceived to be on-ground'. In 2010–11, of the member organisations (including seven CMNs), 11 contributed between $250 and $2000. With CMNs reliant on external funding, this is not a secure position.

Ability to understand, measure, demonstrate outcomes

Ultimately, CMNs, and SWAG as the coordinating mechanism for Victorian CMNs, will be judged by their contribution to biodiversity conservation. They need to have the means of demonstrating success. They do of course meet the reporting requirements of funding bodies, nevertheless, development of measures to show the contribution of community-based biodiversity conservation relative to other approaches is in its infancy. CMNs achieve their goals indirectly – by mobilising people, communities and agencies around biodiversity conservation. Often their success is told in stories rather than in quantifiable numbers (see for example the recent Achievements Report (SWAG 2010)). Examples include the discovery of new populations of threatened orchids by local CMN members, large land purchases for conservation that can be attributed to the CMN presence, and the cooperation between different landholders and government agencies in weed and fox control.

Undertaking adequate monitoring and reporting can be very challenging for community conservation groups due to their limited resources and reliance on volunteers who lack time and may not have relevant qualifications or training. Partnerships with university and government researchers have helped in some activities, such as fox control.

Future directions

Monitoring and reporting

A key role for SWAG is to provide support, guidance and consistency in the way CMNs measure and report on their on-ground and community engagement activities. SWAG is in now in a very good position to support CMNs to do this by helping them establish an evidence base for their work, assisting them to establish local research programs that have measurable inputs and outputs, or running workshops on the latest developments in monitoring and reporting. The Box-Ironbark CMN program recently initiated a 'systematic evidence' project, and it is producing internal reports on current data collection and storage. The program is now developing a simple electronic system for the four Box-Ironbark CMNs to record a more complete and consistent set of information.

Ongoing establishment of SWAG as an independent entity

As SWAG plays a vital role in the success of CMNs, its ongoing development is vital. Currently the organisation is on the path towards becoming a legal entity via incorporation. Incorporation will enable SWAG to seek funding from a wide range of sources, both to support its own operation and for projects that support CMNs. Incorporation will also allow SWAG greater independence in advocacy for CMNs and for independent comment, while allowing DSE, Trust for Nature and Parks Victoria to still be actively involved. After incorporation, SWAG will progress to securing deductible gift recipient and charitable organisation status that will further increase its capacity to secure funding.

Communicating success of CMNs and SWAG

CMNs need to continue to work on effectively communicating the breadth of their activities and their considerable achievements, and SWAG could develop a communication strategy to support them as well as promote itself. This might include supporting individual CMNs to develop their own communication material, providing advice and setting standards around use of CMN logo and name, and developing statewide communication materials such as a yearly report card on CMN achievements. SWAG could also develop a central website for CMNs, and organise events to promote the benefits of the CMN model.

Input into policy process and identifying future landscapes for CMNs

SWAG should continue to proactively influence relevant policy processes where possible. There is now considerable knowledge about the factors influencing successful operation of CMNs. SWAG is currently identifying future landscapes and locations for CMNs based on biodiversity values and community capacity and capability. This will provide greater argument for funding for new CMNs and provide a guide to policy makers on areas they might like to target to make the greatest possible contribution to biodiversity conservation in Victoria.

Conclusion

CMNs are a recognised way of linking landscapes in Victoria. The role and structure of statewide coordination arrangements for CMNs has changed over time and will continue to evolve. Statewide coordination has enhanced the capacity of individual CMNs to operate and seek funding, and also provides a platform for new CMNs to be established. The number of CMNs in Victoria is continuing to increase, with three new CMNs established in 2011. The strength and direction of each CMN varies and, as a whole, CMNs in Victoria are gaining momentum, and receiving funding from an increasing range of sources. Much has been achieved with limited resources, and significant grassroots support, and this shows no signs of changing. Better funding will ensure that this work is built on and is sustainable. This offers lessons for other jurisdictions attempting to coordinate numerous multi-tenure conservation networks.

Acknowledgements

The views expressed in this chapter are those of the authors and not necessarily those of the Department of Sustainability and Environment, the Statewide Advisory Group or any other agency. We thank Kate Shanahan for creating the map.

Biographies

Jim Crosthwaite is an economist by training. From 1988 until 2012 he worked for the Department of Sustainability and Environment on the economic and social dimensions of various biodiversity programs. He managed the CMN program for some years. Jim completed his PhD in 2001 on Farm Businesses and Natural Resource Management. He fits in some consulting work, while also teaching Alexander Technique (www.jimcrosthwaite.blogspot.com), and volunteering for Journeys for Climate Justice, which he co-founded (www.journeysforclimatejustice.org.au).

James Fitzsimons is the Director of Conservation for The Nature Conservancy's Australia Program and an Adjunct Associate Professor at the School of Life and Environmental Sciences, Deakin University. His particular research interests are in the fields of protected area policy, practical conservation planning and mechanisms to integrate conservation outcomes on public and private lands. He has worked in the fields of conservation policy and planning for government environment departments and agencies, and for non-government environment organisations. His PhD researched multi-tenure conservation networks and he has been involved with CMN coordination in Victoria since 2001.

Julia Stanley worked for the Department of Sustainability and Environment on several programs including Conservation Management Networks, and was the executive officer for the Statewide Advisory Group for CMNs from 2010 to 2012. She is passionate about exploring and understanding the deep connection between human communities and environmental conservation and is a member of Journeys for Climate Justice.

Jane Greacen was the inaugural president of the East Gippsland Rainforest CMN, and also chaired SWAG for nearly 2 years. She is a general practitioner and occupational physician and also works in Aboriginal Health in East Gippsland. She has a Master of Public Health and is a senior lecturer with Monash University Regional Clinical School. She has had various careers in the public sector, the most recent being 10 years as CEO of the Rural Workforce Agency, Victoria.

References

Binning CE and Young MD (1997). *Motivating People: Using Management Agreements to Conserve Remnant Vegetation*. Paper 1/97, National Research and Development Program on Rehabilitation, Management and Conservation of Remnant Vegetation. Environment Australia, Canberra.

Commonwealth of Australia (2001). *National Objectives and Targets for Biodiversity Conservation 2001–2005*. Environment Australia, Canberra.

Context Pty Ltd (2008). *Strategic Plan for Conservation Management Networks in Victoria: Working together to protect biodiversity*. Context Pty Ltd, Melbourne. Available: www.dse.vic.gov.au/cmn [Accessed 21 July 2012].

DSE (2010). *Biodiversity is Everyone's Business: Victoria's Biodiversity Strategy 2010–2015. Consultation draft*. Department of Sustainability and Environment, Melbourne.

ECC (2001). *Box-Ironbark Forests and Woodlands Investigation Final Report*. Environment Conservation Council, Melbourne.

EWR (2006). 'Broken Boosey and Wedderburn Conservation Management Networks – Identifying pathways to self-maintenance'. A report for the Department of Sustainability & Environment. EWR, Bendigo.

Fitzsimons JA (2004). The contribution of multi-tenure reserve networks to biodiversity conservation. PhD thesis, School of Ecology and Environment, Deakin University, Melbourne.

Fitzsimons JA and Wescott G (2005). History and attributes of selected Australian multi-tenure reserve networks. *Australian Geographer* **36**, 75–93.

Garbutt S (2002). New conservation initiative for box–ironbark parks. Media release from the Hon Sherryl Garbutt, Minister for Environment and Conservation, 1 October 2002. Available: http://www.legislation.vic.gov.au/domino/Web_Notes/MediaRelArc02.nsf/fce5683f42af57f04a256b36001bd4df/ee5af520937c987bca256c45007f72e5!OpenDocument [Accessed 21 July 2012].

Government of Victoria (2002). *A Community Landscape: Victoria's Box-Ironbark Forests and Woodlands – Victorian Government Response to Environment Conservation Council Box-Ironbark Forests and Woodlands Investigation Final Report*. Victorian Government, Melbourne.

Mansergh I, Cheal D, and Fitzsimons JA (2008). Future landscapes in south-eastern Australia: the role of protected areas and biolinks in adaptation to climate change. *Biodiversity* **9**(3–4), 59–70.

Prober SM, Thiele KR, and Higginson E (2001). The Grassy Box Woodlands Conservation Management Network: Picking up the pieces in fragmented woodlands. *Ecological Management & Restoration* **2**, 179–188.

SWAG (2008). Making it happen: Conservation Management Network Actions from the CMNs Victoria Strategic Plan. Statewide Advisory Group for Victorian Conservation Management Networks, Melbourne. Unpublished.

SWAG (2010). Progress and achievements in implementing the Strategic Plan for the CMNs June 2008 to June 2010. Statewide Advisory Group for Victorian Conservation Management Networks, Melbourne. Unpublished.

Thiele KR and Prober SM (1999). Reserve concepts and conceptual reserves: options for the protection of fragmented ecosystems. In: *Temperate Eucalypt Woodlands in Australia:*

Biology, Conservation, Restoration and Management (Eds RJ Hobbs and CJ Yates) pp. 351–358. Surrey Beatty, Chipping Norton, NSW.

Thiele K and Prober S (2000). Conservation management networks – a model for coordinated protection and management of remnant vegetation. In: *Balancing Conservation and Production in Grassy Landscapes: Proceedings of the Bushcare Grassy Landscapes Conference, Clare, South Australia, 19–21 August 1999* (Eds T Barlow and R Thornburn) pp. 58–63. Environment Australia, Canberra.

Thiele KR, Prober SM, Higginson E, and Bell L (2002). Conservation management networks: a working national framework for their establishment and operation. Discussion paper prepared in conjunction with the National Workshop on Conservation Management Networks, Canberra, 5–6 March 2001.

VEAC (2008). *River Red Gum Forests Investigation Final Report.* Victorian Environmental Assessment Council, Melbourne.

VEAC (2010). *Remnant Native Vegetation Investigation Discussion Paper for Public Comment.* Victorian Environmental Assessment Council, Melbourne.

VEAC (2011). *Remnant Native Vegetation Investigation Final Recommendations.* Victorian Environmental Assessment Council, Melbourne.

Victorian Government (2009). *Securing Our Natural Future: A White Paper for Land and Biodiversity at a Time of Climate Change.* Department of Sustainability and Environment, Melbourne.

Appendices

Appendix 21.1. Membership of the Statewide Advisory Group (2008–present) and the earlier CMN Statewide Stakeholders' Group (2004–2005)

	CMN Statewide Stakeholders' Group	Statewide Advisory Group
Individual CMNs		x
CMN facilitators		x
Trust for Nature ^	x	x
Goulburn Broken Catchment Management Authority	x	x
North Central Catchment Management Authority	x	x
Other Catchment Management Authorities *		x
Victorian National Parks Association	x	
Greening Australia Victoria Inc.	x	
Parks Victoria	x	x
The Wilderness Society	x	
Victorian Farmers Federation	x	
North West Region, DSE	x	x
Resources & Regional Services Division, DSE	x	
Biodiversity & Natural Resources Division, DSE	x	x
Conservation & Recreation Division, DSE	x	

	CMN Statewide Stakeholders' Group	Statewide Advisory Group
Land & Catchments Division, DSE	x	x
Peter Wilcock, Caring for our Country program, Australian Government Department of Sustainability, Environment, Water, Population & Communities #		x
James Fitzsimons, The Nature Conservancy #		x

* Other Catchment Management Authorities are welcome to attend
^ The CEO of Trust for Nature is currently the chair of the Statewide Advisory Group
# SWAG sought the expertise of these individuals rather than their organisations

Appendix 21.2. Directions in the *Strategic Plan for Conservation Management Networks in Victoria* (Context Pty Ltd 2008)

1 Articulate a clear role for CMNs in achieving improved biodiversity outcomes at a federal, state and local scale.
2 Undertake necessary planning to support achievement of local biodiversity objectives.
3 CMNs to nurture and formalise relationships with key partners to achieve improved biodiversity outcomes.
4 Support CMNs to engage/empower a diverse cross-section of community in biodiversity conservation at a landscape scale.
5 Implement programs to increase the number, quality and connectivity of sites within the CMN network.
6 Demonstrate the effectiveness of CMNs in delivering biodiversity outcomes at a landscape level.
7 Implement appropriate governance and explore sustainable resourcing for existing and new CMNs.
8 Improve linkages between CMNs at a state and national level.

Appendix 21.3. History of forming CMNs in Victoria

CMNs formed before the Strategic Plan was prepared		
CMN	**When established**	**Process leading to establishment & leading founders**
Gippsland Plains	1999	Initiated by Trust for Nature, supported by public workshops (see Chapter 11).
Broken Boosey	2003	Created under the Victorian Government's Box-Ironbark Implementation Program, with DSE funding Goulburn–Broken CMA as manager. An appointed facilitator invited landholders and groups to join. The fox baiting and Bush Stone-curlew projects rallied people to become involved.
Wedderburn	2003	Created under the Victorian Government's Box-Ironbark Implementation Program, with DSE as manager. An appointed facilitator invited landholders, agencies and groups to join (see Chapter 10).
Northern Plains	2004	Created by North Central CMA and supported by Trust for Nature and DSE. Established as a means of coordinating NRM works across land tenure and halting the decline in grassland habitat.

(Continued)

Appendix 21.3. (*Continued*)

East Gippsland Rainforest	2006	Created at a public meeting, supported by East Gippsland CMA and Trust for Nature. There was a particular concern about the conservation of highly fragmented rainforest. Landcare members, and others with rainforest on their land, were integral right from the start.
CMNs formed during discussions leading to the Strategic Plan		
Mid-Loddon	2007	Created by DSE and the Mid-Loddon Landcare Network, it focuses on restoring the health of public Box-Ironbark forests and both private and public Box Woodlands, to increase quality of habitat for Swift Parrots, Bush Stone-curlews and Brush-tailed Phascogales. The CMN meets jointly with the Upper Spring Landcare Group and involves all Landcare Network members and a wide range of environmental groups, public land managers and agency representatives.
Whroo Goldfields	2007	Created by Goulburn–Broken CMA with support from DSE, the facilitator of Broken–Boosey CMN expanded their role and invited a large number of groups and individuals to the first meeting of the new committee. Fox baiting and nest box programs attracted people to the CMN.
CMNs formed after SWAG was established		
Longwood Plains	2008	The CMN provided a means of formalising, and gaining wider recognition for, a group that had been planning and implementing activities to increase protection and connectivity since 2002 under the local Biodiversity Action Planning process, with support from Trust for Nature.
CMNs formed after SWAG was established		
Victorian Volcanic Plains	2009	Trust for Nature appointed a facilitator, and invited many groups to join a steering committee. The facilitator was employed using funds obtained by Arthur Rylah Institute (DSE) as part of a larger Caring for our Country grant for the Victorian Volcanic Plains.
Kara Kara	2010	Many community groups led by Buloke and Southern Grampians Landcare Network came together in public meetings to plan the CMN, after success in winning a Caring for our Country grant.
Strathbogie Ranges	2011	The CMN developed on the basis of three things – a long-active Landcare community with an interest in biodiversity; a recent tender program leading to conservation of wetlands and bogs that attracted much interest; and a $20 000 grant obtained by the Goulburn-Broken CMA to set up the CMN. 35 people attended the first 'let's do something' meeting.
Otway Community Conservation Network	2011	A facilitator was appointed to set up the Network, after DSE and local partners won a Caring for our Country grant focused on environmental weeds. The facilitator invited a large number of groups to join.
Lower Goulburn	2011	A member of Broken–Boosey CMN, who was their representative on SWAG, invited many groups to the public meetings that led to the formation of the CMN.

22

Reconnecting Natural Northland: New Zealand's Large Landscape Initiative

Nick Edgar

Description of the framework

This chapter describes a New Zealand large landscape restoration initiative – Reconnecting Natural Northland. In 2008, a philanthropic trust, the Tindall Foundation, initiated a 'large landscape' conversation with a selection of environmental groups in New Zealand and Australia. Study tours to examine ecological restoration projects were undertaken to Australia by representatives of the Tindall Foundation, the NZ Landcare Trust and WWF-New Zealand (WWF-NZ). A wider group of New Zealand stakeholders was then consulted and consensus was reached that a large landscape initiative in New Zealand could deliver substantial benefits to biodiversity condition, resilience to climate change, sustainable land management, and water quality, as well as potentially delivering social and economic benefits.

The Tindall Foundation, being a sponsor and funder of a range of community-based environmental initiatives in New Zealand, was acutely conscious of the gap in support of conservation on private land outside the public conservation estate. The Foundation was also aware of one of the difficulties of being able to strategically target funding to large, high priority issues and landscapes given that the grants provided by so many funding agencies are directed at local communities for smaller, locally-based projects. The problem with funding a range of smaller ecological restoration projects across the country is the perception that they lead to numerous small-scale initiatives that do not seem to be well coordinated or integrated across the landscape and that they lack a clear cohesiveness. It became apparent that there were some larger-scale landscape projects and initiatives underway in both Australia and the United States (and elsewhere) that might be able to satisfy this need to take ecological restoration to a larger more integrated scale in New Zealand.

The challenge New Zealand faces is to ensure that while maximising land productivity we simultaneously recognise and provide for the natural systems and processes that make us clean, green, and special in terms of our unique biodiversity and landscapes. The successful introduction of landscape-scale restoration projects in New Zealand has the potential to be a fundamental step in addressing issues of biodiversity decline at a landscape level. Landscape-scale restoration has the potential to provide a solid base for building sustainable projects that incorporate resilience to climate change in land management practices and community-based activities.

This New Zealand Large Landscape Initiative (LLI) has the overall aim to increase the extent of habitat, improve the health, connectivity, quality of habitat and the sustainable land management of a candidate (pilot) large landscape. This chapter describes the journey taken in selecting this candidate landscape in Northland and what the Reconnecting Natural Northland initiative intends to do. A fundamental element of what sets this New Zealand LLI apart from other international approaches has been the strategic planning approach to landscape selection and the development of a series of landscape selection criteria that incorporate both biophysical and socio-economic characteristics. Also of interest is the work that has been undertaken to examine how the project's scale and governance arrangements will influence its implementation. There has also been a strong focus on incorporating new enterprises into the landscape to assist the initiative to become financially self-sustaining in the long term.

When was the framework established and why?

An initial driver for this project was a loose desire by the Tindall Foundation to help 'green' New Zealand's hosting of the 2011 Rugby World Cup through indigenous forest establishment. After some discussion it was recognised that although this event could be a useful motivator for specific environmental projects, there were more significant challenges to achieving durable, long-term ecological restoration outcomes that needed to be addressed before tackling one particular event-related effort. Consideration was then given to what else might be done.

The fundamental question being asked was not whether New Zealand is doing enough at the moment, but rather 'what could be done to promote a step-change improvement in New Zealand's environmental outcomes?' Initial discussion and scoping work indicated that there seemed to be no well-resourced, nationwide organisation in New Zealand acting in a manner that answered this vaguely formed question.

Many of the country's landcare and environmental restoration groups appeared to be fragmented and often focused on financial survival (due to funding challenges), without the resources or cohesiveness to pull together a sector-wide effort and achieve results on a nationally significant scale. Given the significant decline of New Zealand's native biodiversity due to historic land clearance and associated habitat loss, the impact of introduced weeds and animal pests, and the vulnerability of ecosystem resilience to the future impacts of climate change there is a clear need for enhancing ecological restoration efforts (Ministry for the Environment 2007). This opportunity to improve the integration and cohesiveness of individual, fragmentary efforts to protect and enhance biodiversity and ecosystem services was seen as a possible way of leading to a step-change improvement in environmental outcomes. This opportunity was loosely recognised as 'connecting the dots', in essence, an initiative to enhance the collective effort of an array of individual ecological restoration projects within a landscape.

An initial meeting was undertaken by the Tindall Foundation in December 2008 to examine Australian approaches to this issue. This was followed by a study visit to examine large-scale landscape restoration initiatives in Australia (e.g. Gondwana Link, Habitat 141°, and the Great Eastern Ranges – Chapters 3, 8, 13, respectively). This visit focused in more detail on how these large landscape projects were initiated, managed and funded. After this study visit was completed, several New Zealand environmental organisations and stakeholders were then consulted and a series of dialogue meetings and conversations were held to determine if a large landscape restoration project could be implemented in New Zealand. A key meeting of potential project partners was organised in February 2010.

A consensus was reached at this meeting that a large landscape initiative in New Zealand could deliver substantial improvements in biodiversity condition, resilience to climate change, sustainable land management, and water quality, as well as potentially delivering social and economic benefits. The primary aim, and what the project steering group were hoping to inject into New Zealand's toolbox of responses to environmental issues, were the various large landscape-scale methodologies that have been adopted in Australia and internationally (Handford *et al.* 2011).

Achievements and developments to date

As a first stage in advancing the LLI, the project steering group decided to undertake work to identify a suitable pilot landscape. This work was intended to identify a landscape where it might be possible to support the implementation of landscape-scale restoration concepts, expand thinking around how this approach could work in New Zealand and demonstrate its application. Deciding which pilot landscape to work in became a pivotal challenge for this initiative and also its key point of difference from other New Zealand approaches.

The project steering group identified two key phases of work that would be required to progress the LLI in relation to a selected landscape. The first phase of work was described as the preparatory phase and would involve a national assessment of potential New Zealand landscapes for the LLI and then the selection of one of these as the candidate landscape. The second phase of work was described as the implementation phase and would involve: securing implementation funding for the LLI; establishing a governance framework for the selected landscape; recruiting staff to undertake project management and on-ground operational work; stakeholder engagement in the landscape; and establishing a monitoring and evaluation framework to track project outcomes.

This section will largely focus on describing the achievements of the preparatory phase of the LLI as the decision to fund the first 5-year implementation phase has only recently been made. The preparatory phase of the LLI involved two non-government organisations – the NZ Landcare Trust and the WWF-NZ – applying for funding from the Tindall Foundation to conduct a piece of work to inform the further development and selection of a candidate landscape. A project steering group made the decision to tender this work through a request for proposal process from research providers and private environmental consultancy firms. A series of work streams was identified as part of the request for proposal process. Table 22.1 outlines these work streams and deliverables.

The critical component of the work was to develop a set of criteria for the selection of a landscape based on a series of both ecological/biodiversity and socio-economic characteristics. Rather than identifying potential landscapes based on the project steering group's existing knowledge (or other expert knowledge), it was decided that developing a series of criteria for identifying suitable landscapes would lead to a more objective consideration of

Table 22.1. Process for undertaking candidate landscape selection for New Zealand's Large Landscape Initiative

Workstream
Workstream 1 – Developing a set of criteria for selection of a landscape considering: • Ecological/biodiversity characteristics • Economic and social characteristics • Other factors such as replicability, sustainability, synergies with other initiatives etc.
Workstream 2 – Learning from international examples of landscape-scale conservation/restoration. This will involve conducting a literature review, reviewing metadata and visits/interviews with participants.
Workstream 3 – Exploring opportunities to monetise restoration work, exploring models to help make conservation/restoration economically sustainable.
Workstream 4 – Identification of the most appropriate 'tools' and methods to identify priority sites/actions in the selected landscape.
Workstream 5 – Identification of potential governance models and organisational entities for delivery of the landscape project in the selected landscape.
Workstream 6 – Using agreed criteria, to identify a shortlist of candidate landscapes in New Zealand which meet the criteria.
Workstream 7 – Select one landscape, from the shortlist as a proposed candidate landscape.

landscape options. It would allow both ecological/biodiversity and socio-economic landscape characteristics to be compared between different landscapes. By further adding a scoring (weighting) system to each characteristic it was then possible to compare scores between landscapes for particular characteristics. A total score was also developed for each landscape for overall comparative purposes.

The ecological/biodiversity selection criteria incorporated such fundamental characteristics as: sufficient core natural vegetation cover; diversity of species and ecosystems; presence of rare or threatened species; potential large-scale connections; degraded ecosystems (which had not passed a permanent loss threshold); impacts from pests and habitat loss evident but manageable; and the matrix of land type and land use provides opportunities for restoration. Socio-economic selection criteria incorporated such fundamental characteristics as: sufficient cohesion and logic in the combination of natural landscape and human communities; a (restoration) vision can be reasonably easily developed and articulated; potential to generate economic benefits to tourism and primary production sectors; likelihood that ecosystem services could generate income in the landscape; land ownership likely to support long-term restoration and land managers likely to have the capacity to be involved; and iwi (Maori) and wider community relationships and capacity to sustain restoration over the long term.

From an initial national assessment of 20 potential landscapes, a short list of eight candidate landscapes was scored using these landscape selection criteria (Figure 22.1). The landscape that scored highest was Northland – from Kaipara Harbour in the south through to North Cape, the northernmost extent of New Zealand's North Island. Northland was chosen because it ranked highly on so many of the landscape selection criteria. Northland has: a large coherent landscape; a good quantity, quality and pattern of indigenous habitats; a high diversity of species and ecosystems; a high proportion of unique, rare and

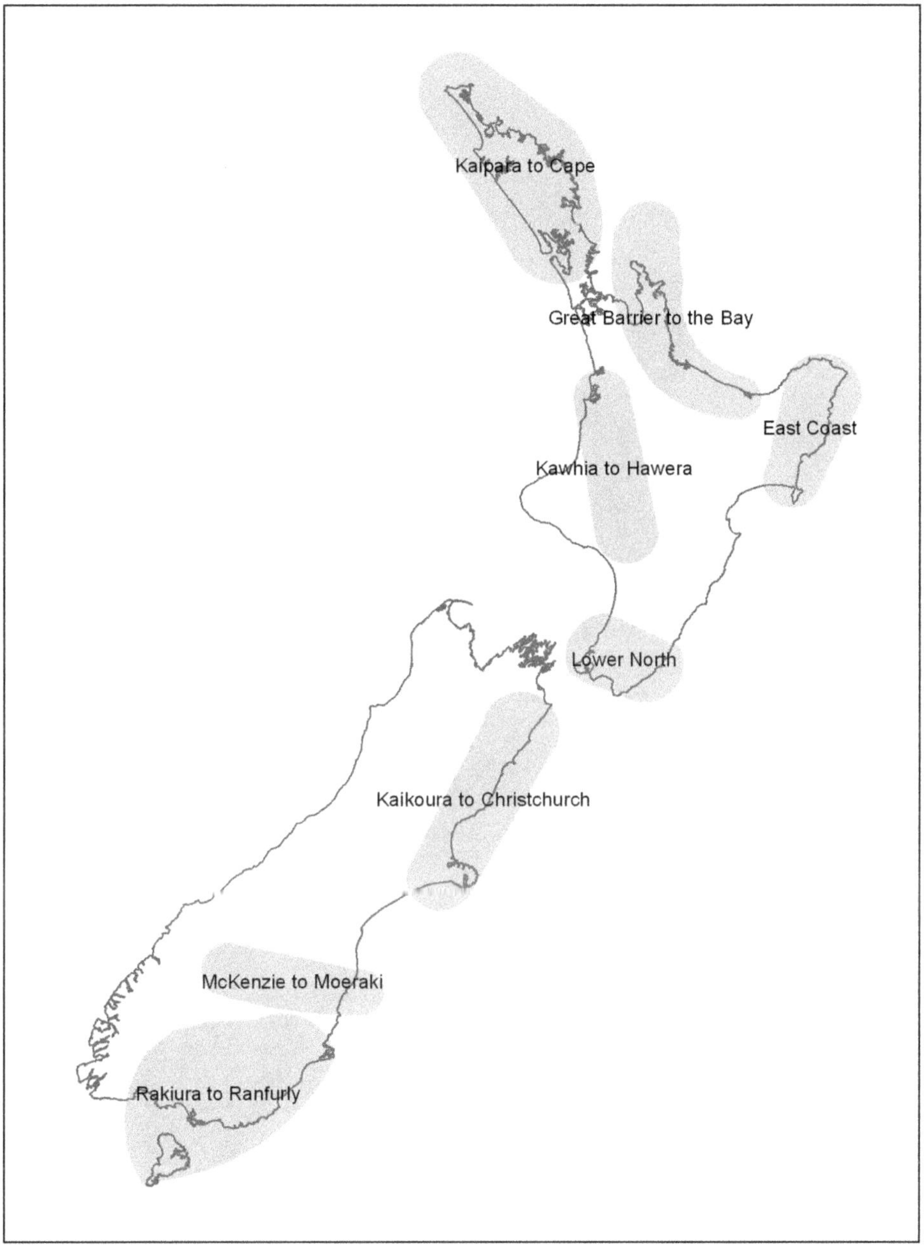

Figure 22.1 Short list of eight candidate landscapes for New Zealand's Large Landscape Initiative.

threatened species; a good mix of land use (Figure 22.2) and a community receptive to sustainable land management; and the potential to generate economic and social benefits from a large-scale ecological restoration initiative.

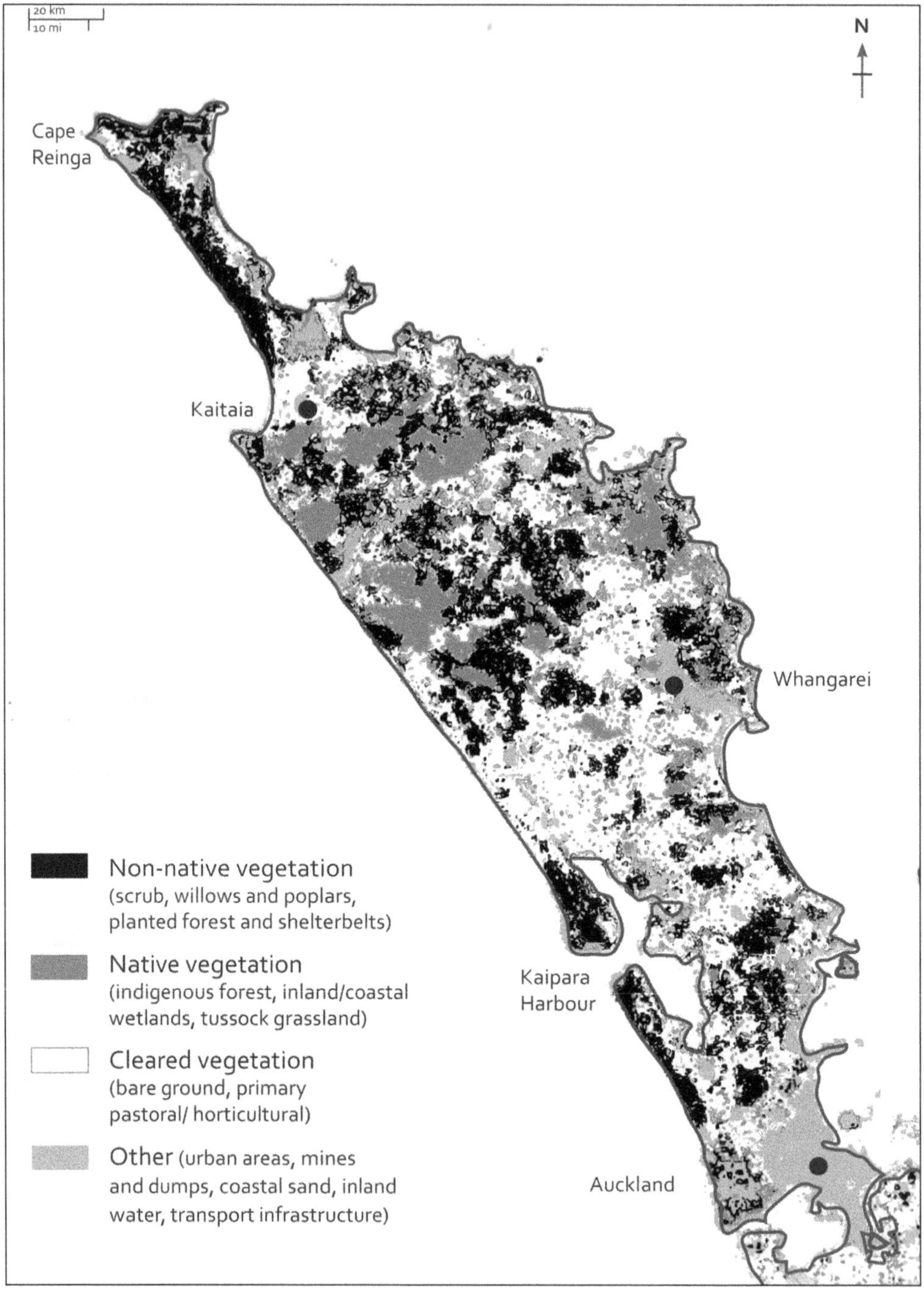

Figure 22.2 Land cover in the Northland Region.

Major limitations experienced in developing or implementing the framework

Several limitations and challenges were experienced in undertaking the preparatory phase of the Large Landscape Initiative (LLI) and in determining a candidate landscape for restoration as outlined below.

Candidate landscape selection

The LLI set out to select a potential candidate landscape where a large landscape restoration initiative could be supported. Landscape selection followed a strategic approach, where key biophysical and socio-economic characteristics formed the basis of the selection criteria for choosing a landscape. The aims of the criteria were 'to help select landscapes where landscape-scale ecological restoration was of high value through high potential to retain and increase biological diversity across a landscape, and had a high likelihood of success' (Handford *et al.* 2011). The process of determining which selection criteria were fundamental to achieving project success involved case study analysis and review of existing New Zealand and international landscape restoration initiatives.

One of the principle limitations of the LLI framework that has been developed has been the lack of a suitable body of knowledge and analysis of which factors have a key role in the success of landscape restoration, particularly in New Zealand. Despite a growing body of knowledge on governance arrangements for managing collective resources and evaluating critical success factors for collaborative action (Ostrom and Nagendra 2006; Hearnshaw *et al.* 2011), there is a gap in the analysis and evaluation of the outcomes of landscape restoration initiatives, both in New Zealand and globally. One of the potential benefits of the LLI framework is the work that has been undertaken to examine which biophysical and socio-economic characteristics can have a significant impact on landscape restoration outcomes.

Landscape scale

Working at a landscape scale involves moving away from simply working on the management of discrete sites. However, the question remains as to how to define the area within which landscape-scale ecological restoration will take place. How large does the area need to be to maintain ecological processes and be able to respond to natural or human disturbances and still maintain its resilience?

One of the key limitations of the framework was being able to adequately define the scale at which the LLI would be undertaken. In New Zealand, we generally lack methods for determining the size of areas, such as reserves, to maintain ecological processes that incorporate natural or human disturbance. Despite a growing body of research on fragmentation and connectivity, core areas and stepping stones, and restoration and buffer areas, the study of ecological processes in New Zealand ecosystems is not well advanced for many systems (Handford *et al.* 2011). Research relating to the movements of 'keystone' species and the minimum size of the habitats they require to be self-sustaining, is still required for many species (Handford *et al.* 2011).

The ultimate scale of the project and the finer delineation of project boundaries will occur as Reconnecting Natural Northland works with local communities to define the size and scope of restoration and sustainable land management activities across the landscape. Fundamentally, this will involve developing a community-owned vision for the landscape restoration initiative.

Monetising the Large Landscape Initiative

Economically sustaining restoration work is essential if landscape-scale restoration is to be achieved over the long term. Information on practical approaches to economically sustaining conservation/restoration work was gathered from a range of sources, including the review of case study projects (Handford *et al.* 2011). There is a range of potential opportunities to monetise and economically sustain conservation/restoration work. They include: sustainable forestry and carbon farming; honey production; landscape product branding; biodiversity offsets and credits; ecotourism and job creation.

Analysis was undertaken of the current readiness of these approaches for adoption and the strengths and weaknesses inherent in each opportunity (Handford *et al.* 2011). There is clearly considerable potential to develop some of these opportunities into revenue generating enterprises in Northland. However, there are several limitations associated with these opportunities. For example, the lack of clearly defined New Zealand legislation and policies for managing carbon within the Emissions Trading Scheme combined with volatility in the price of carbon. Honey production and ecotourism will require the development of new business arrangements for farmers and landowners and the acquisition of new skills and capabilities to take advantage of non-traditional farm-based enterprises. Landscape product branding and biodiversity offsets/credits are relatively new concepts with considerable development and trialling still required to determine their efficacy.

The reality is that many of these monetising opportunities that could support self-sufficiency in resourcing large landscape restoration are in their infancy in New Zealand and their applicability is far from firmly established.

Monitoring and evaluating outcomes

The application to the Tindall Foundation for the first 5-year phase of project implementation funding described the initiative as 'audacious, visionary, long-term and large scale'. It is anticipated that the LLI will be implemented over a 30–50 year period and hopefully longer. The LLI focus of interest is large, the whole of the Northland Region being a little over 5% of New Zealand's land area (13 940 km^2). Given the potential duration of the project and its geographical extent, only some broad overall project outcomes have been signalled at this time.

There has been much criticism of the lack of evaluation of the outcomes of natural resource management to date (Conley and Moote 2003; Edgar 2009). This applies not only to land and water management outcomes but equally to biodiversity and conservation outcomes. There is also a need to evaluate the socio-economic and cultural outcomes of large-scale ecological restoration initiatives.

Clearly a monitoring, reporting and evaluative framework will need to be developed at the commencement of the implementation phase of the LLI. This framework will need to incorporate both the intended duration of the project and its scale. It will need to include indicators and metrics that allow reporting against broad biophysical and socio-economic outcomes. It will also need to be put in place rapidly so that there is baseline information by which to analyse the impact of the LLI over time. Importantly, the indicators and metrics chosen will need to reflect where the initial focus of restoration is undertaken in the Northland landscape and be adaptive to changes in the priorities and work focus areas over time.

Future directions for developing and implementing the framework

The Tindall Foundation requested that the NZ Landcare Trust and the WWF-NZ prepare an application for funding the first 5-year implementation phase of the LLI in the Northland landscape. In February 2012, a funding application entitled 'Reconnecting Natural Northland' was submitted. This application was successful and the Tindall Foundation has indicated that they will provide significant financial support, up to 50% of the NZD$3.1 M cost of the project over the next 5 years.

Reconnecting Natural Northland will begin by identifying key native ecosystems and their threats at a landscape level. The project will work with the community to develop a landscape vision and identify and map potential opportunities for restoring connectivity through developing buffer zones, habitat corridors and stepping stones. Early implementation phase activities will include the following: (i) recruiting staff to project manage and implement Reconnecting Natural Northland on the ground; (ii) identifying conservation priorities and actions using conservation management planning and GIS mapping tools; (iii) identifying and supporting specific on the ground 'demonstration' restoration and land management projects; and (iv) scoping specific income-generating opportunities to assist the initiative to become self-sustaining.

In tandem with the physical restoration focus there will also be a need to develop the people dimension of the project. This will include: development and agreement on a long-term plan and vision for the landscape with stakeholders; identification and engagement with key stakeholders to sell this vision and begin to create a core governance group for Reconnecting Natural Northland; identification and support for project champions in the landscape; and building relationships and gaining the support of iwi, land management agencies, national and local government, community conservation groups, landowners and farmers, and business interests.

Reconnecting Natural Northland will need to develop a monitoring and evaluation plan including agreeing how the outcomes of the initiative will be measured, monitored and reported on to assess the initiative's progress and effectiveness. This will be dependent on the early agreement of some clear project goals and outcomes and indicators to assess progress against. An adaptive management approach will be adopted, developing a feedback loop so that lessons learned from reviewing the project's progress at regular intervals can be used to review and refine delivery accordingly.

It may be possible to transfer successful lessons for project implementation of Reconnecting Natural Northland to some of the other New Zealand landscapes that were identified as part of the landscape selection process. Clearly, initiating further large landscape initiatives in New Zealand will require resources and a community willing to take up the challenge.

Conclusions

The project represents a desire to undertake a large-scale biodiversity restoration initiative in the Northland landscape. The process of selecting the candidate large landscape has been by way of strategic planning and evaluation, as opposed to a direct reflection of the resident community's desire to restore landscape values. This strategic approach to landscape selection is the initiative's key point of difference from other New Zealand projects that have often been initiated by grassroots, bottom-up, community-driven processes. The initiative is about to embark on a phase of relationship building, both in the Northland region with resource management agencies and the local community, and with national agencies including the New Zealand government, non-government organisations and the corporate sector.

The challenge and the future promise of this initiative lies in working with the Northland community to articulate and then have them take ownership of a unique vision and approach to reconnecting natural Northland. That vision will need to resonate strongly with local communities and in a way that affirms the community's role in the driving seat of the project. A project that captures the hearts, minds and hands of the local community, making them firmly believe they can make a lasting positive difference to the Northland landscape.

Acknowledgements

The author would like to acknowledge the support, mentorship and resourcing provided by the Tindall Foundation, the supportive partnership that has developed between the NZ Landcare Trust, the WWF-NZ, and other members of the project steering group, and the excellent work undertaken by PLaCE Consulting in developing recommendations on candidate large landscape selection.

Biography

Nick Edgar has a background in sustainable land and water management with a specific focus on community-led catchment management initiatives. Dr Edgar has worked with a range of New Zealand and international organisations to encourage landowners and community groups to enhance their use of Best Management Practices and mitigation technologies to manage the impact of land use intensification on water resources. He is the CEO of the NZ Landcare Trust.

References

Conley A and Moote MA (2003). Evaluating collaborative natural resource management. *Society & Natural Resources* **16**, 371–386.

Edgar NB (2009). Icon lakes in New Zealand: managing the tension between land development and water resource protection. *Society & Natural Resources* **22**, 1–11.

Handford PA, Green W, Anstey C, Buchan D, and Pratt JP (2011). *Large Landscape Initiative: Criteria and Assessment of Potential Pilot Landscapes.* PLaCE Consulting, Paekakariki.

Hearnshaw EJS, Holmes SN, Yeates JJ, Karl DD, Schollum AC, and Simms MN (2011). *Collective Action Success in New Zealand.* Ministry for the Environment, Wellington.

Ministry for the Environment (2007). *Environment New Zealand 2007.* Ministry for the Environment, Wellington.

Ostrom E and Nagendra H (2006). Insights on linking forests, trees and people from the air, on the ground, and in the laboratory. *Proceedings of the National Academy of Sciences of the United States of America* **103**, 19224–19231.

SECTION 4.
BROAD THEMES

23

Connectivity conservation principles for Australia's National Wildlife Corridors

Brendan Mackey, Hugh P. Possingham and Simon Ferrier

Introduction

Welcome to the Anthropocene – the epoch in Earth's history when humans became a planetary force. Such is the power and scale of our technology and consumption, the decisions we take today will determine the course of Earth's environment in the coming century, the extent to which planetary boundaries are breached, and how much of living nature – the outcome of billions of years of evolutionary, ecological and biological processes – will remain and in what condition (Rockström *et al.* 2009). The time has come when society needs to think big and scale-up its responses to match the magnitude of the challenge we face in keeping nature functional and diverse in the 21st century. This point was made clear in a recent OECD (2012) report '… the sheer scale of economic and population growth has overwhelmed the pace of progress in curbing environmental pressures. Progress on an incremental, piecemeal, business-as-usual basis in the coming decades will not be enough … urgent and holistic action is needed now.' Connectivity conservation has emerged out of many decades of conservation science and practice as a response to the biodiversity challenge of the Anthropocene (Crooks and Sanjayan 2006). Its scientific origins date back to theories of island biogeography and metapopulation ecology where the roles of connectivity and habitat quality are combined (MacArthur and Wilson 1967). It reframes conventional conservation planning and management in terms of goals that are based on the long-term needs of species and ecosystems, recognising that conservation outcomes are required on a landscape-wide basis, inclusive of many tenures and sectors, and acknowledging that we must consider factors at multiple scales.

The *National Wildlife Corridors Plan* is the Australian Government's latest and most targeted policy response to the call for connectivity conservation (DSEWPC 2012; see also Chapter 17). The broad objective of National Wildlife Corridors is to contribute to the long-term persistence of biodiversity in the face of multiple threats, including human-forced climate change. Consistent with the definition provided by the international Convention on Biological Diversity, 'biodiversity' is defined here as the many different kinds of plant, animal, bacterial and fungal species that occur in Australia, the genetic diversity found within species, the different ecological communities that arise, the diversity of ecosystems and the habitats they provide, and the underlying ecological and evolutionary processes.

Connectivity conservation and the *National Wildlife Corridors Plan*

The scientific foundation for the Australian Government's *National Wildlife Corridors Plan* is based on the emerging body of theory and practice called 'connectivity conservation' (Crooks and Sanjayan 2006), which builds upon the existing foundations of conservation science including: the equilibrium theory of island biogeography; the study of metapopulation dynamics; macroecology (the emergent statistical properties of ecological datasets); and knowledge of evolutionary, historic and environmental biogeography. Connectivity conservation is a logical extension of these foundations which is necessary given the scale of the current biodiversity extinction crisis and the potential synergistic impacts of climate change (Bennett 2003; Mackey *et al.* 2010).

Connectivity conservation extends biological conservation to more explicitly consider the different scales over which species and ecosystems occur, as well as the scales at which important ecological and evolutionary processes operate. From connectivity conservation thinking arose the concept of a large scale 'corridor' that provides a planning and management framework for maintaining and re-establishing natural connections and related biodiversity assets in a coordinated way across multiple land tenures that occur in a catchment, bioregion, water basin or continent. Connectivity conservation therefore encompasses scales larger than those usually associated with the more established term 'landscape ecology'. The term 'wildlife corridor' has been used in a narrower context by some researchers, such as to refer to narrow bands of native vegetation connecting core habitat areas (e.g. Lindenmayer and Nix 1993). Here we follow the Australian Government's preference to refer to connectivity conservation corridors as 'wildlife corridors', that is, containing a range of elements including: narrow habitat corridors that link remnant vegetation patches; networks of habitat patches that serve as stepping stones for animals; conservation management regimes on farms that make the landscape more biologically permeable to the movement of animals; and ecological connections that maintain watershed hydrological flows.

Connectivity can be thought of in terms of 'structure' and 'function'. Structural connectivity refers to the spatial arrangement of different types of habitats or habitat patches in the landscape and is measured by analysing landscape pattern without any explicit reference to the movement of a particular species or the flow of a specific ecological process (Lindenmayer and Fischer 2006). Conversely, functional connectivity is broader and it focuses on the movement through landscapes of particular species and ecological processes such as environmental water flows. The 'connectivity' part of connectivity conservation therefore refers to various kinds of connections including:

- the structural configuration of habitats or habitat patches in a landscape mosaic;
- the permeability of a landscape mosaic for dispersal and movement of a specific species;
- the presence or absence of barriers or impediments to the natural flux of water, nutrients, or fire experienced in a landscape;
- landscape permeability with respect to meta-population dynamics; and
- gene flows associated with micro- and macro-evolutionary processes (Mackey *et al.* 2010).

Central to systematic conservation planning is the need to identify priorities for taking action, in particular with respect to the resources available for a given period (Margules and Sarkar 2007). While connectivity conservation does not change this planning requirement, it does highlight the need for prioritised short-term actions to be framed by an understanding of the long-term conservation goals for National Wildlife

Corridor initiatives. These long-terms goals will vary depending on the landscape's ecology, condition, and threats. However, put simply, (as with all conservation planning) we should ask of each corridor the questions: 'what does nature need and how can her needs be met?' The answers to these questions provide the framework for identifying targets and timetables, and thus the basis for prioritising conservation actions from year to year.

Complementarity between Australia's National Reserve System and the *National Wildlife Corridors Plan* is essential. It is axiomatic that a representative sample and well-managed suite of dedicated core protected areas should, in the long term, function as the backbone of wildlife corridors as they provide the 'anchors in the landscape' around which connectivity conservation management is planned. Generally, core protected areas contain relatively intact ecosystems and often represent the 'the best of what's left'. Ideally, at a regional scale, core areas should represent all major ecosystem types and be sufficiently large to have the capacity to largely 'self-manage' through natural processes that include the dispersal of biota and their propagules, natural selection and evolution, and biotic regulation of local biogeochemical and water cycles. However, the reality is that many currently do not possess these ideal characteristics and one important reason for the establishment of wildlife corridors is to help ensure their viability into the future. In addition, there will always be the need for some kind of conservation management intervention that cross protected area boundaries, even in large intact areas, given that these extensive areas have been subject to, among other perturbations, unsustainable land use and impacts from introduced species.

Scientific principles of connectivity conservation

The *National Wildlife Corridors Plan* (hereafter, the Plan) is a framework and mechanism for implementing connectivity conservation across the Australian continent. Implementation of the Plan must be tailored to the ecological, historical and socio-economic characteristics of each bioregion. Nonetheless, general principles can be identified that are grounded in conservation science and practice, and which can be used to guide how the Plan is implemented and what it means in practice in a given landscape. The following connectivity conservation principles were developed by the expert science panel supporting the National Wildlife Corridors Plan Advisory Group drawing upon, among other sources, Soulé *et al.* (2004) and Mackey *et al.* (2010). The focus here is on terrestrial ecosystems and biodiversity inclusive of wetlands, estuaries and coastal areas.

1. Connectivity conservation recognises the importance of the landscape matrix between areas of intact native vegetation, including protected areas

The matrix around intact native vegetation needs to be managed to facilitate natural levels of connectivity, including biological permeability between more intact native vegetation (Craig *et al.* 2000). The matrix can have biodiversity values that are absent from formal protected areas and those values should be identified and conserved. Matrix management may also be essential for many large-scale evolutionary and ecological processes, including ecological fire management. Where native vegetation exists as remnant patches, special conservation management may be required in the adjacent matrix to provide a buffer zone.

2. Connectivity conservation deals with biodiversity conservation at multiple spatial scales

That said, it places a particular emphasis on the medium to large scale, including connectivity across the entire continent. Planning needs to be mindful of the spatial scale of relevant ecological and evolutionary processes, as well as the spatial scale of socio-economic systems. There are multiple types of connectivity that exist at different scales and perform different functions (see principles below).

3. Prioritisation and implementation of actions within National Wildlife Corridors should be underpinned by explicit conservation objectives

These objectives should define as precisely as possible the long-term goals, milestone targets towards those goals and intended outcomes. The specific goals, targets and outcomes of any given corridor initiative will depend on the specifics of the region's biodiversity, land use history, current state of environment, extant and future threatening processes, and socio-economic drivers, capacities and opportunities.

4. Achieving conservation outcomes in National Wildlife Corridors will require active management targeting key threats

Australia's biodiversity faces an array of threatening processes including habitat destruction and degradation, introduced plants (weeds) and animal species (such as foxes, cats, camels, water buffalo) altered fire regimes and hydrological regimes, disease and human-forced rapid climate change. Land management interventions are needed to eliminate or contain the impacts of these threatening processes on species and ecosystems and in some cases prevent their introduction.

5. Prevention is usually better than cure when managing vegetation within National Wildlife Corridors

In extensively cleared landscapes, priority should generally be given to protecting and improving the quality of existing native vegetation over revegetation. In landscapes that have not been extensively cleared but suffer habitat degradation, restoration of this degraded habitat should not be at the expense of protecting and managing vegetation in good condition that provides high quality habitat. However, in many highly fragmented southern Australian landscapes, large-scale revegetation will be essential.

6. Ecological restoration and rehabilitation may be required to re-establish connectivity within National Wildlife Corridors in more fragmented landscapes

In landscapes subject to broad-scale clearing, re-establishing connectivity should focus on restoration of remnant vegetation. In rehabilitating previously cleared land, the taxonomic composition of plantings requires careful consideration. Wherever possible, species that characterise the original native ecosystem (or an analogue ecosystem given current landscape condition) should be used. This will optimise benefits for biodiversity as well as carbon, as the restored or rehabilitated ecosystem will be more resilient and have more stable carbon stocks.

7. Planning of National Wildlife Corridors needs to consider multiple ecological processes, operating at multiple spatial and temporal scales.

These processes include (Soulé *et al.* 2004; Mackey *et al.* 2007):

> *Trophic relations and functional interactions* – especially for (i) highly interactive species critical to top-down predator control; and (ii) ecosystem functions such as seed dispersal and pollination critical to reproduction;

Hydro-ecological processes – including (i) the interplay between ground and surface water resources and phenomenon such as ground-water dependent ecosystems; (ii) environmental flows for aquatic and floodplain ecosystems; and (iii) the relationship between catchment health and habitat productivity;
Dispersive species – especially species that exhibit large-scale movements including migratory, nomadic and eruptive species. For example, around half (364) of Australia's land and freshwater birds are dispersive (and therefore are not obligate residents) and conservation of these dispersive species requires consideration of habitat needs in multiple bioregions and landscapes (Gilmore *et al.* 2007);
Meta-population dynamics – especially for mobile species and animal species with large home ranges; many species occur within a landscape as a network of populations which are connected through the dispersal of juveniles and other displaced individuals; these interactions between populations can contribute to maintenance of genetic diversity and the replenishment of extirpated populations; movement between populations of a species requires consideration of connectivity at scales commensurate with their mobility;
Ecological refuges – micro-habitats often function as networks that provide species with refuge from threatening processes and extreme climatic events (especially fire and drought) (e.g. Morton *et al.* 1995; Mac Nally *et al.* 2000); also relevant are habitat locations that might provide refuge for some species in the face of climate change (Vos *et al.* 2008).

8. Planning of National Wildlife Corridors needs to consider evolutionary processes, both past and ongoing

Evolution is, put simply, inherited genetic changes in an interbreeding population. Macro-evolutionary changes can lead to the emergence of new species. However, just as important to the survival of existing species is ongoing micro-evolution that can result in adaptations to local conditions by a population and the transmission of adaptive traits between populations. The genetic legacy of evolutionary history is also revealed in genetic differences at the population level and concentrations of ancient, narrow range endemics can represent refugia for interesting and useful genetic diversity. Also of importance are zones of potential future speciation.

9. Accommodating anticipated impacts of climate change within National Wildlife Corridors will require multiple forms of connectivity implemented at multiple scales

For many naturally dispersive species, their natural adaptation response will involve large-scale movement such as migration to previously uninhabited locations. Other species with more limited mobility may be forced to migrate as climate change impacts intensify in coming decades, albeit at slower rates. Some species may be able to persist without moving, through evolutionary, physiological and/or behavioural changes – i.e. through in situ local adaptation. Connectivity planning should identify potential elevation and latitudinal gradients that facilitate dispersal and migration along thermal gradients. This planning needs to consider major differences in dispersal ability and adaptation responses (e.g. migration versus in situ local adaptation), between different species and functional groups. Another relevant factor is the steepness in climatic gradients to which species will have to respond.

10. The continental system of National Wildlife Corridors should include (i.e. 'represent') as many biodiversity elements as possible

The main components of biodiversity (genetic-population, species, community, functional) are apparent at multiple scales with a useful distinction made between 'α' and 'β' diversity at local and regional scales, respectively. Maintaining biodiversity at all scales enhances the resilience and adaptive capacities of ecosystems and demands a whole of landscape approach to conservation. Good planning will ensure that proper consideration is given to the protection and restoration of habitats that are otherwise poorly conserved or disproportionately cleared and degraded.

11. A resilient National Wildlife Corridor will include redundancy in the representation of elements

A conservation plan that is robust to catastrophes including climate-related disasters will be designed to capture multiple examples of biodiversity elements that during normal conditions may be considered 'redundant' in terms of representativeness or ecosystem function. Building redundancy into conservation planning is a good insurance policy in the face of catastrophes that can for example extirpate an entire population of a species. Having replicates of the locations where species and their habitat occur, and functional redundancy between species, contributes to the resilience of ecosystems to external impacts and threatening processes.

12. National Wildlife Corridors should be managed adaptively

As new knowledge comes to hand we need to have the flexibility to alter actions to accommodate that new information. Furthermore, some actions will have the dual role of delivering outcomes and information that can be used for adaptation.

In addition to the above science-based principles, two additional principles are worth noting that relate to issues of implementation.

13. Different conservation management instruments will be needed depending on landscape context

Relevant contextual information includes the geographic scale of the corridor and local social, legal and economic circumstances. Conservation management instruments range from land acquisition and tenure change, voluntary change of land use by private landowners, stewardship agreements, tax-concessions and other incentives that facilitate outcomes across the landscape.

14. A connectivity conservation approach demands partnerships with multiple stakeholders and action across land tenures

Established cooperative initiatives such as Landcare (see Chapter 19) provide a good model of the level of cooperation than can be achieved in addressing shared land management problems. Successful implementation of connectivity conservation initiatives have been shown to require (i) a shared vision; (ii) strategic long-term planning; and (iii) innovative partnerships.

Lessons learnt to date in applying these principles

The *National Wildlife Corridors Plan* builds upon a set of existing connectivity conservation initiatives in Australia being implemented by consortiums of land owners and managers and non-government organisations with some supported by state government agencies and local government. These initiatives have yielded important insights into what is critical to successful implementation of connectivity conservation corridors.

One important lesson is that the Plan cannot be implemented through a 'one size fits all' approach. Rather, the goals, targets and actions must be tailored to the specific ecological and social characteristics of each region. One of the landscape characteristics that fundamentally frame connectivity conservation initiative is the condition of the native vegetation cover as it has been impacted by land use. The native vegetation cover of the Australian continent has been cleared and degraded to varying degrees by land-use activities over the last 200 years. At one end of this spectrum, the land cover is transformed into a largely exotic cover (e.g. cropping) or an artificial land surface. However, a range of vegetation conditions can be recognised: (i) land that has been *transformed* from a native to non-native vegetation cover; (ii) land that has been largely cleared but where *remnant* patches of native vegetation remain; (iii) land that largely retains its native vegetation cover but has been extensively *modified* by land use; and (iv) land whose native vegetation is basically *intact* (Thackway and Lesslie 2006). We can also recognise (v) *peri-urban and coastal development zones* that warrant recognition as these areas bear the brunt of Australia's growing human population.

Different conservation strategies are needed in each of these categories, as is cross-tenure conservation management in order to protect and restore biodiversity on a landscape-wide basis. On land that has been heavily cleared and fragmented, there is a need for large-scale ecological restoration and rehabilitation. In intact country, the option remains to maintain natural connectivity. In all cases, wildlife corridors provide a vehicle for promoting biodiversity conservation in an integrated way through a combination of protected areas and complementary conservation management on other tenures.

Wildlife corridors should also be framed by scientifically based long-term conservation goals. Central to systematic conservation planning is the need to identify priorities for taking action, in particular with respect to the resources available for a given period. While connectivity conservation does not change this planning requirement, it does highlight the need for such prioritised short-term actions to be framed by an understanding of the long-term conservation goals for the wildlife corridor. These long-term goals will vary depending on the landscape's ecology, condition, and threats. The answers to these questions, coupled with prudent return-on-investment analysis, provide the framework for identifying targets and timetables, and thus the basis for prioritising conservation actions from year to year.

Of particular interest to land managers working in landscapes that have been largely cleared but contain remnant patches of native vegetation, is connectivity in terms of the structural configuration of habitats patches in the landscape matrix at the paddock and farm scale. For example, the following guidelines are based on a review of 80 connectivity studies and provide insights into how to link small, fragmented remnants of native vegetation that may be useful for mammals and birds living in woodland and forest ecosystems (Doerr *et al.* 2010a, 2010b):

Step 1: Evaluate the landscape context to see if connectivity management and restoration is appropriate for the ecosystem in question. There is a trade-off to consider between connecting patches and improving patch size and quality.

Step 2: Determine which types of structural connectivity are likely to be most effective given your goals (facilitate movement between patches or provide additional habitat) and local environment (temperate or tropical), e.g. stepping stones (scattered trees, shrubs, etc.), vegetated linear connections. Selection may vary with landscape context (e.g. tropical versus temperate) and vegetation/habitat quality may be an important consideration.

Step 3: Given the information above on effectiveness of connectivity options, select which option to pursue by also considering the other benefits desired in the landscape, such as shelter for stock, riparian restoration, and wind breaks, compared to the costs involved.

Step 4: Monitor the results of connectivity management actions and be prepared to adjust management activities based on data from this ongoing monitoring as well as any recommendations coming from new research, i.e. an adaptive management approach to connectivity projects is essential.

Future directions and resilience thinking

We use the term 'resilience thinking' to refer to the application of resilience theory to problems of conservation planning and management, following the work of Walker and Salt (2006) who applied it to natural resource management more broadly.

Wildlife corridor plans provide an opportunity to advance conservation planning and management in innovative ways. Resilience thinking is increasingly seen as one innovation that can guide conservation actions so that they better reflect ecological realities, leading to improved conservation outcomes (Walker and Salt 2006). However, understanding the implications of resilience thinking for wildlife corridor plans (that is, how would a plan be different if it were designed in terms of resilience thinking?) remains an ongoing challenge for researchers and practitioners alike.

Ecosystem resilience can be understood by reference to Holling's adaptive cycle (Walker *et al.* 2004). Ecosystems are conceived as being dynamic and comprising four phases: (1) *r* – growth; (2) *K* – conservation; (3) *omega* – release; and (4) *alpha* – reorganisation. Three kinds of resilience can be recognised. First, a system can be considered resilient if it is able to absorb the impact of external perturbations and remain stable in the conservation phase. Second, a system can also be resilient if, when a perturbation causes a shift from the conservative to the release phase, it is able to reorganise and grow back into basically the same kind of system. Third, if the resilience of system 'A' is breached, then it can be transformed into a different kind of system (Walker *et al.* 2004). Adaptive capacity is perhaps most obviously brought into play when external forcing requires a transformation change. However, adaptive capacity also refers to mechanisms that enable an ecosystem to maintain a stable conservation phase or to 'bounce back' following reorganisation and growth.

The resilience of an ecosystem to changing environmental conditions is determined by biodiversity at multiple scales including: (i) the taxonomic and functional diversity of species, including microorganisms; (ii) the genetic variability within species (i.e. the diversity of genetic traits within populations of species); and (iii) the regional pool of species and ecosystems. The diversity of species and functional types found at local and regional scales provides an ecosystem with options in the face of changing circumstances including climate change. Resilience is also influenced by the size of ecosystems – generally, the larger and less fragmented, the better – and by the condition and character of the surrounding landscape (Thompson *et al.* 2009). Wildlife corridors can help protect and restore biodiversity at multiple scales, including β (regional) diversity, thereby contributing to the adaptive capacity and resilience of landscape ecosystems.

Conclusion

In this chapter we have drawn together connectivity conservation issues relevant to Australia's circumstances and biodiversity challenges. We believe that the *National Wildlife Corridors Plan* has the potential to deliver outcomes for biodiversity that are more integrated, durable and cost effective. However, in implementing the Plan those parties responsible for the on-ground delivery of projects operating under its auspices, and other initiatives undertaken independently, will need to keep a close eye on the scientific principles enunciated here, the lessons learnt to date, and focus on monitoring outcomes in the long term. More generally, Australia will need to invest considerably more in conservation, and on a sustained basis, to ensure the many willing hands that support connectivity conservation have the resources they need to help realise the Plan's potential.

Biographies

Brendan Mackey is a Professor at Griffith University and Director of the Griffith Climate Change Response Program. He is a member of the Australian Climate Commission science advisory group and serves on the Council of the International Union for Conservation of Nature (IUCN). Brendan's research is currently focused on climate change science, ecosystem-based mitigation and adaptation, and the integration of biodiversity into regional assessments.

Hugh Possingham is a Professor at The University of Queensland from where he directs two national research centres. He is a member of The Wentworth Group and has led several scientific consensus statements that delivered outcomes on: land clearing, stock routes and marine conservation. During his 12 years in Brisbane he has coauthored over 200 refereed scientific papers and supervised 39 PhD students. Hugh sits on 16 boards and committees external to The University of Queensland and suffers from obsessive bird-watching.

Simon Ferrier is a Senior Principal Research Scientist with CSIRO Ecosystem Sciences, based in Canberra. Up until 4 years ago he worked with the NSW Department of Environment, Climate Change and Water. He has more than 30 years of experience in researching, developing and applying quantitative spatial analysis and modelling approaches to biodiversity assessment and conservation planning at regional, continental and global scales.

References

Bennett AF (2003). *Linkages in the Landscape: The Role of Corridors and Connectivity in Wildlife Conservation.* IUCN, Gland, Switzerland and Cambridge, UK.

Craig JL, Mitchell N, and Saunders DA (Eds) (2000). *Nature Conservation 5: Conservation in Production Environments – Managing the Matrix.* Surrey Beatty & Sons, Chipping Norton, NSW.

Crooks KR and Sanjayan M (Eds) (2006). *Connectivity Conservation.* Cambridge University Press, Cambridge.

Doerr VAJ, Doerr ED, and Davies MJ (2010a). Does structural connectivity facilitate dispersal of native species in Australia's fragmented terrestrial landscapes? CEE review 08-007 (SR44). Collaboration for Environmental Evidence, Bangor. Available: www.environmental evidence.org/SR44.html [Accessed 5 June 2012].

Doerr VAJ, Doerr ED, and Davies MJ (2010b). Guidelines for connectivity management and restoration in Australia. Collaboration for Environmental Evidence. Available: http://www.environmental evidence.org/Documents/ManagementguidelinesSR44.pdf [Accessed 5 June 2012].

DSEWPC (2012). *National Wildlife Corridors Plan: A Framework for Landscape-scale Conservation.* Department of Sustainability, Environment, Water, Population and Communities, Canberra.

Gilmore S, Mackey B, and Berry SB (2007). The extent of dispersive movements in Australian vertebrate animals, possible causes, and some implications for conservation. *Pacific Conservation Biology* **13**, 93–103.

Lindenmayer DB and Fischer J (2006). *Habitat Fragmentation and Landscape Change: An Ecological and Conservation Synthesis.* Island Press, Washington, DC.

Lindenmayer DB and Nix HA (1993). Ecological principles for the design of wildlife corridors. *Conservation Biology* 7, 627–631.

Mac Nally R, Soderquist TT, and Tzaros C (2000). The conservation value of mesic gullies in dry forest landscapes: avian assemblages in the box-ironbark ecosystem of southern Australia. *Biological Conservation* **93**, 293–302.

MacArthur RH and Wilson EO (1967). *The Theory of Island Biogeography.* Princeton University Press, Princeton.

Mackey BG, Soulé ME, Nix HA, Recher HF, Lesslie RG, Williams JE, Woinarski JCZR, Hobbs J, and Possingham HP (2007). Towards a scientific framework for the WildCountry project. In: *Key Topics and Perspectives in Landscape Ecology.* (Eds J Wu and RJ Hobbs) pp. 192–213. Cambridge University Press, Cambridge.

Mackey B, Watson J, and Worboys GL (2010). *Connectivity Conservation and the Great Eastern Ranges Corridor.* Department of Environment, Climate Change and Water, Sydney.

Margules CR and Sarkar S (2007). *Systematic Conservation Planning.* Cambridge University Press, Cambridge.

Morton SR, Short J, and Barker RD (1995). *Refugia for Biological Diversity in Arid and Semiarid Australia.* Biodiversity Series, Paper No. 4. Department of Environment, Sport and Territories, Canberra.

OECD (2012). *Environmental Outlook to 2050: The Consequences of Inaction.* Organisation for Economic Co-operation and Development, Paris.

Rockström J, Steffen W, Noone K, Persson Å, Chapin FS, III, Lambin E, Lenton TM, Scheffer M, Folke C, Schellnhuber H, Nykvist B, De Wit CA, Hughes T, van der Leeuw S, Rodhe H, Sörlin S, Snyder PK, Costanza R, Svedin U, Falkenmark M, Karlberg L, Corell RW, Fabry VJ, Hansen J, Walker BH, Liverman D, Richardson K, Crutzen C, and Foley J (2009). A safe operating space for humanity. *Nature* **461**, 472–475.

Soulé ME, Mackey BG, Recher HF, Williams JE, Woinarski JCZ, Driscoll D, Dennison WC, and Jones ME, (2004). The role of connectivity in Australian conservation. *Pacific Conservation Biology* **10**, 266–279.

Thackway R and Lesslie R (2006). Reporting vegetation condition using the Vegetation Assets, States and Transitions (VAST) framework. *Ecological Management & Restoration* 7, S53–S62.

Thompson I, Mackey B, McNulty S, and Mosseler A (2009). *Forest Resilience, Biodiversity, and Climate Change. A Synthesis of the Biodiversity/Resilience/Stability Relationship in Forest Ecosystems.* Technical Series no. 43, Secretariat of the Convention on Biological Diversity, Montreal.

Vos CC, Berry P, Opdam P, Baveco H, Nijhof B, O'Hanley J, Bell C, and Kuipers H (2008). Adapting landscapes to climate change: examples of climate-proof ecosystem networks and priority adaptation zones. *Journal of Applied Ecology* **45**, 1722–1731.

Walker BH and Salt D (2006). *Resilience Thinking: Sustaining Ecosystems and People in a Changing World*. Island Press, Washington, DC.

Walker B, Holling CS, Carpenter SR, and Kinzig A (2004). Resilience, adaptability and transformability in social–ecological systems. *Ecology and Society* **9**(2), 5. http://www.ecologyandsociety.org/vol9/iss2/art5

24

Social aspects: linking the people *and* their landscapes

Judy Lambert

Introduction

The interested observer need delve only a very short distance back into Australian history to find nature conservation focused on threatened species and the national parks and other protected areas that provided those species with their primary refuge. Only in the past decade has that focus broadened.

Through a mix of natural area protection, restoration and landscape rehabilitation, large-scale habitat connectivity has received increasing attention in both policy and practice in recent years. Initially focused on biodiversity conservation and the creation of habitat corridors, connectivity projects now encompass the notion of maintaining resilience in the face of human-induced climate change.

Landscape restoration is a relatively new science in Australia. However, in recent years Williams (2008), Whitten *et al.* (2011), Worboys and Pulsford (2011) and others have highlighted a diverse range of large-scale connectivity projects. As Williams (2008) reports, the impetus and conduct of these projects and the policy that underpins them 'spans the intuitive to the systematic'.

No matter what the motivation for these landscape-scale connectivity projects, each involves actively managed linkages across land of different tenures, managed by people with different needs, expectations and values. The success of connectivity projects is critically dependent upon successful collaboration between landholders and managers responsible for both public and private land and in almost all instances community-based non-government organisations provide both the catalyst and the 'glue' that holds participants together.

Complex systems

As Fischer and Lindenmayer (2007) highlight in their synthesis paper, the modified landscapes across which large-scale connectivity projects are being developed are complex systems in which human activities and land uses interact with both individual species and the natural habitats that remain.

Ecologically, landscape connectivity varies across space, scale and time such that each situation is to an extent unique (Whitten *et al.* 2011).

To that naturally occurring complexity can be added the human dimensions of landscape management. Jurisdictional and institutional aspects of landscape management are inextricably linked with socio-economic aspects. However, since Wyborn (Chapter 26) addresses governance arrangements relating to landscape linkages and Whitten *et al.* (Chapter 25) address socio-economic perspectives, the remainder of this chapter focuses on the human interactions important in linking Australia's landscapes.

The human complexity: different people, different values, different needs

The land involved in connectivity conservation projects focuses around core areas of existing natural vegetation in which ecological processes are largely intact. That land may vary from a remnant patch in a farm back paddock within a highly fragmented landscape to community-managed Indigenous land or a remote wilderness area in a national park. Beyond these core areas, the restoration and rehabilitation needed to build connectivity occur in an even more complex array of land management priorities in rapidly changing landscapes.

Those responsible for managing the different parcels of land contributing to connectivity conservation will bring to a project widely differing needs, expectations and values. At a government level, national policy makers and those responsible for funding allocations might see the priorities very differently from either the state-based national parks ranger whose focus is on nature conservation and biodiversity or the extension officer advising rural landholders on optimising their agricultural production. Staff from regional natural resource management (NRM) bodies bring to connectivity conservation a focus centred more on regional outcomes. With more than 50% of Australia privately owned or leased, primarily for food and fibre production (Australian Bureau of Statistics 2011), all are interacting with and seeking the participation of the landholders managing primarily for their rural production livelihoods.

During more than two decades of work at the interface between scientists, rural landholders, government agencies and the community-based conservation sector, I have frequently observed the rural landholders/farmers treated as one entity in the complex human mix that is an essential part of connectivity conservation across Australia.

However, landholders differ in many ways. Whether on a family farm, leasing property or managing for a corporate owner; whether raising livestock in an extensive grazing enterprise or intensively as in dairying or a piggery, cropping, growing fruit or vegetables, or experimenting with 'sunrise' products; whether getting started, educating children, developing a mid-life second interest, or nearing retirement; whether a new settler recently moved to farming or a fourth-generation farmer; and whether rich or poorer – all are significant factors influencing farming decisions. Vanclay (2011) captures the socio-cultural diversity of farming in his development of social principles guiding the adoption of NRM in agriculture.

Different temperaments

As adoption of new approaches to land management has increasingly become recognised as a social process, several studies have examined the role of personality or temperament in influencing the decisions made by individual farmers or farming families.

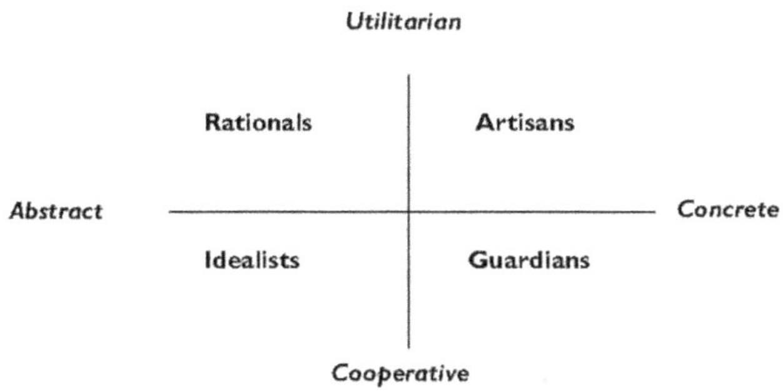

Figure 24.1 Keirsey's four temperaments (Adapted from Keirsey 1987).

Using the Myer-Briggs personality indicator, Shrapnel and Davie (2001) showed Queensland graziers had a much stronger tendency to introversion and discomfort with group situations than did the wider community. Perhaps more useful when considering interactions between rural landholders, scientists, government agency staff and others needing to work together in connectivity conservation are the measures of temperament identified by Keirsey (1987). Based on whether individuals value concrete, reality-based thinking or abstract, ideas-based thinking and the extent to which they value cooperative, socially approved action as compared with utilitarian, pragmatic results-based action, Keirsey identifies four dominant temperaments that generally remain with us throughout our lifetime (Figure 24.1).

As Strachan (2011) highlights, relatively higher numbers of people working in rural industries have a 'Guardian' or traditionalist temperament. Guardians value 'social stability, security, right order, loyalty, industry and belonging'. They are seen as 'cautious, careful, steady-paced and reliable'. By contrast, research scientists more often fit Keirsey's 'Rationalist' or visionary temperament, favouring intuitive abstract thinking and utilitarian results-based action.

Different ways of 'knowing'

Not only do different individuals have different temperaments that influence their decision making, but people from different backgrounds and life experiences also have different ways of knowing. Educator and social change researcher Professor Valerie Brown identifies five different types of knowledge influencing decision making in Western cultures (Brown 2008). Only when brought together in a collective process, do these forms of knowledge result in the wisdom likely to maximise success in tackling the complex problems that beset today's world, including connectivity conservation.

Brown (2008) describes the five knowledges (or ways of knowing) as:

1 individual knowledge
2 local knowledge
3 specialised knowledge
4 strategic knowledge; and
5 holistic knowledge.

As described in Figure 24.2, each of these has a different origin and content and is derived from different sources of evidence. Those for whom a different kind of knowledge is dominant often treat other types of knowledge with some mistrust or dismiss it altogether.

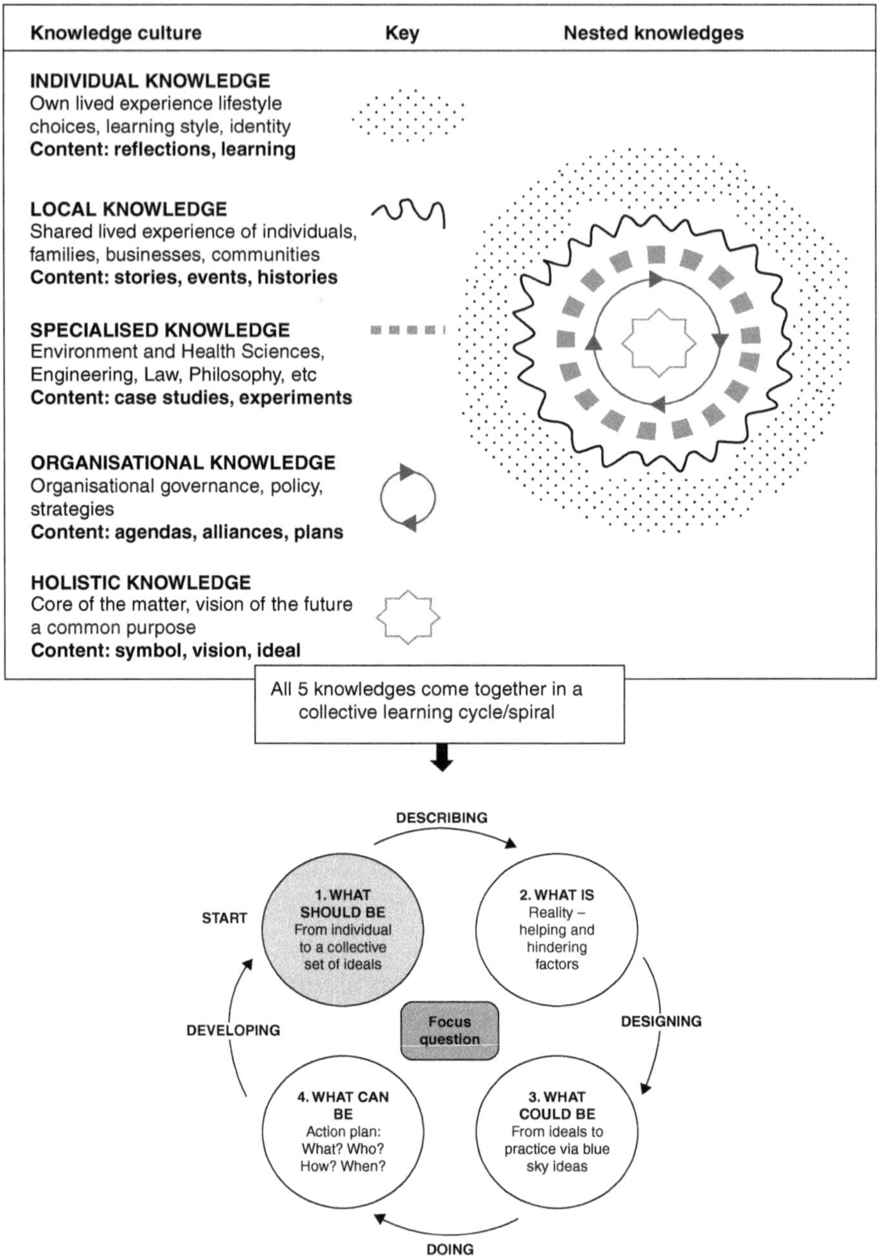

Figure 24.2 The knowledge cultures of Western decision making as they are synthesised to a collective learning cycle/spiral (Adapted from Brown and Lambert 2012).

In a connectivity conservation project, people from each of the knowledge cultures bring to the process their individual knowledge, shaped as it is by their own individual life experiences. Rural producers, lifestyle settlers, community-based environmentalists, Indigenous people and others living or participating in a local community bring their shared experiences and local knowledge in all its forms. Ecologists, district agronomists, land use planners, bankers and others trained in scientific and technical subjects bring specialist technical knowledge, while staff in government agencies and in business bring the administrative, governance, policy and program knowledge described by Brown (2008) as strategic knowledge. Because of the scale at which most connectivity conservation projects operate, it is important that the process includes some holistic thinkers able to synthesise and focus on the 'bigger picture' resulting from the bringing together of what are often smaller, more localised initiatives.

Rarely will the diversity of participants needing to be involved come together naturally and interact comfortably without specific invitation, hosting and skilled facilitation. An independent facilitator with some knowledge of NRM and a focus on building shared understanding can do much to strengthen connectivity and shared commitment among the people involved.

Partnerships and collaboration

Connectivity conservation places strong emphasis on integrated approaches to land use planning and management. This necessitates people from different backgrounds working together. Wyborn (2011) captures the framework for connectivity initiatives in three basic concepts: 'a system of interconnected properties; strategic management to achieve conservation objectives; and collaboration between multiple landowners and managers'.

However, collaborations and partnerships do not simply happen. When people from different backgrounds and cultures – sometimes even within different government agencies, but more particularly across sectors – come together in partnership, the success of that partnership depends in significant part on:

- mutual respect of knowledge, skills, opinions, needs and expectations;
- trust built on effective communication, honesty, openness and transparency;
- cooperative effort; and
- shared power (Elix *et al.* 2005).

Too often within existing programs anticipating 'partnerships', insufficient time, effort and resources are directed to doing the groundwork needed to ensure that these pillars of success are in place and mutually agreed. In the haste often necessary to meet a funding deadline, newly formed alliances assume common language, misinterpret each other's needs and values, or fail to clearly define roles and responsibilities of each participant. While some aspects can be addressed through formal contracts and governance arrangements, others require more informal processes to determine whether a partnership is right for the groups involved. Crawford *et al.* (2007) describe this as 'active negotiation of learning roles between farmers, researchers and advisors' – a process that 'requires a unique form of facilitation to support the learning environment and use of learning tools'.

Even once a project is underway, continuing efforts to maintain these aspects of the relationships between partners are necessary. Staff turnover is often a key factor in undermining established relationships; changes in funding priorities within a participating organisation might shift the balance in cooperative effort; or the interpersonal dynamic may simply change as relationships mature.

The importance of a shared vision

Only in recent years has the importance of shared understanding been highlighted in complex multi-partner teams working for connectivity outcomes. Crawford *et al.* (2007), in a study of the factors influencing the success of 'learning partnerships' in the dairy industry, highlighted the importance of 'negotiating goals, roles and responsibility between commercial farms, research and extension'. Both Lindenmayer *et al.* (2008), in their checklist for ecological management of landscapes for conservation, and Worboys and Pulsford (2011) in their guide to connectivity conservation, take this further and place a shared vision at the centre of the process.

The importance of developing a shared vision, understood and mutually agreed by the diversity of partners in complex landscape connectivity projects, is reinforced by the extensive body of work done by Brown *et al.* (2010) on collective learning for transformational change. While the learning cycle used in such projects is the well-known adult learning cycle developed by Kolb (1984), practical experience in a suite of complex collective learning projects stresses the importance of starting not with *what is now*, but with *what should be* – the vision for the future.

Making relevant information accessible

Even when participants have a shared vision, the task of locating the most useful information to help achieve that vision is often daunting. Although landscape restoration and connectivity are relatively new sciences, the volume of literature is rapidly escalating. Layered over that is the growth in knowledge and understanding of the factors influencing successful partnerships between sometimes disparate individuals and organisations.

As Lovett *et al.* (2008) and many others have identified, those on the ground (whether rural landholders or the government and community-based staff who are the knowledge brokers) are often time poor and feel pressured to deliver outcomes in unrealistically short timeframes. Theirs is a need for the 'right' information at the right time and in a context that has local meaning. Research findings are scattered across different publications and focal points providing useful syntheses of information are relatively few. Staff turnover is often sufficiently high that it impacts adversely on the trusted relationships that enable community practitioners to access the scientists. Those scientists are themselves under pressure to produce more peer-reviewed publications that influence their career prospects but may not be widely read outside academic circles, rather than spending time interpreting and sharing their knowledge.

In seeking to address these needs Lovett *et al.* (2008) recommended:

- development of an 'information hub' on landscape restoration (or connectivity);
- investment in the development of distance education on relevant topics;
- a national database of landscape restoration projects; and
- working with scientists and science communicators to develop easily accessible on-line information sessions.

However, these are initiatives that largely remain to be addressed and, in the meantime, the shared learning that comes of collaborative processes can make some contribution.

Beyond collaboration: the collective learning approach

As Fischer *et al.* (2006) observe in their development of guiding principles for ecosystem function and resilience in commodity production landscapes, 'Management approaches that solve all ecological and economic problems at once do not exist'.

The issues faced are *wicked problems* as defined by Rittel and Webber (1973). They are problems that defy complete definition and have no final solution. Brown *et al.* (2010, p. 4) identify that the solutions to 'wicked problems' 'are not true or false or good or bad, but the best that can be done at the time'. Because these challenges are born of the society in which they occur, they need all the wisdom and knowledge available to achieve that solution. They also require a new approach that goes beyond our traditional collaboration to a collective effort that develops transformational change.

These are new approaches, not yet well accommodated in government programs and the funding application processes that they sustain.

Merely requiring 'partnerships' does not necessarily enable the shared vision, commitment, understanding and respect needed for success. The building of partnerships among the diversity of interests in landscape connectivity requires specific investment of time and resources. One useful framework to guide this investment is the collective learning spiral developed by Brown (2010).

In her discussion of the collective learning spiral, Brown (2010) highlights some of the differences between this cycle aimed at achieving transformational change in complex situations and our usual approaches to partnership or collaboration. Major differences from our usual approach include:

- the bringing together of people representing each of the five types of knowledge (individual, community, specialised or technical, organisational and holistic) and providing each with equal respect and value;
- the emphasis on beginning with the development of a shared vision (rather than beginning with the current situation that is seen as fact);
- moving through the facts to a creative new preferred situation, then determining what can actually be done;
- incorporating points of reflection (the connectors between the stages) throughout the learning cycle; and
- the recognition of change as something for celebration, rather than something to be feared in complex 'wicked' problems.

Conclusions

While connectivity conservation should always be underpinned by sound ecological science, there is now a substantial body of literature that makes clear the importance of social considerations to the success of these initiatives. Key social factors for success begin with the importance of developing a shared vision, which provides a point of reference throughout the collaborations that are required for projects that are striving to enhance connectivity in the landscape. Managed as a collective process, the development of that shared vision does much to build mutual understanding among the participants.

Given the diversity of people who need to be involved in large-scale landscape connectivity projects (both those who make decisions and those who will be affected by those decisions), conscious effort is required to ensure that all views are heard and respected. Recognising that the people involved will bring with them different temperaments, different preferred ways of learning and different ways of 'knowing', building of trust will take time and commitment.

The biggest pitfall for coordinators in developing conservation connectivity initiatives usually arise from assumptions of shared understanding. Too often the partners in a landscape connectivity project assume that all attach the same purpose to their efforts and the same meaning to their words (even the words of a 'shared' vision). Large, and ultimately very successful projects may even risk self-destruction if these differences are not recognised and addressed early (Lovett *et al.* 2008; Wyborn 2011; J. Lambert, facilitation experience; see also Fitzsimons and Wescott 2007). Because of the central role that trust and shared understanding play in projects that link landscapes across tenure and land use, the lack of continuity that so often arises from discontinuity of government programs and funding cycles has serious adverse impacts. New personnel, changed program terms of reference and the need to reshape projects to enable their survival all lead to a need to rebuild trust and shared understanding among the key participants. Associated with this may come a loss of spirit and enthusiasm, especially among those who have made larger commitments in a voluntary capacity. Strong and inclusive leadership at all levels is repeatedly identified as important in addressing these problems.

These are challenges that are relevant across all landscapes. However, the large numbers and diversity of likely participants in fragmented commodity production landscapes will likely see the challenges exacerbated. They are, however, the places where the opportunities for shared celebration and rewards for those who do persist will likely be greatest.

Acknowledgements

To the hundreds of committed people (farmers and graziers, government agency staff, research scientists, primary industry bodies and others in local communities) with whom I have worked over the past 30 years, I am deeply grateful. Emeritus Professor Valerie Brown has taught me new ways to view the interactions between people and to understand the importance of encompassing their different knowledge cultures. Without my colleague and former business partner Jane Elix and my long-time and very dedicated research assistant Heather Pearce I would probably not have embarked on the extremely rewarding post-research, post-policy phase of my career which has enabled me to contribute in some small ways to landscape connectivity and the understanding of the importance of involving different people in achieving connectivity outcomes. This paper has benefitted significantly from valuable comments provided by colleagues Karen Alexander and Carina Wyborn. I appreciate the time and effort they put in.

Biography

Judy Lambert is trained in social, environmental and health sciences and in business administration, and has a particular interest in the connections between the environmental and social aspects of sustainable living. She has more than 25 years' experience working at the interface between farm production, biodiversity conservation and natural resource management. In a career spanning research, community sector advocacy, government policy and consulting, Judy is committed to a future in which integrated management becomes a reality in ensuring our unique Australian landscapes are sustained in a changing climate.

References

Australian Bureau of Statistics (2011). Land Management and Farming in Australia, 2009–10. Available: www.abs.gov.au/ausstats/abs@.nsf/Products/4627.0-2009-10 [Accessed April 2012].

Brown VA (2008). *Leonardo's Vision: A Guide to Collective Thinking and Action.* Sense Publishers, Rotterdam, The Netherlands.

Brown VA (2010). Collective inquiry and its wicked problems. In: *Tackling Wicked Problems Through the Transdisciplinary Imagination.* (Eds VA Brown, JA Harris and JY Russell) pp. 61–81. Earthscan, London.

Brown VA, Deane P, Harris JA, and Russell JY (2010). Introduction. In: *Tackling Wicked Problems Through the Transdisciplinary Imagination.* (Eds VA Brown, JA Harris and JY Russell) pp. 1–15. Earthscan, London.

Brown VA, and Lambert JA (2012). *Collective Learning for Transformational Change: A Guide to Collaborative Action.* Earthscan/Routledge, London.

Crawford A, Nettle R, Paine M, and Kabore C (2007). Farms and learning partnerships in farming systems projects: A response to the challenges of complexity in agricultural innovation. *Journal of Agricultural Education and Extension* **13**, 191–207.

Elix J, Lambert J, and Williams M (2005). *Working Together – Collaborate Now: Avoid the Rush.* Training manual prepared for Australian Government National Conservation Incentives Forum, La Trobe University, Bundoora, 6–8 July 2005.

Fischer J and Lindenmayer DB (2007). Landscape modification and habitat fragmentation: a synthesis. *Global Ecology and Biogeography* **16**, 265–280.

Fischer J, Lindenmayer DB, and Manning AD (2006). Biodiversity, ecosystem function, and resilience: ten guiding principles for commodity production landscapes. *Frontiers in Ecology and the Environment* **4**, 80–86.

Fitzsimons JA and Wescott G (2007). Perceptions and attitudes of land managers in multi-tenure reserve networks and the implications for conservation. *Journal of Environmental Management* **84**, 38–48.

Keirsey D (1987). *Portraits of Temperament.* Prometheus Nemesis Book Co, Del Mar, CA.

Kolb DA (1984). *Experiential Learning: Experience as the Source of Learning and Development.* Prentice-Hall, New Jersey.

Lindenmayer D, Hobbs RJ, Montague-Drake R, Alexandra J, Bennett A, Burgman M, Cale P, Calhoun A, Cramer V, Cullen P, Driscoll D, Fahrig L, Fischer J, Franklin J, Haila Y, Hunter M, Gibbons P, Lake S, Luck G, MacGregor C, McIntyre S, Mac Nally R, Manning A, Miller J, Mooney H, Noss R, Possingham H, Saunders D, Schmiegelow F, Scott M, Simberloff D, Sisk T, Tabor G, Walker B, Wiens J, Woinarski J, and Zavaleta E (2008). A checklist for ecological management of landscapes for conservation. *Ecology Letters* **11**, 78–91.

Lovett S, Lambert J, Williams J, and Price P (2008). *Restoring Landscapes with Confidence – an Evaluation of the Science, its Methods and Application. Final Report.* Land & Water Australia, Canberra.

Rittel H and Webber M (1973). Dilemmas in the general theory of planning. *Policy Sciences* **4**, 155–169.

Shrapnel M and Davie J (2001). The influence of personality in determining farmer responsiveness to risk. *Journal of Agricultural Education and Extension* **7**, 167–178.

Strachan R (2011). Myer Briggs Type Indicator preferences by industry and implications for extension. In: *Shaping Change: Natural Resource Management, Agriculture and the Role of Extension.* (Eds J Jennings, R Packham and D Woodside) pp. 171–174. Australasia-Pacific Extension Network, Australia.

Vanclay F (2011). Social principles for agricultural extension for the promotion of natural resource management. In: *Shaping Change: Natural Resource Management, Agriculture and the Role of Extension.* (Eds J Jennings, R Packham and D Woodside) pp. 62–71. Australasia-Pacific Extension Network, Australia.

Whitten S, Freudenberger D, Wyborn C, Doerr V, Doerr E, and Langston A (2011). 'A compendium of existing and planned Australian wildlife corridor projects and initiatives, and case study analysis of operational experience'. A report to the Australian Government Department of Sustainability, Environment, Water, Population and Communities. CSIRO Ecosystem Sciences, Canberra.

Williams J (2008). 'Restoring landscapes with confidence: state of knowledge discussion paper on landscape restoration science in Australia'. Report prepared for Land & Water Australia, Canberra.

Worboys GL and Pulsford I (2011). 'Connectivity conservation in Australian landscapes'. A report to the Australian Government Department of Sustainability, Environment, Water, Population and Communities on behalf of the State of the Environment 2011 Committee, Canberra.

Wyborn C (2011). Landscape scale ecological connectivity: Australian survey and rehearsals. *Pacific Conservation Biology* **17**, 121–131.

25

Socio-economic issues in establishing and successful operation of landscape-scale connectivity conservation initiatives

Stuart M. Whitten, Hannah Parris, Veronica A.J. Doerr and Erik D. Doerr

Introduction

The attraction of connectivity conservation initiatives, regardless of whether they are led by governmental or non-governmental organisations, is the assumption that a relatively small spatially or cost-effectively targeted investment will yield a large pay-off for conservation and potentially for ecosystem services, with consequent pay-offs to individual and community wellbeing. In this chapter we discuss this assumption, taking into account potential biophysical thresholds. We describe the potential for secondary pay-offs in ecosystem service flows and individual and community benefits, as well as the indirect economic benefits of connectivity conservation initiatives. The critical role of inclusive planning, implementation and management processes in delivering these outcomes is discussed. Finally, we assess the potential for new or additional wider ecosystem benefits to be protected, created, or enhanced.

Delivering these benefits of connectivity conservation will require policy mechanisms able to better deliver the full set of activities necessary to support connectivity conservation initiatives (e.g. protecting and enlarging core areas, managing and restoring multiple types of linkages, buffering and matrix management; see Mackey *et al.* 2010). While some mechanisms are already available or can be modified from existing approaches, we identify examples of new approaches that will be required.

The nature of the costs and benefits from connectivity conservation initiatives

Social and economic costs and benefits

Given the competing demands on public and private conservation budgets, the amount of resources (financial, social, physical and natural capitals) will likely always be relatively scarce compared to the potential opportunities for investment in connectivity conservation-related activities (including land use change). Resource availability influences the range and extent of additional ecological and socio-economic outcomes (over and above the *status quo* outcomes) that one may expect from a connectivity conservation initiative – which may be more limited than has been argued to date. Understanding the nature of the costs and benefits available from investment in connectivity conservation helps to focus the allocation of these scarce resources towards their most effective use.

Given the likelihood of a relative scarcity of resources compared to the investment opportunity available, we suggest that the following parameters are likely to characterise future major landscape-scale conservation initiatives in Australia.

- Management of structural connectivity (i.e. establishing new, or improving existing physical links in the landscape) to deliver functional connectivity objectives (i.e. facilitating landscape-scale species movement or ecological processes) will involve a range of land uses, many of which will continue to support a mix of agricultural and conservation outcomes;
- Where possible, targeting protection and maintenance of existing assets over creation of new assets could keep the implementation costs to improve landscape-scale outcomes relatively modest;
- The overall opportunity costs would be relatively small (as measured by the value of lost agricultural production), because while there may be significant land use change for some (relatively few) individual landholders (reflecting change at the 'paddock' and 'property' scale in targeted high priority areas), change is unlikely at the scale of a regional economy (e.g. within a catchment area).
- The costs in the targeted high priority areas where land use is changed are likely to be high at the point of conversion and involve significant ongoing management costs in targeted areas. Actual costs by site will vary, for example revegetation/restoration activities will be more costly than reducing agricultural intensity;
- Connectivity investments may lead to significant loss of option values (that is, loss of opportunity for future economic development) particularly in peri-urban or more intensive agricultural and forestry settings. However, this loss in option value may be relatively small, given that existing legal restrictions, such as vegetation clearance laws, may have restricted economic development regardless of connectivity actions; and
- There are likely to be significant monetary and non-monetary transaction costs (i.e. the costs of establishing and maintaining connectivity institutions) associated with establishing and investing in connectivity conservation initiatives, although a focus on policy and implementation design can greatly reduce these costs.

While investments in connectivity conservation may appear substantial (in the tens to hundreds of millions per year) the relative scale of funding flowing into communities for connectivity projects is generally small compared to the size of existing regional economies (in the billions per year). For example, URS (2010) estimate the costs of a major landscape restoration project at less than 0.5% of regional agricultural income. However, the targeted nature of investment is likely to provide an important contribution to the local and regional environmental sector by providing funds for land restoration or conservation activities. More significant are the likely benefits generated through social capital within the local communities. These are the value of networks and relationships among individuals which are key elements in building and strengthening community resilience and community adaptation capacity (Pahl-Wostl 2006). Enhanced social capital generates flow-on economic benefits as networks and the resultant communication and information sharing have been shown to facilitate relatively low cost transfer of economic innovation (Pannell *et al.* 2006).

Connectivity conservation initiatives also have the potential to take account of the distinctive socio-cultural characteristics that influence Indigenous engagement in, and historical marginalisation from, natural resource management (Hill and Williams 2009). Investment through Indigenous Protected Areas and similar arrangements where Indigenous communities conduct their own planning and management priorities for natural resource management (NRM), on their own terms, can be important in overcoming Indigenous marginalisation (Lane and Williams 2009).

Conservation benefits

The focus of connectivity conservation in Australia is on managing and recreating metapopulations through protecting, managing and buffering relatively high quality 'patches' or areas of ecological communities, as well as linking those patches so the populations they contain can function like a larger population (Doerr *et al.* 2011). Investments in linkages are focused on protecting or restoring key elements of structural connectivity in landscapes with the objective of delivering functional connectivity (movement of individuals and genes) between patches (Doerr *et al.* 2010). Given the relatively scarce resources for connectivity investment, the success of this approach is dependent on the pay-off from spatially coordinating these different types of investments. That is, we assume a set of spatially coordinated investments will deliver a proportionately larger conservation outcome than a larger investment which does not coordinate activities. This assumption is likely to be valid in many settings, but in others a threshold investment may be required in order to sufficiently functionally connect landscapes to impact on conservation outcomes (see Figure 25.1). For example, in some moderately intact or variegated landscapes, there are

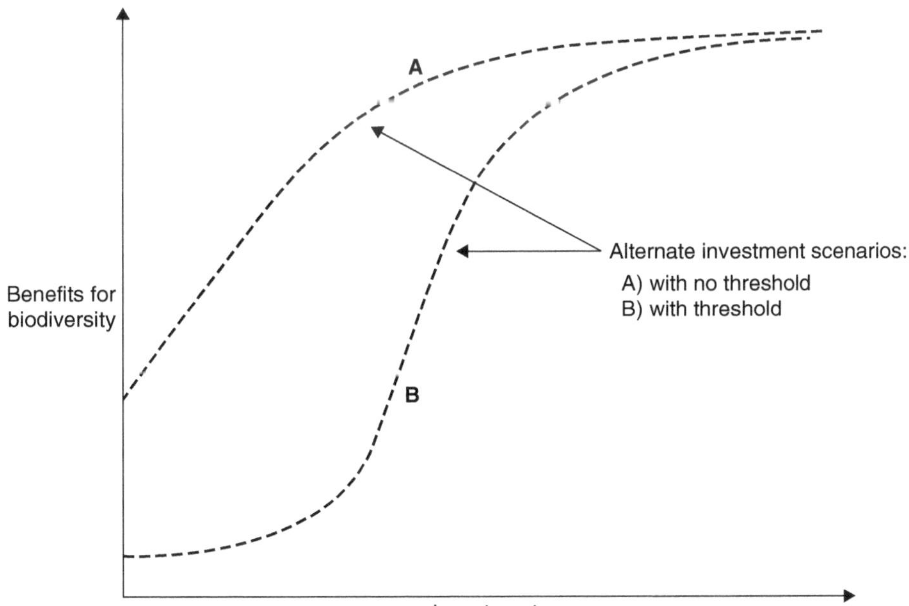

Figure 25.1 Potential connectivity conservation investment biodiversity benefit relationships. Curve A may apply when the scale of management is small and/or the landscape is still relatively intact. Curve B may apply when the desired scale of investment is large and/or the landscape has been more heavily modified and fragmented.

already moderately large patches so a very small investment in just a few connections linking them could make their constituent populations viable. In contrast, in more relictual landscapes where only small patches remain, a threshold level of investment would be needed to restore the many connections necessary to make the constituent populations viable in the long term. We also note that the principle of declining marginal benefit (declining pay-offs from additional investment) would apply at some point in both cases as illustrated in Figure 25.1. It would be prudent in effectively targeting investment to identify where and to what extent thresholds and declining marginal benefits are present.

Potential for wider ecosystem service benefits

Some researchers have reported on the potential ecosystem service benefits (monetary and non-monetary) to the community from connectivity conservation initiatives (e.g. Pulsford *et al.* 2010). The scale of these benefits is dependent on the extent to which ecosystem service benefits are generated from connectivity conservation initiatives *and* the extent to which these services deliver conservation and socio-economic benefits to communities over and above more conventional conservation strategies. The difficulty is in assessing the relationship between investment levels and changes to ecosystem service outcomes.[i] Benefits are likely to fall into four general classes (loosely following Millennium Ecosystem Assessment 2005).

1 Benefits to primary industries (food and fibre production, water-related services, pollination and local climate regulation) within the initiative boundary. These benefits directly generate economic return to farmers and businesses (Zhang *et al.* 2007). There may also be costs if land use change leads to increased fire, weed or pest animal threats (see Mooney and Hobbs 2000 on weeds for example) or a reduction in the critical mass of industry in the region.
2 Use benefits (water-related services, recreation and aesthetic, and education in particular) both within and beyond the initiative boundary. These benefits support healthy and vibrant communities and human wellbeing.
3 A range of non-use benefits (sense of place, Indigenous and non-indigenous cultural heritage, and spiritual values) which contribute to a wider cultural richness and human wellbeing.
4 Strengthening resilience of ecosystems by supporting retention of functional redundancies (or 'option value'), rather than loss of biodiversity assets *per se*.

While connectivity conservation investments will help secure initiatives that already generate substantive ecosystem service benefits, the potential for new or additional benefits from investment is less certain. Evidence suggests that where spatial processes are important there will also be response thresholds in individual ecosystem services that are similar to curve B in Figure 25.1, but potentially at different scales (see for example Samways *et al.* 2010). Where responses exhibit threshold effects, three conditions would need to be met for investments in connectivity conservation initiatives to deliver substantive ecosystem service benefits beyond those already delivered: (i) benefits are generated from the same activities; (ii) at the same locations in the landscape; and (iii) the scale of investment is sufficient to deliver threshold benefits. Therefore, while wider ecosystem benefits *may* be protected, created, or enhanced through investment in connectivity conservation; it is unlikely there would be a substantive enhancement to ecosystem services or new benefits to income-producing activities, such as agriculture, tourism or to urban communities. As we argue for biodiversity benefits, the case for both protection and enhancement of ecosystem co-benefits should be evidence based.

Benefit and costs of operating at a landscape scale

Connectivity conservation initiatives, as a vehicle that operates across a range of existing institutional scales at which there are already existing conservation organisations, should be expected to deliver benefits over and above the capacity or remit of existing activities. Conversely, connectivity conservation initiatives may also introduce new types of risks into the Australian NRM sector, which will need to be managed, and ideally minimised, through careful institutional design and policy selection. These benefits and costs would typically be assessed via changes to delivery costs or effectiveness when compared against alternative approaches.

Potential delivery benefits from a connectivity conservation focus

Delivery benefits centre on the conceptualisation of initiatives as simultaneously being a 'broad vision' for connectivity conservation objectives, a management framework, and a network necessary for building strategic relationships between potential project partners (Chester 2006).

As a 'broad vision', prior experience identifies the importance of creating – and communicating – an ambitious, even romantic, goal for biodiversity conservation (Worboys *et al.* 2010). The sense of tangibly contributing to regional and continental-scale connectivity motivates individuals, groups and communities to participate as donors and volunteers. This motivation to participate acts to reduce the transaction costs associated with 'selling' any particular on-ground activity related to the initiative (because individuals are already committed to the vision of what that activity is trying to achieve). It can also help in managing friction that may arise should communities or individuals be required to curtail their income-generating, recreational or other social activities in order to implement initiative activities. However, there remains a tension between the scales at which institutions operate and the necessary scale of physical intervention for on-ground works (see discussion below).

Connectivity conservation offers a management framework within which to coordinate biodiversity conservation activities across scales within the multiple spatial, temporal, jurisdictional, institutional, community, knowledge and ownership domains. Connectivity conservation initiatives can assist in overcoming institutional fragmentation, knowledge uncertainty, and contested interests across a landscape (Pert *et al.* 2010); often through the use of participatory tools that foster integration of science, local and Indigenous knowledge, and recognise the diversity of human values associated with the environment (Hill *et al.* 2010).

Initiatives achieve this primarily via an emergent 'boundary' organisation, commonly employing a partnership model (Whitten *et al.* 2011). Boundary organisations are embedded within a network of diverse stakeholders, each of which have different levels of involvement, interest and participation in, or contribution to, the initiative (Hill *et al.* 2010). The boundary organisation concept allows aggregation of needs across on-ground conservation groups and communities to levels that are able to engage with large, cross-scale institutional players who may otherwise not have the capacity to form relationships with small-scale community organisations. Thus, the partnership model tends to be more inclusive of community groups, non-government organisations and other interested stakeholders in high level internal decision making. The partnership model is likely to generate a more conducive atmosphere, supporting adaptive capacity, flexible delivery and potential for organisational and operational transformation (Gooch and Warburton 2009).

Potential risks and costs

Connectivity conservation initiatives may be exposed to several costs and risks that are not present or as prevalent at existing Australian NRM institutional scales of operation. These result from organisational or institutional risks and uncertainty in science or implementation.

Organisational or institutional risks arise from:

1. *Proliferation of potentially overlapping organisations* adding to complexity, operational costs, and the costs (monetary and non-monetary) of community participants (preparation, meetings, negotiations, etc.). If the participant pool is low (not an unreasonable assumption in sparsely populated areas) then volunteer burnout or high volunteer turnover increase costs in developing and maintaining social capital and organisational knowledge.
2. *Barriers to implementation* posed by institutional constraints, financial limitations or other factors, that may prevent key linkages and access to the scale benefits of connectivity conservation initiatives.
3. *Fragmented implementation and poor delivery.* Even with the best biophysical and spatial planning, the translation of connectivity conservation concepts into on-ground works is often diluted from the original plan due to a misalignment between opportunities and objectives (Whitten *et al.* 2011). Misalignment may arise from:
   - Inadequate spatial coordination of activities due to either expense and time in targeting specific areas and specific landholders, or continuing gaps in activities due to reliance on *ad hoc*, opportunistic funding; and
   - Diverse initiative partners whose priorities may not cover, or who may be unable to deliver, a full suite of connectivity conservation activities.
4. *The risk of conflict.* Landscape-scale initiatives could undermine conservation imperatives by making explicit the differences in values and perspectives on ecological assets in a landscape (Prager 2010). For example, differences may arise between conservationists and farmers on whether to focus connectivity outcomes on promoting biodiversity or ecosystems services that explicitly support production.

Uncertainties in science or implementation derive from either the underpinning conservation science or its application in connectivity conservation initiatives. Issues include:

1. *The science of landscape ecology* is relatively new and relevant questions for management have not necessarily been thoroughly addressed. For example, it is still unclear which structural elements between 'patches' (e.g. corridors, scattered trees, etc.) truly provide functional connectivity (but see Doerr *et al.* 2010, 2011, and Gilbert-Norton *et al.* 2010 for emerging general principles). Parris *et al.* (2011) provide an indication of the level of scientific research available for differing connectivity elements and the risks associated with incomplete or incorrectly applied research leading to poor outcomes.
2. *The risk of perverse or unanticipated outcomes from enhancing connectivity* such as enhancing invasive species mobility, increased feral or nuisance species populations, fire propagation or unanticipated interactions between species leading to undesirable biodiversity outcomes. None of these issues has received much research attention so the likelihood and management of these risks is often unclear.
3. *Limited ability to apply the often new and more complex landscape science including limited skills or resources to access and translate scientific literature into practical and pragmatic management approaches.*

4 *Scale/investment mismatches* from thresholds meaning little or no biodiversity response from small or incorrectly targeted investments. Long time lags to biodiversity benefits but immediate costs also complicate otherwise positive social feedback loops in supporting connectivity conservation.

Emerging investment and institutional support needs at the landscape scale

Connectivity conservation envisages a range of activities, some of which are familiar components of existing NRM schemes, and others that are entirely new, or that may be implemented in different ways and locations or supported by different investment tools. A key need is the development of landscape-scale metrics that are able to describe the relative benefits of different activities specifically to landscape-scale rather than just site-scale outcomes. Without such metrics it is difficult to identify which investment tools or institutions are appropriate, or to evaluate the performance of such investments at scale.

The investment tools and other institutional support underpinning these activities will, in some cases also require attention at a larger scale than the NRM region or community (Cash *et al.* 2006). For example, some land tenure institutions (legal frameworks for determining ownership and use rights for land) are intended to promote development and food security and may inadvertently work against the conservation objectives of the initiative.

New supporting institutions

While there are many existing conservation tools that can support connectivity conservation, it is likely that additional or modified legal tools will be needed to support conservation connectivity. Three examples of supporting institutional structures that are likely to be of particular importance at scale are:

1 *Extension of formal protection to linkage areas and not just core 'patches' via legislative recognition* acts as a signal of the importance of connectivity, a first legal option for protection, and can aid in access to some funding.
2 *More flexible covenanting or similar legal protection approaches* covering buffers and agricultural matrix attributes. These would assist in supporting long-term investment in assets such as paddock trees, and preventing threats such as fertiliser or chemical use near environmental assets.
3 *Greater use of connectivity biodiversity protection zones* assist in preventing undesirable land use change and may be particularly useful in conjunction with offsets or other mechanisms in signalling preferable locations for beneficial investment.

New and emerging investment tools

Connectivity conservation has two major differences to existing investments in biodiversity and landscape sustainability. First, investments need to be coordinated carefully from paddock to regional and landscape-scale due to the interaction between activities at different scales. Second, new investors need to be identified to deliver the step change in investment required. While many of the necessary investment tools have yet to emerge some preliminary examples include:

- *Landscape approaches to conservation tenders*: designing tenders to encourage and facilitate cooperation between landholders while maintaining cost-minimising designs can support coordinated landscape investment (Reeson *et al.* 2011). Assessing the contribution of each tender (or a set of tenders) to a defined landscape objective allows the set that most cost-effectively delivers towards the landscape outcome to be selected.

- *New tools for de-intensification and supporting management* in buffering and matrix management areas will be needed to support connectivity conservation initiatives. An important difference from existing schemes will be a focus on managing a mix of production and conservation activities within a land use matrix to reduce threats to connectivity, and to appropriately address the relatively large opportunity costs of reduced production (as opposed to restoration costs) for landholders in targeted high priority areas.
- *Investigation of complementary sources of income* may be particularly important to sustaining management in matrix management areas. Several initiatives are exploring potential activities including Sandalwood plantations (Gondwana Link), and bush foods in northern Australia.
- *The emergence of environmental markets, including carbon bio-sequestration and biodiversity offsets.* Such markets may support both complementary matrix management and restoration activities (see Whitten *et al.* 2011 for suggestions). Other related market activities such as revolving funds will continue to play an expanded role in connectivity conservation initiatives.
- A related area of interest is in *environmental services markets* including restoration and fire management activities for the carbon market, particularly by Indigenous communities (e.g. Fitzsimons *et al.* 2012).
- *Philanthropic contributions* offer one source of funds to expand investment: many connectivity conservation initiatives have received significant philanthropic contributions that have supported their initial development, planning and ongoing investment.

Conclusions

Connectivity conservation is predicated on the assumption that targeted investment in spatially coordinating patch protection and the restoration/management of functional links between patches deliver major biodiversity benefits. The same assumption suggests opportunity costs are likely to be low. However, restoration, ongoing management and transaction costs of implementation may be significant. Initiatives are likely to deliver a range of wider benefits through social capital in the environment sector, and protection of ecosystem services. They are less likely to deliver a major increase in ecosystem services. They also face several risks in delivering on their potential due to the need to build investment at scale (particularly if a threshold investment is needed to achieve benefits), the organisational challenge of landscape-scale conservation, scientific and implementation uncertainties, and misalignment of goals and opportunities. To deliver on their potential, new institutional supporting structures and new investment tools will be required, some of which are already emerging.

Acknowledgements

This chapter is largely drawn from: Parris *et al.* (2011). 'An overview of key socio-economic factors, principles and guidelines in wildlife corridor planning and implementation'.

A report for the Australian Government Department of Sustainability, Environment, Water, Population and Communities, CSIRO Ecosystem Sciences.

Biographies

Stuart Whitten is a team leader in CSIRO's Social and Economic Sciences Program and is also a stream leader in their Biodiversity Theme: Building Resilient Australian Biodiversity Assets. Dr Whitten is a specialist in market-based instruments such as conservation 'auctions', including the consideration of mechanisms to deliver coordinated outcomes across landscapes.

Hannah Parris is with CSIRO's Ecosystem Sciences Division. She is an expert in institutional analysis and design for natural resource management, and the application of economic policies to sustainability issues. She has recently completed her PhD research on institutional design and analysis of large-scale natural resource commons.

Veronica Doerr is a Senior Research Scientist with CSIRO's Ecosystem Sciences Division. Dr Doerr researches ways to design and manage landscapes so they might best support biodiversity alongside other land uses. She makes use of genetic and behavioural data as well as process-based modelling approaches to reveal how the behaviour and ultimately persistence of organisms depend on the characteristics of the landscapes they live in.

Erik Doerr is a Research Scientist with CSIRO's Sustainable Ecosystems Division. Dr Doerr researches the problem of habitat fragmentation. He uses both genetic and behavioural data to identify the precise landscape elements that animals use during dispersal, and applies that information to help design more effective habitat restoration plans.

Endnote

i. In practice investors should make explicit assumptions, measure the responses that do occur, and modify management accordingly.

References

Cash DW, Adger WN, Berkes F, Garden P, Lebel L, Olsson L, Pritchard L, and Young O (2006). Scale and cross-scale dynamics: governance and information in a multilevel world. *Ecology and Society* **11**(2), art. 8. http://www.ecologyandsociety.org/vol11/iss2/art8/

Chester C (2006). *Conservation Across Borders: Biodiversity in an Interdependent World*. Island Press, Washington, DC.

Doerr VAJ, Doerr ED, and Davies MJ (2010). *Systematic Review #44: Does structural connectivity facilitate dispersal of native species in Australia's fragmented terrestrial landscapes?* Collaboration for Environmental Evidence, Bangor.

Doerr VAJ, Barrett T, and Doerr ED (2011). Connectivity, dispersal behaviour and conservation under climate change: a response to Hodgson *et al. Journal of Applied Ecology* **48**, 143–147.

Fitzsimons J, Russell-Smith J, James G, Vigilante T, Lipsett-Moore G, Morrison J, and Looker M (2012). Insights into the biodiversity and social benchmarking components of the Northern Australian fire management and carbon abatement programmes. *Ecological Management & Restoration* **13**, 51–57.

Gilbert-Norton L, Wilson R, Stevens JR, and Beard KH (2010). A meta-analytic review of corridor effectiveness. *Conservation Biology* **24**, 660–668.

Gooch M and Warburton J (2009). Building and managing resilience in community-based NRM groups: An Australian case study. *Society & Natural Resources* **22**, 158–171.

Hill R and Williams LJ (2009). Indigenous natural resource management: overcoming marginalisation produced in Australia's current NRM model. In: *Contested Country: Local and Regional Environmental Management in Australia.* (Eds MB Lane, C Robinson and BM Taylor) pp. 161–178. CSIRO Publishing, Melbourne.

Hill R, Williams KJ, Pert PL, Robinson CJ, Dale AP, Westcott DA, Grace RA, and O'Malley T (2010). Adaptive community-based biodiversity conservation in Australia's tropical rainforests. *Environmental Conservation* **37**, 73–82.

Lane MB and Williams LJ (2009). The Natural Heritage Trust and Indigenous lands: the trials and tribulations of 'new technologies of governance'. *Australian Geographer* **40**, 85–107.

Mackey BG, Watson J, and Worboys GL (2010). 'Connectivity conservation and the Great Eastern Ranges corridor'. An independent report to the Interstate Agency Working Group (Alps to Atherton Connectivity Conservation Working Group) convened under the Environment Heritage and Protection Council/Natural Resource Management Ministerial Council, ANU Enterprises Pty Ltd.

Millennium Ecosystem Assessment (2005). *Ecosystems and Human Well-being: Synthesis.* Island Press, Washington, DC.

Mooney HA and Hobbs RJ (Eds) (2000). *Invasive Species in a Changing World.* Island Press, Washington, DC.

Pahl-Wostl C (2006). The importance of social learning in restoring the multifunctionality of rivers and floodplains. *Ecology and Society* **11**, art. 10. http://www.ecologyandsociety.org/vol11/iss1/art10/

Pannell DJ, Marshall GR, Barr N, Curtis A, Vanclay F, and Wilkinson R (2006). Understanding and promoting adoption of conservation practices by rural landholders. *Australian Journal of Experimental Agriculture* **46**, 1407–1424.

Parris H, Whitten SM, Wyborn C, Hill R, and Freudenberger D (2011). 'An overview of key socio-economic factors, principles and guidelines in wildlife "corridor" planning and implementation'. Report to the Department of Sustainability, Environment, Water, Population and Communities. CSIRO, Canberra.

Pert PL, Hill R, Williams KJ, Harding EK, O'Malley T, Grace RA, Dale AP, Bohnet I, and Butler JRLA (2010). Scenarios for community-based approaches to biodiversity conservation: a case study from the Wet Tropics, Queensland, Australia. *Australian Geographer* **41**, 285–306.

Prager K (2010). Local and regional partnerships in natural resource management: the challenge of bridging institutional levels. *Environmental Management* **46**, 711–724.

Pulsford I, Worboys GL, and Howling G (2010). Australian Alps to Atherton connectivity conservation corridor. In: *Connectivity Conservation Management: A Global Guide.* (Eds GL Worboys, WL Francis and M Lockwood) pp. 97–105. Earthscan, Washington, DC.

Reeson AF, Rodriguez LC, Whitten SM, Williams K, Nolles K, Windle J, and Rolfe J (2011). Adapting auctions for the provision of ecosystem services at the landscape scale. *Ecological Economics* **70**, 1621–1627.

Samways M, Bazelet C, and Pryke J (2010). Provision of ecosystem services by large scale corridors and ecological networks. *Biodiversity and Conservation* **19**, 2949–2962.

URS (2010). A natural investment project: a regional development case study. A report prepared for Victoria Naturally Alliance. URS, Melbourne. Available: http://victorianaturally.org.au/documents/File/rpt_naturalinvestmentproject.pdf [Accessed 1 July 2012].
Whitten SM, Freudenberger D, Wyborn C, Doerr V, and Doerr E (2011). 'A compendium of existing and planned Australian wildlife corridor projects and initiatives, and case study analysis of operational experience'. Report to the Department of Sustainability, Environment, Water, Population and Communities. CSIRO, Canberra.
Worboys G, Francis W, and Lockwood M (Eds) (2010). *Connectivity Conservation Management: A Global Guide.* Earthscan. Washington, DC.
Zhang W, Ricketts TH, Kremen C, Carney K, and Swinton SM (2007). Ecosystem services and dis-services to agriculture. *Ecological Economics* **64**, 253–260.

26

Collaboration across scales: the governance challenges of linking landscapes

Carina Wyborn

Introduction

Designing collaborative governance arrangements remains a key challenge of any attempt to 'link Australian landscapes'. This task is complicated by the desire to initiate landscape-scale change through mobilising 'grassroots' efforts. While a great diversity of governance arrangements exist, all must address the challenge of aligning diverse players to coordinate land management across local to regional to supra-regional scales. The range of actors involved in connectivity conservation complicates governance – public, private, civil society and individual landholders may be involved in connectivity governance. Thus governance arrangements should enable individuals or organisations to operate according to their own motivations, mandates, and in the case of organisational actors, their own governance. Drawing on theory and practice in emerging Australian initiatives, this chapter (i) outlines the challenges posed by connectivity conservation governance; (ii) provides an overview of governance attributes to highlight the diverse mechanisms available to address the practical challenges inherent to connectivity governance; and (iii) concludes with some general guidelines or principles to consider when establishing a connectivity initiative. Connectivity governance should be viewed as an experiment to be adapted for the particular social and ecological landscape of an initiative. Questions about governance will remain ever present for practitioners of connectivity conservation; consequently, governance mechanisms must be flexible and adaptive to changes in knowledge and context.

'Governance' is widely used in conservation and natural resource management (NRM), but is a slippery term encompassing many different notions of collective decision making. Governance refers to the structures and processes used by groups to organise and achieve their goals: it is the space where actors negotiate decisions and determine a path of action (Lebel *et al.* 2006; Lemos and Agrawal 2006). The idea of governance applies to the internal mechanisms of a single entity, public or private, but can also relate to interactions, collaborations or networks among actors. Governance includes both formal rules – laws, regulations, negotiation, mediation, conflict resolution, elections, public consultations and informal interactions – and norms and principles that shape decision making.

While specifics should be negotiated in context, governance tasks can broadly be grouped into four areas: (i) maintain internal and external communication; (ii) set priorities for strategic goals; (iii) obtain financial resources; and (iv) provide general mechanisms for accountability (Mitchell and Shortell 2000).

A wealth of literature and practice outlines the principles of good environmental governance (see, for example, Folke *et al.* 2005; Lebel *et al.* 2006; Lemos and Agrawal 2006; de Loe *et al.* 2009; Lockwood *et al.* 2009). These authors cite trust, integrity, inclusivity, transparency, accountability, flexibility, reciprocity and communication as foundations of good governance and collaboration. This chapter moves beyond a general discussion of environmental governance to focus on the specific aspects and challenges associated with connectivity governance. In particular, it focuses on the challenges presented by two key dimensions that define the connectivity conservation approach: inclusion of multiple stakeholders in governance and the alignment of efforts across local to large landscape scales.

Nature of the governance challenge

As Ryan *et al.* (2010) identify, the central governance challenge faced in NRM and conservation in Australia is the coordination of many actions, undertaken by diverse players, across multiple scales. The emerging initiatives formed under the banner of 'connectivity conservation' represent one approach to achieve this alignment and coordination. To realise their goals of enhancing ecological connectivity and landscape function, these initiatives promote landscape-scale behaviour change through coordination and collaboration of diverse actors. These efforts require internal alignment with the characteristics of the partners and external alignment with the context and needs of specific landscapes and communities. While the approach requires landscape-scale behaviour change, a landscape is the sum of many different parcels of land and widespread behaviour change requires a critical mass of individuals or organisations to alter management at the property scale. The geographic scope of these initiatives presents practical challenges in engaging with individual property managers. Connectivity initiatives therefore tend to focus on aligning the efforts of pre-existing groups or agencies, assuming that individuals will be reached through these partners. In order for collaborations to succeed, diverse expectations and motivations of players need to be integrated in a coordinated fashion that enables each group to work towards a common vision in line with their core values and operations (Wondolleck and Yaffee 2000; see also Chapter 24). Connectivity governance thus requires a structure of 'meta governance' whereby the alignment and coordination does not undermine the individual and autonomous efforts of their constituents. For organisational partners this involves respecting their individual governance and in the case of individual landholders, connectivity governance in Australia is yet to include regulatory measures or requirements to force behaviour change or alter property rights regimes.

The complexity of the governance challenge is confounded by the social and ecological diversity across large landscapes. This applies both to the landscape within a particular initiative but also presents challenges for understanding and experimenting with governance arrangements across initiatives. What 'works' in one place is unlikely to be easily transferred elsewhere. Moreover, governance is an evolving process: a workable arrangement for the start-up phase of an initiative is unlikely to remain viable as collaborations solidify and start to attract large sums of money. Connectivity initiatives are guided by long-term (50–100 year) visions for landscape change. Operating across these timeframes requires flexible and adaptive governance arrangements in order to remain relevant in the face of changes to social demographics, political and economic drivers, scientific knowledge and ecological processes. The evolution of governance arrangements has already been seen in several initiatives (e.g. Fitzsimons and Wescott 2008; other chapters in this volume).

Attributes of connectivity governance

The following overview of governance attributes illustrates the multitude of ways in which different initiatives have approached the challenges of multiple stakeholders and cross-scale governance. With a focus on aspects of governance specific to connectivity conservation,[i] the following five dimensions are considered: (i) partnership composition; (ii) the nature of collaboration; (iii) approach to up-scaling; (iv) delivery mechanisms; and (v) funding (see Table 26.1). Given the diversity and complexity of existing arrangements it is not possible to group the various initiatives into ideal types or categories. The intention here is not to prescribe particular approaches or arrangements, rather it is to illustrate the different governance attributes to be considered and outline different mechanisms available to those seeking to understand or establish connectivity governance. This discussion complements the work of Lockwood (2010), which illustrates broader institutional characteristics of connectivity governance.

Partnership composition

The heterogeneity of organisations is the key characteristic of partnership composition. Connectivity initiatives in Australia encompass a broad range of partners across the public-private spectrum including local stewardship groups, non-government organisations (NGOs), various government agencies (NRM bodies, national park agencies, state government environment departments), philanthropic groups and corporate entities (Wyborn 2011). The complexity of governance increases with a greater diversity of sectors and number of players present in a collaboration. Within that, the presence or absence of government agencies fundamentally shapes the nature of an initiative as the financial and institutional power of government can challenge initiatives that are rhetorically grassroots. Although most Australian initiatives include some form of partnership with government, the role varies. NGOs are the driving players in many initiatives (e.g. Gondwana Link (Chapter 3), Habitat 141° (Chapter 8), and the Tasmanian Midlandscapes (Chapter 9), while

Table 26.1. Connectivity conservation governance attributes

Governance attribute	Determinants
Partnership composition	• Heterogeneity and number of partners • Catalysing organisation
Nature of collaboration	• Formality of structures • Collaborative processes
Approach to up-scaling	• Existence of hierarchy • Relationship across scales
Delivery mechanisms	• Direct or indirect influence on land use change • Staffing and hosting arrangements
Funding	• Sources • Management of funds

other efforts were initiated by government agencies that have maintained a significant presence throughout the evolution of the initiatives (e.g. NatureLinks (Chapter 6) and the Great Eastern Ranges Initiative (see Chapter 13)) (Whitten *et al.* 2011). The presence of government agencies can provide vital access to financial, informational and logistical resources. However, these must be balanced with the (perceived or real) potential for government dominance (see also Chapter 21).

Nature of collaboration

Collaboration comes in many forms, is used to meet various ends and can be underpinned by a diversity of structures. Sabatier *et al.* (2005) distinguish between three types of collaboration: collaborative engagement, partnerships, and superagencies. Engagement tends to be a short-term process focused on determining different perspectives on a particular issue, partnerships are focused on building a longer-term structure of collaboration while maintaining the autonomy of the players involved, while superagencies are created as fully amalgamated institutions operating on collaborative principles to determine and administer actions. With a greater range of government agencies focused on land management – the Forest Service, Fish and Wildlife Service, the National Parks Service and the Bureau of Land Management – the United States has created several superagencies to support and coordinate cooperative management between these agencies to overcome jurisdictional and institutional fragmentation (see Heikkila and Gerlak 2005).

The dominant approach in Australia is partnerships and we are unlikely to see an emergence of superagencies here. However, collaborative engagement processes could perform an important function in connectivity governance. Habitat 141° conducted an engagement process to gather comments and feedback from prospective partner groups on a proposed governance model. This followed a typical process of circulating a draft for comment before a partnership meeting where issues were negotiated. Given the practical challenges of directly consulting with individual land managers across a large landscape, engagement processes could provide an important mechanism for providing input into areas that require consultation with those affected by an initiative (i.e. strategic planning processes or proposed changes to future governance). The Great Eastern Ranges Initiative has adopted a two-tiered approach to governance, with a lead partners group overseeing the whole landscape and six regional partnerships across the initiative. The lead partners group comprises NGO leaders from the regional partnerships. The governance of the regional partnerships was developed through community forums to form local governance structures that were tailored to local requirements (Chapters 14 and 15).

The formality of underlying organisational structures often changes as the project evolves. At the outset, structures tend to be informal – working groups or steering committees – with partners operating under a memorandum of understanding (MOU). It is easier for partners to enter into or withdraw from an MOU without a legally binding structure, enabling greater flexibility for partners. Without a formally recognised legal structure in place, receiving funds is difficult – an issue that can drive collaborations towards formal structures, such as an Incorporated Association (Inc.) or a Company Limited by Guarantee (CLG). Regardless of the structure, specifics should be tailored to the particular purpose and context, and legal advice should be sought in early discussions about formalising governance structures. Discussion of legal structures should also include the more informal aspects of governance: creating a shared understanding of appropriate behaviour of partners, the norms and social practices of how the collaboration operates and how partners work together.

Approach to up-scaling

Successfully operating at scale requires explicit consideration of how to build on small successes to generate landscape-scale change, as well as the mechanisms to ensure cross-scale communication. Dividing a region into smaller planning units is a common strategy for operating across a large landscape. For example, the Great Eastern Ranges Initiative, which covers the alpine region of south-eastern Australia to the tropics in the north-east, currently comprises six, largely autonomous partnerships; as one conservation plan to cover this gamut of ecosystems or communities would not make social or ecological sense. This polycentric approach distributes decision-making authority throughout a disbursed network (McGinnis 1999). Not all partners or groups will be working across the entire region so these smaller planning units operate on a more logical scale enabling planning to be tailored to, and negotiated within, the specific context. Proponents of a polycentric approach claim these arrangements are more responsive to local knowledge and aspirations, facilitate learning and flexibility in the face of internal or external change (Marshall 2009; Ostrom 2010).

Distributed decision making across the landscape fulfils part of the promise of connectivity conservation; however, it does not necessarily enable coordination and communication across scales. The question of how best to do this remains largely unresolved, yet there is some indication that this function can be performed by 'cross scale brokers' – people who act as conduits between different scales of decision making and action (Ernstson *et al.* 2010). Habitat 141° has approached this through creating nine zones operating within particular regions and a council which, when established, will be comprised of representatives from the partner participating in one or more of the zones. This model operates on the principle of subsidiarity whereby decision making is devolved to the scale relevant to tackling the problem (see Marshall 2008). Essentially this means decisions relevant to one zone will be determined in that context, while cross-regional issues are negotiated at the council level. For example a fox baiting program within one region is discussed by that zone, while the strategic direction of Habitat 141° is negotiated by the council. The connection between the zones and the council currently relies on partners who participate in both; it is yet to be seen whether this enables adequate alignment from the zone scale up to the whole landscape.

Delivery mechanisms

Several mechanisms are available to influence land-use change: (i) provision of information through field days, workshops, publications, websites or brochures; (ii) volunteer coordination; (iii) field surveys and monitoring; (iv) covenants, stewardship grants and other market-based instruments; (v) purchase of land; or (vi) services for managing invasive species, weeds or fire (Whitten *et al.* 2011). The primary purpose of many connectivity initiatives is often not to deliver programs themselves, rather to facilitate collaboration between partners. This is undertaken by a small coordinating entity or secretariat with the primary function of acting as a conduit between partners, promoting collaboration and attracting resources (Parris *et al.* 2011). The Yellowstone to Yukon Conservation Initiative, a leading international example of connectivity conservation, plays a similar role through catalysing or supporting regional-scale collaboration in 12 priority areas across the landscape (Chester 2006; Locke 2010). Habitat 141° uses the metaphor of 'the gluing not doing' to describe its role in the landscape. Brokering between partners (be they individual landholders or organisations) and between partners and individual land managers is an important function of the initiatives. For example, the Kosciuszko to Coast partnership of the Great Eastern Ranges Initiative produced an information leaflet for landholders outlining the different covenanting programs or grants available through the partners. The partnership facilitator operates as a broker to landholders for the various conservation instruments and programs offered by the partners.

As the central purpose of these initiatives is to coordinate diverse partners across the landscape, the importance of coordinator cannot be overstated. In less formalised governance arrangements, staff members of the partners undertake work that supports the initiative within their existing roles. A progression towards formalisation is the employment of a coordinator hosted by one of the partner organisations. By using the existing structures of the partners, this arrangement reduces transaction costs. However, it can create a perceived lack of independence and an additional cost burden for the host organisation. Progressing towards formalised governance often accompanies a step towards independently employing staff in an independent office.

Funding

A primary focus of connectivity initiatives is attracting resources for conservation, restoration and natural resource management. Diverse funding sources are sought to provide flexibility in funding procurement to support the different aspects and operations of a connectivity initiative. Four primary funding areas are state and federal government programs, philanthropic or corporate donations, emerging environmental markets and in-kind contributions from partners and landholders. Initiatives yet to establish formal governance structures that enable the procurement of funds face challenges in establishing mechanisms for transparency and accountability in financial transactions. A common approach that reduces associated transaction costs is to utilise the administrative structures of partners – however, this faces (perceived or real) challenges of independence and ownership. Like staffing arrangements, the progression towards more formalised governance is often paired with or driven by a need to seek out and maintain independent funding. Regardless of the structure, finding and securing funding to support coordination and governance is likely to remain a significant challenge into the future.

The sometimes-divergent ideological positions of partner organisations can, at times, present issues for receiving corporate donations. Both WildEyre (part of South Australia's NatureLinks program) and Gondwana Link have dealt with circumstances where individual partners were unable to have their brand associated with corporate entities seeking to contribute to the broader initiative. Given the diversity of partners involved, it is likely to be difficult to reach some kind of consensus on issues such as this. In the above cases a strategy was adopted where the sponsored projects were branded with the partners happy to receive the funds rather than under the broader initiative. This is one of many issues highlighting a tension between the creation of an independent entity and the founding partners. This challenge is also faced in questions about the ability of different partners to undertake environmental advocacy campaigns, as it is difficult for government agencies to be associated, through the collaboration, with advocacy that seeks to change government policy. The intention of a connectivity initiative is to foster landscape-scale change. To do this, initiatives will need to create lasting collaborations capable of delivering outcomes over and above that of the individual partners. Part of this involves freeing the initiative from restrictions of the partners, enabling the whole to become more than the sum of the parts. These issues underscore the importance of flexible governance arrangements that allow partners to opt in and out of particular projects.

General principles

This overview highlights some of the complexity associated with connectivity governance. While specific details must be developed for the social and ecological context of a region, some general principles can be identified and divided into three themes: communication, collaboration and coordination. While not a guarantee, considering these elements in the establishment and evolution of governance should enable connectivity initiatives to manage the challenge of multi-stakeholder collaboration across multiple scales.

Communication

- Communicate clearly from the outset the expectations and contributions of different parties;
- Establish and maintain mechanisms for communication between different scales of decision making as well as between those planning and implementing projects or initiatives;
- Utilise various forms of media (newsletters, email, blogs, wikis, websites) to maintain communication among partners and to communicate externally, BUT remember the importance of face-to-face contact between partners.

Collaboration

- Agree on a vision to guide the initiative and, through that collaborative process, generate a shared understanding of how partners agree to work towards that vision;
- Display leadership, either through an individual or a collective, to support and empower participation in the collaborative effort;
- Consider adopting a form of collaborative leadership where there are multiple leaders across various levels undertaking leadership roles;
- Be patient, willing to compromise and appreciative of alternative perspectives;
- Recognise that building and developing relationships and trust takes time and concerted effort but pays off in the long term;
- Acknowledge and seek to ameliorate financial and institutional power imbalances between partners;
- Build collaborations between research institutions, the coordinating entity and those undertaking planning and implementation.

Coordination

- Clearly define the roles and responsibilities of different partner organisations and of the coordinating entity;
- Establish flexibility within the governance structures to enable groups to opt in or out of particular projects or initiatives;
- Maintain linkages from the local to large landscape scales through communication (face-to-face, hard copy and electronic);
- Establish links between decision-making mechanisms at different scales;
- Facilitate, where necessary, communication and coordination between different partners operating at the same scale;
- Dedicate resources and time to coordination and governance, find a core funding base that will outlive short-term project cycles;
- Demonstrate action at multiple scales and, where possible, at multiple locations across a project – seeing and participating in tangible successes will inspire further actions.

Conclusion

Connectivity conservation was inspired largely by a desire to improve ecological connectivity at large landscape scales. However, achieving this requires social and institutional connectivity across scales, tenures and contexts. The heterogeneity of partners and the spatial scale of these initiatives require new modes of governance that span multiple scales and diverse interests. Despite the complexity inherent to these arrangements, the initiatives across Australia show promise for addressing such governance challenges. While not a comprehensive breakdown of all aspects of connectivity governance, the typology outlined in this chapter highlights some of the important attributes to incorporate in the analysis or formation of connectivity governance. Considering these attributes in combination with the principles of communication, collaboration and coordination is a first step towards a continual evolution in the practice and understanding of connectivity governance.

Acknowledgements

This chapter is based on research funded by a Land and Water Australia PhD scholarship and a top-up scholarship from the CSIRO Climate Adaptation Flagship. The ideas reflect the negotiations of the Habitat 141° Governance Working Group and were developed in discussion with Stuart Whitten, Hannah Parris, David Freudenberger, Stephen Dovers, Lorrae van Kerkhoff and Mike Dunlop.

Biography

Carina Wyborn is a PhD student from the Australian National University in Canberra. Her doctoral research is examining collaborative governance in large landscape connectivity conservation projects. The research involves two case studies, one in Australia (Habitat 141°) and one in North America (Yellowstone to Yukon). She teaches in the Human Ecology program at the Fenner School of Environment and Society and holds a PhD scholarship from Land and Water Australia and a top-up scholarship from the CSIRO Climate Adaptation Flagship.

Endnote

i. For a discussion of the broader dimensions of environmental governance see Folke *et al.* (2005), Lemos and Agrawal (2006), de Loe *et al.* (2009), Lockwood *et al.* (2009) and Ryan *et al.* (2010).

References

Chester C (2006). *Conservation Across Borders: Biodiversity in an Interdependent World.* Island Press, Washington DC.

de Loe RC, Armitage D, Plummer R, Davidson S, and Moraru L (2009). 'From Government to Governance: A state-of-the-art review of environmental governance'. Final report prepared for the Alberta Environment, Environmental Stewardship, Environmental Relations. Rob de Loe Consulting Services, Guelph.

Ernstson H, Barthel S, Andersson E, and Borgström ST (2010). Scale-crossing brokers and network governance of urban ecosystem services: the case of Stockholm. *Ecology and Society* **15**, art. 28. http://www.ecologyandsociety.org/vol15/iss4/art28/

Fitzsimons JA and Wescott G (2008). Evolving governance arrangements in multi-tenure reserve networks. *Environmental Conservation* **35**, 5–7.

Folke C, Hahn T, Olsson P, and Norberg J (2005). Adaptive governance of social-ecological systems. *Annual Review of Environment and Resources* **30**, 441–473.

Heikkila T and Gerlak AK (2005). The formation of large-scale collaborative resource management institutions: Clarifying the roles of stakeholders, science, and institutions. *Policy Studies Journal* **33**, 583–612.

Lebel L, Anderies J, Campbell B, Folke C, Hatfield-Dodds S, Hughes T, and Wilson J (2006). Governance and the capacity to manage resilience in regional social-ecological systems. *Ecology and Society* **11**, art. 19. Available: www.ecologyandsociety.org/vol11/iss1/art19/

Lemos MC and Agrawal A (2006). Environmental governance. *Annual Review of Environment and Resources* **31**, 297–325.

Locke H (2010). Yellowstone to Yukon Initiative. In: *Connectivity Conservation Management: A Global Guide.* (Eds GL Worboys, W Francis and M Lockwood) pp. 161–181. Earthscan, London.

Lockwood M (2010). Scoping the territory: considerations for connectivity conservation managers. In: *Connectivity Conservation Management: A Global Guide.* (Eds G Worboys, W Francis and M Lockwood) pp. 34–52. Earthscan, London.

Lockwood M, Davidson J, Curtis A, Stratford E, and Griffith R (2009). Multi-level environmental governance: lessons from Australian natural resource management. *Australian Geographer* **40**, 169–186.

Marshall GR (2008). Nesting, subsidiarity, and community-based environmental governance beyond the local level. *International Journal of the Commons* **2**, 75–97.

Marshall GR (2009). Polycentricity, reciprocity, and farmer adoption of conservation practices under community-based governance. *Ecological Economics* **68**, 1507–1520.

McGinnis MD (1999). *Polycentric Governance and Development: Readings from the Workshop in Political Theory and Policy Analysis.* University of Michigan Press, Ann Arbor.

Mitchell SM and Shortell SM (2000). The governance and management of effective community health partnerships: a typology for research, policy, and practice. *Milbank Quarterly* **78**, 241–289.

Ostrom E (2010). Polycentric systems for coping with collective action and global environmental change. *Global Environmental Change* **20**, 550– 557.

Parris H, Whitten SM, Wyborn C, Hill R, and Freudenberger D (2011). 'An overview of key socio-economic factors, principles and guidelines in wildlife "corridor" planning and implementation'. Report for the Australian Government Department of Sustainability, Environment, Water, Population and Communities. CSIRO Ecosystem Sciences, Canberra.

Ryan S, Broderick K, Sneddon Y, and Andrews K (2010). *Australia's NRM Governance System: Foundations and Principles for Meeting Future Challenges.* Australian Regional NRM Chairs, Canberra.

Sabatier P, Focht W, Lubell M, Trachtenberg Z, Vedlitz A, and Matlock M (Eds) (2005). *Swimming Upstream: Collaborative Approaches to Water Management.* MIT Press, Cambridge, Massachusetts.

Whitten SM, Freudenberger D, Wyborn C, Doerr V, and Doerr E (2011). 'A compendium of existing and planned Australian wildlife corridor projects and initiatives, and case study analysis of operational experience'. Report for the Australian Government Department of Sustainability, Environment, Water, Population and Communities. CSIRO Ecosystem Sciences, Canberra.

Wondolleck JM and Yaffee SL (2000). *Making Collaboration Work: Lessons from Innovation in Natural Resource Management.* Island Press, Washington, DC.

Wyborn C (2011). Landscape scale ecological connectivity: Australian survey and rehearsals. *Pacific Conservation Biology* **17**, 121–131.

27

The importance of interdisciplinary research in conservation networks: lessons from south-eastern Australia

James Fitzsimons and Geoff Wescott

Introduction

Approaches to establishing real, on-ground connectivity conservation networks in Australia have varied over the years. Despite their recent proliferation, there has been little detailed comparison of different network models or differences between landscapes. Here we outline the importance of multi-disciplinary research into different approaches by comparing three multi-tenure conservation networks (herein 'networks') in south-eastern Australia (Bookmark Biosphere Reserve, Grassy Box Woodlands Conservation Management Network [CMN] and Gippsland Plains CMN). Further details on the current status of each of these three networks are outlined in Chapters 7, 10 and 12.

While there has been an increase in discussion about the benefits or otherwise of multi-tenure conservation networks, the amount of actual research on operating examples of these models is limited. This chapter reflects on the lessons acquired from earlier research into these networks and discusses with this practical insight the difference between rhetoric and performance in this vital area. In particular, we discuss some of the ecological, social, governance and legal aspects of these networks.

The role of these networks in protecting biodiversity was evaluated by: (i) analysing the spatial configuration (size, shape, connectivity) of the networks and their individual components; (ii) evaluating the contribution of the networks (in real terms and in reporting procedures) to biodiversity conservation objectives; (iii) analysing the influence of the attitudes and perceptions of land managers on the functionality of these networks; and (iv) evaluating the influence of coordinating bodies on network functionality. Finally, we highlight some of the limitations that these particular networks faced in achieving biodiversity conservation outcomes and the implications for current practice.

Major findings

The Bookmark (now Riverland) Biosphere Reserve located in the Murray Mallee of South Australia, the Gippsland Plains CMN on the eastern Gippsland Plains of Victoria, and the Grassy Box Woodlands CMN across the inland slopes of New South Wales (NSW) were case studies for this research, which was undertaken between 1999 and 2004 (Fitzsimons 2004; Fitzsimons and Wescott 2005). The spatial configuration of individual components (i.e. participating public and private lands) within networks and contribution to protecting under-reserved ecosystems were measured using spatial analysis techniques within a geographic information system (see Fitzsimons and Wescott 2008a, 2008b). The attitudes and perceptions of landowners and managers within the networks were obtained using questionnaires (see Fitzsimons and Wescott 2007) and aspects of network governance, goals and operations via questionnaires and discussions with network coordinators (see Fitzsimons and Wescott 2008c).

The sizes of individual parcels of conservation lands varied markedly between the three networks. However, within each network public reserves were on average larger than private conservation lands (Fitzsimons and Wescott 2008a). In each of the three networks, non-government organisations (NGOs) purchased at least one large private reserve (Fitzsimons and Wescott 2005). Although levels of physical connectivity varied between networks, Bookmark Biosphere Reserve and Gippsland Plains CMN showed greater similarity to each other than to the Grassy Box Woodlands CMN. The findings raise important questions about how different the Biosphere Reserve (BR) and Conservation Management Network models are. However, it also provided evidence that multi-tenure reserve networks could be applied in a variety of different landscapes – i.e. one of the networks (Grassy Box Woodlands CMN), was set in a heavily cleared and fragmented landscape and comprised of mostly isolated components separated by lengthy distances, stretching across the entire inland slopes of NSW; conversely, the bulk of land within the Bookmark BR was within semi-arid rangeland and the majority of components adjoined another parcel within the biosphere reserve.

All networks, and particularly those components outside the public reserve estate, contributed to enhancing the protection of ecosystems that were unrepresented or underrepresented in the protected area estate, although the extent of this contribution varied between networks (Fitzsimons and Wescott 2008b). Trade-offs between reserve design efficiency and a contribution to a comprehensive, adequate and representative reserve system were evident between networks. Bookmark BR was characterised by high connectivity, and strong reserve design integrity but made a lower contribution to protecting ecosystems under-represented in the reserve system, whereas the opposite was evident in the Grassy Box Woodlands CMN (which protected mostly remnants of nationally endangered and under-represented vegetation communities).

Different landholders and managers of properties involved in multi-tenure reserve networks perceived the aims of their network in different ways (Fitzsimons and Wescott 2007). This has the potential to cause uncertainty and may lead to conflict if decisions made in the name of the network are contrary to an individual manager's or participant's expectations. The level of communication among land managers and landowners varied significantly between networks. Geographic spread and property connectivity appeared to have a bearing on this. Over 88% of managers considered their involvement in multi-tenure conservation networks to be a positive or very positive experience. Although land managers considered insufficient time and finances to be the major limitation towards achieving network objectives, almost all other perceived limitations and suggested improvements related to issues that can be influenced by the networks' coordinating body. The majority (80%) of private land managers within these networks indicated a willingness to have their land included in a national reserve system of conservation lands (Fitzsimons and Wescott 2007).

The results of the research also highlighted important differences in aims, structure, authority and reporting mechanisms between multi-tenure reserve networks (Fitzsimons and Wescott 2008c). These differences are based on several factors including: (i) the legislative/institutional arrangements of the jurisdiction; (ii) the strength/attractiveness of voluntary conservation mechanisms; (iii) the level of commitment/knowledge of ecological processes and mechanisms; (iv) the densities of population in the focus region; and (v) the physical nature of the landscape (e.g. amount of subdivision/remnant vegetation). The changing nature of the network coordination arrangements, evidenced in two of the three networks at the time of the research, suggests an organic fluid evolution of network structures is likely, which contrasts with the desire for legalistic and administrative rigidity promoted by government agencies (Fitzsimons and Wescott 2008c).

In summary, all the networks studied contribute in varying degrees to biodiversity conservation. The key factors influencing the current and potential contribution that such networks make are: (i) the aims, directions and restrictions set by or imposed upon the coordinating body; and (ii) the biophysical nature of the surrounding bioregion and historical land use and tenure pattern. Although the successful operation of such 'multi-tenure' networks ultimately relies on the willing participation of private landholders, ongoing institutional support is likely to be required for maintaining networks in the longer term. Considering that networks are increasingly formed outside of the influence of government institutions, this presents a significant challenge for effective coordination of conservation activities.

Implications of findings

The high level of willingness from private landowners to participate in a national reserve system of conservation lands provides an important stimulus to strengthen the coordination of activities between public and private conservation lands at a national level. Importantly, these networks provided a mechanism for reporting on the protection of these ecosystems outside of the public protected area estate, thus providing input into national, state and regional biodiversity conservation planning decisions.

The rise of NGOs which purchase land for conservation is having an increasing, and potentially positive, influence on efforts to achieve cross-tenure biodiversity conservation, in Australia and other parts of the world. Their influence is multi-faceted, ranging from direct land purchase to network coordination to on-ground management to the raising of a network's profile by association. The NGOs are aware of the real and potential role they play in seeding the establishment of multi-tenure conservation networks (e.g. Bentley *et al.* 2002; Cowell and Williams 2006; Chapter 18).

Local communities are increasingly forming networks with different interpretations and applications of existing network models (see also Chapter 21). This situation has affinities to Mason's (1987, p. 22) observation that despite a lack of a formal 'greenline' parks program in the United States, there were 'numerous examples of greenline-type parks throughout (that) country'.

Limitations to achieving conservation objectives

The results of this study indicate that there are several factors that could limit the potential of multi-tenure conservation networks to contribute to national and regional biodiversity conservation objectives. Some of these limitations are discussed in greater detail in Chapter 28, which considers a broader range of initiatives from across the country.

The definition of a conservation network

Bennett and Lowe (2002) suggest that in southern Australia there is a wide recognition of the need for conservation networks but their implementation is not straightforward. The multi-tenure conservation networks studied in this research alone have shown there are several different interpretations of the term 'network' and these may influence establishment, operation and outcomes. For example, the term 'conservation networks' could mean:

- a physical entity of parcels of land managed for conservation objectives (e.g. reserve system);
- a group of managers undertaking on-ground conservation management actions; or
- a communicative (social network) structure designed for the flow of information.

To some degree Conservation Management Networks and Biosphere Reserves and larger-scale networks (such as corridor initiatives) aim to incorporate all of these definitions. However, while all are related to some extent, there are obvious important differences between a physical entity and a communication structure which often gets lost in definition and understanding of a 'network'. The varying and disparate responses of land managers in this study to the questions of their perception of the aims of their respective 'networks' highlights this potential confusion. The two most common broad responses were 'protection of conservation areas' and 'education of land managers/information source/involvement of community'. While some managers mentioned both, many from the same network stated one or the other. Defining what a 'network' means in a particular context (or indeed any other description for connectivity conservation) must be done early so as not to confuse, and potentially disillusion, participants.

The Biosphere Reserve and Conservation Management Network models

Major differences between the two types of networks lie in the significance attached to a site. The CMN model, in theory, recognises all sites equally, although with an increased emphasis on those of a higher conservation significance, higher quality, and higher level of protection, whereas the biosphere reserve model has a defined framework which is broadly expected to be followed (internationally at least). The terms 'core', 'buffer' and 'transition' zones within biosphere reserves have the potential to diminish the recognition of smaller private properties actively conserving biodiversity. For example, the placement of South Australian National Trust reserves within the 'transition zone' in Bookmark BR implies that these reserves are merely carrying out activities that do not impact negatively on the 'core'.

Unlike other international designations which are included under the Federal *Environment Protection and Biodiversity Conservation Act 1999* (such as World Heritage, Ramsar wetlands), legal boundaries for biosphere reserves as a whole are not well defined and there is no real penalty (internationally or nationally) for breach of objectives. Although the application of stricter national or international legal compliance requirements may deter the involvement of some private landholders within biosphere reserves, it should be recognised that without individually negotiated binding agreements, long-term protection cannot be assured. For example, little of the private land within Bookmark has legally binding agreements so if that network were to collapse, individual remnants would not necessarily be protected in the long term.

Transparency in participation criteria, goals and targets

The criteria for a participant's 'entry' into a network is another important determinant of how conservation networks operate and are perceived. All networks had some difficulty rigidly defining which properties were actually formally considered part of the network. This suggests more clearly defined entry requirements such as minimum standards for protection mechanisms, management intent, quality and size all need to be clarified if networks are to be considered in biodiversity conservation auditing. Further, this requirement will become increasingly important for justifying outcomes resulting from public funding for such networks.

The ability for networks to assess biodiversity conservation efforts

Elsewhere, we have shown that conservation networks could be used as a mechanism for accounting for biodiversity protection and management across a range of tenures (Fitzsimons and Wescott 2005, 2007, 2008a, 2008b). However, this potential is not being realised at present due to broader institutional obstacles relating to the lack of data collected on various private land conservation initiatives (e.g. Fitzsimons and Carr 2007), combined with a lack of a centralised body to collate outcomes.

Conservation networks will evolve based on both their vision and the reality of implementing that vision. For example, during the course of this research the Gippsland Plains CMN has moved from the local (Perry River) to the regional (Gippsland Plains) in its geographic scope. While expansion in sites or area within networks is generally straightforward, shrinking or splitting a network if so desired maybe less clear. For example, although it was originally envisaged that different vegetation communities within a similar geographic range could comprise separate CMNs (e.g. White Box Woodlands, Yellow Box Woodlands, Fuzzy Box Woodlands on the inland slopes of NSW; Prober *et al.* 2001) it is unlikely that this would be feasible or practical on the ground. This could have involved splitting a reserve/remnant that may contain part White Box Woodland and part Yellow Box Woodland into two separate CMNs.

However, there is likely to be a critical number (if not geographical size) of sites a network can effectively support. The larger a network gets the potential for smaller sites to get 'lost in the system' increases (as does the alienation of individual participants). Striking a balance between the importance of social cohesion of participants and optimal scales of operation for biodiversity conservation is a major challenge but likely to be essential if networks are to operate effectively. The realisation of larger, subcontinental-scale networks may require the formation of numerous smaller subregional networks, as has occurred in the Great Eastern Ranges Initiative (Chapters 13, 14 and 15).

Long-term financial security for operational and stewardship sustainability

Considering the scale and range of tenures over which they operate, multi-tenure conservation networks serve to highlight deficiencies in funding for conservation at a landscape scale. The need for improved resourcing for biodiversity conservation activities is required on several levels. Current resourcing arrangements that are typified by short-term funding and a reliance on externally provided grant-based resources (e.g. Jennings and Moore 2000; Robins and Kanowski 2011) run counter to this. A socially skilled and respected extension officer/coordinator with credibility in the region and a well-developed regional action plan are considered prerequisites for multi-tenure conservation networks to succeed (Edwards and Traill 2002; Stoll-Kleemann and O'Riordan 2002). With this level of importance comes the risk that if key individuals were to move, the network could easily collapse. Without adequate job security and appropriate remuneration, particularly in more remote locations, the risk of this occurring is likely to increase. In addition, while funding for land acquisition and stewardship agreements has increased substantially since this research was undertaken, it is likely to be well below what is required in many landscapes (Possingham *et al.* 2002).

Conclusions

This research has shown that these types of multi-tenure conservation networks can make a significant contribution to biodiversity conservation and fill a sizeable gap in coordinating conservation management across a variety of tenures. This contribution varies between networks with an apparent 'trade-off' between reserve design efficiency and the contribution they make to enhancing the comprehensiveness, adequacy and representativeness of the protected area system.

The key factors influencing the current and potential future contribution that such networks make are:

1. the aims, directions and restrictions set by or imposed upon the coordinating body;
2. the biophysical nature of the landscape the network operates within, and historical land use and tenure patterns;
3. the financial and other resources that are available and the willingness of landholders to participate; and
4. the capacity and skills available to undertake adequate monitoring evaluation and reporting on progress.

Although understanding of the networks' role and aims was mixed between participants, the overwhelming majority of landowners and managers considered their involvement in the network to be a positive experience.

Multi-tenure reserve networks offer a tangible means of implementing bioregional planning and conservation principles. However, they also provide a prime example of the tension between encouraging community involvement in nature conservation and ensuring limited resources are used in the highest priority areas as determined by scientific research. Ultimately both aspirations will have to be accommodated if these models are to become accepted by landholders, governments and the community at large.

Although the successful operation of multi-tenure reserve networks ultimately relies on the willing participation of private landholders, ongoing continuous institutional support is likely to be required for maintaining networks in the longer term. Considering that networks are increasingly formed outside of the influence of government institutions, this presents a significant challenge for effective coordinated conservation in the future. Possibly the most interesting outcome of the independent research outlined in this chapter (which was carried out several years ago) and the other chapters in this book (reporting on the situation in 2012) is that the challenges identified in our study are still reported by practitioners across the entire country today.

Acknowledgements

James Fitzsimons was supported at the time of this research by a Deakin University Postgraduate Research Scholarship and during the preparation of this chapter by The Nature Conservancy.

Biographies

James Fitzsimons is the Director of Conservation for The Nature Conservancy's Australia Program and an adjunct associate professor at the School of Life and Environmental

Sciences, Deakin University. His particular research interests are in the fields of protected area policy, practical conservation planning and mechanisms to integrate conservation outcomes on public and private lands. He has worked in the fields of conservation policy and planning for government environment departments and agencies, and for non-government environment organisations. He has a PhD in multi-tenure conservation networks and has been involved with CMN coordination in Victoria since 2001.

Geoff Wescott is Associate Professor of Environment at Deakin University's Melbourne Campus. He is a former Chair of the Victorian National Parks Advisory Council, Deputy Chair of Parks Victoria and Chair of the Marine and Coastal Community Network. He is currently Vice President of the Australian Coastal Society and a member of Zoos Victoria's Board of Directors. He is also a member of the Wedderburn Conservation Management Network in Victoria.

References

Bennett AF and Lowe KW (2002). From design to implementation: insights from ecological networks in southern Australia. Paper presented at the Society for Conservation Biology 16th Annual Meeting July 14–July 19 2002. Canterbury, UK. Available: http://www.conbio.org/Activities/Meetings/2002/abstracts/symposia/sympnine.cfm

Bentley J, Cowell S, Fitzsimons J, Cramer V, Hopkins K, and Jungalwalla S (2002). Dunnies, smokewater, salt and other science. *Bush Heritage News* **Spring**, 2–4.

Cowell S and Williams C (2006). Conservation through buyer diversity: a key role for not-for-profit land-holding organizations in Australia. *Ecological Management & Restoration* 7, 5–20.

Edwards R and Traill B (2002). Getting beyond field days: targeting extension to protect threatened ecosystems on private land. *Ecological Management & Restoration* **3**, 229–231.

Fitzsimons JA (2004). The contribution of multi-tenure reserve networks to biodiversity conservation. PhD thesis, School of Ecology & Environment, Deakin University, Melbourne.

Fitzsimons J and Carr B (2007). 'Evaluation of the effectiveness of conservation covenanting programs in delivering biodiversity conservation outcomes'. Report for the Australian Government's Department of Environment and Water Resources. Bush Heritage Australia, Melbourne.

Fitzsimons JA and Wescott G (2005). History and attributes of selected Australian multi-tenure reserve networks. *Australian Geographer* **36**, 75–93.

Fitzsimons JA and Wescott G (2007). Perceptions and attitudes of land managers in multi-tenure reserve networks and the implications for conservation. *Journal of Environmental Management* **84**, 38–48.

Fitzsimons JA and Wescott G (2008a). The role of multi-tenure reserve networks in improving reserve design and connectivity. *Landscape and Urban Planning* **85**, 163–173.

Fitzsimons JA and Wescott G (2008b). Ecosystem conservation in multi-tenure reserve networks: The contribution of land outside of publicly protected areas. *Pacific Conservation Biology* **14**, 250–262.

Fitzsimons JA and Wescott G (2008c). Evolving governance arrangements in multi-tenure reserve networks. *Environmental Conservation* **35**, 5–7.

Jennings SF and Moore SA (2000). The rhetoric behind regionalization in Australian natural resource management: myth, reality and moving forward. *Journal of Environmental Policy and Planning* **2**, 177–191.

Mason RJ (1987). Greenline parks: an accommodation of metropolitan and rural interests? *Operational Geographer* **13**, 22–24.

Possingham H, Ryan S, Baxter B, and Morton S (2002). 'Setting biodiversity priorities'. A paper prepared as part of the activities of the working group producing the report 'Sustaining our Natural Systems and Biodiversity' for the Prime Minister's Science, Engineering and Innovation Council in 2002. Department of Education, Science and Training, Canberra.

Prober SM, Thiele KR, and Higginson E (2001). The Grassy Box Woodlands Conservation Management Network: picking up the pieces in fragmented woodlands. *Ecological Management & Restoration* **2**, 179–188.

Robins L and Kanowski P (2011). 'Crying for our Country': Eight ways in which 'Caring for our Country' has undermined Australia's regional model for natural resource management. *Australasian Journal of Environmental Management* **18**, 88–108.

Stoll-Kleemann S and O'Riordan T (2002). From participation to partnership in biodiversity protection: experience from Germany and South Africa. *Society & Natural Resources* **15**, 161–177.

SECTION 5. SYNTHESIS

28

Challenges and opportunities for linking Australia's landscapes: a synthesis

James Fitzsimons, Ian Pulsford and Geoff Wescott

Introduction

The continent of Australia has been confronting threats to the conservation of its unique landscapes, flora and fauna (and the ecological processes that support them) ever since European settlement almost 230 years ago. The synergistic impact of human-induced climate change has heightened the threat and made the need for widespread holistic action ever more urgent (Steffen *et al.* 2009). The latest evolving major response to these threats has been the application of the concept of connectivity conservation – or 'linking landscapes' – which gave rise to the need for this book. In this synthesis chapter we have attempted to extract the 'big messages' – the repeated recurring themes as well as highlighting points of difference – that the authors of the preceding chapters have outlined from their extensive experience. These are summarised below.

Scale and breadth

The scale of the operation of a multi-tenure connectivity conservation initiative was a topic discussed by many authors. Many of the larger corridor initiatives highlighted that they considered a large scale of operation to be important for ecological function (e.g. Bradby, Chapter 3; Bridges, Chapter 5; Pulsford *et al.*, Chapter 13; and see also Worboys and Mackey, Chapter 2) and for creating an inspiring vision (see below). However, operating at such a large scale was also recognised to have significant challenges. This included issues of coordination, governance (Carr, Chapter 8; Wyborn, Chapter 26) and communication (McLeish *et al.*, Chapter 12; Beyer and Baker, Chapter 15; Fitzsimons and Wescott, Chapter 27).

Almost all of the larger corridor initiatives that operated at continental or subcontinental scale divided their total area into smaller 'operating units'; landscape zones that reflected similarities in the ecological or social attributes (Bradby, Chapter 3; Bridges, Chapter 5; Gates and Kondylas, Chapter 6; Carr, Chapter 8; Pulsford *et al.*, Chapter 13). It was rare that larger initiatives actively 'operated' across their entire area and activities were often focused on high priority operating units within. These smaller units were often given their own names and identities, for example 'Slopes to Summit' and 'Border Ranges' in the Great Eastern Ranges Initiative (Pulsford *et al.*, Chapter 13; Spooner *et al.*, Chapter 14;

Beyer and Baker, Chapter 15), and 'FitzStirling' in Gondwana Link (Bradby, Chapter 3). 'Place' is important, as is effective communication, and both contribute to the important element of social connectedness, which will often only happen at a more local scale. Ensuring that that the planning process includes some holistic thinkers able to synthesise and focus on the 'bigger picture' resulting from the bringing together of what are often smaller, more localised initiatives is important (Lambert, Chapter 24).

Interestingly, these smaller units often operated at a similar scale to other 'stand alone' initiatives such as the various Victorian Conservation Management Networks (CMNs) and Tasmania's Midlandscapes.

This begs the question as to whether some models (e.g. CMNs) can become appropriate on-ground mechanisms (or building blocks) for larger initiatives, e.g. wildlife corridor initiatives. Indeed it raises a bigger issue about the real and perceived differences between 'wildlife corridors', 'conservation management networks', 'biosphere reserves' and other similar initiatives under different names, considering all have a multi-tenure focus and seek to improve or protect habitat connectivity and landscape function within a matrix of conservation and other land.

The importance of a shared and guiding vision

Regional landscape conservation networks (such as CMNs), as well as the larger corridor initiatives, inspire people, help create conservation communities and provide opportunities where 'individuals make a difference' (Worboys and Mackey, Chapter 2). Many of the case studies in this book recognise that a primary requirement for the establishment and ongoing success is the need to provide 'stories' that are compelling to each stakeholder group or individual (Bridges, Chapter 5). This narrative can provide a powerful basis for engagement and help to drive the evolving process of governance and decision making.

A key part of this narrative is summarised in the form of a shared vision statement that describes a desired future landscape or biodiversity condition and seeks the involvement of people in achieving this outcome (Lambert, Chapter 24; Wyborn, 26). The vision is combined with several goals and usually supported by a plan that identifies aspirational targets and priorities for investment. Most authors in Section 2 recognised the paramount importance of a clear guiding vision statement in harnessing community support and engagement and for providing an overarching direction for the long term.

In some cases, such as within the large-scale corridors (the Great Eastern Ranges Initiative, Gondwana Link, NatureLinks of South Australia), each nested regional partnership/network or project area has supplemented the larger vision with their own supporting guiding vision for the local or regional landscape. While the vision is an important part of any initiative's success, it may not be static and can evolve over time and with increased ecological community understanding and feedback (Pulsford *et al.*, Chapter 13), provided it brings its communities along with the evolution of the initiative.

Social and economic factors are critical elements of success

Conservation management is required across tenures in order to protect and restore biodiversity on a landscape-wide basis and different connectivity conservation strategies are needed in different Australian landscapes. This is because the type of conservation actions required in cleared and fragmented forest and woodland landscapes can be different to the contiguous habitats for arid and northern Australia or peri-urban areas. In each of these landscapes the social willingness and economic circumstances of landholders and communities require quite different solutions and mixes of programs. On the ground, Territory Eco-link (Bridges, Chapter 5) encourages the inclusion of Indigenous Protected Areas and pastoral properties, while many Victorian CMNs seek to establish private protected areas or encourage landholders to sign in-perpetuity conservation agreements (Morison and Murphy, Chapter 10; Edwards and Fox, Chapter 11). These regional differences are recognised by the *National Wildlife Corridors Plan* (Zammit, Chapter 17).

A key challenge for some initiatives is to be able to maintain partner and community interest once initial funding sources are discontinued or grant applications are unsuccessful due to changing priorities of funding bodies (e.g. Mackenzie and Fitzsimons, Chapter 7). There can be mismatch between partner and community expectations when a vision is not matched by sufficient funding to make an impact quickly enough (Sanders, Chapter 4; Carr, Chapter 8; Spooner *et al.*, Chapter 14). Loss of social capital is a real threat to the sustainability of the network (Mackenzie and Fitzsimons, Chapter 8) and this is where the strength and commitment to leadership over long periods is a key factor.

It appears that programs built on trust and openness will be more resilient in hard times and hence it is worth investing the time and effort early in development to build trust (Bradby, Chapter 3; Cowell *et al.*, Chapter 9; Edwards and Fox, Chapter 11; Freudenberger *et al.*, Chapter 16; Lambert, Chapter 21).

Another critical element would appear to be the achievement of visible and tangible on-ground results early (Bradby, Chapter 3; Edwards and Fox, Chapter 11; Pulsford *et al.*, Chapter 13; Edgar, Chapter 22). However, several authors (Cowell *et al.*, Chapter 9; Edwards and Fox, Chapter 11; Spooner *et al.*, Chapter 14; Beyer and Baker, Chapter 15; Freudenberger *et al.*, Chapter 16; Crosthwaite *et al.*, Chapter 21) stressed the importance of building on past conservation programs and activities that were in operation before the formation of a connectivity/landscape initiative (although see Morison and Murphy, Chapter 10). These past activities are likely to have built important social networks and shared visions.

Finally, it is critical to maintain a core group of volunteers (Walker, Chapter 19) to ensure the continuity of the program if a facilitator, or equivalent government officer, is lost through leave, forced redundancies or loss of grants.

Leadership and the notion of 'champions'

The key role of a dedicated 'facilitator' was highlighted by several authors as an essential element of success for connectivity initiatives (Carr, Chapter 8; Morison and Murphy, Chapter 10; Edwards and Fox, Chapter 11; Spooner *et al.*, Chapter 14; Wyborn, Chapter 26). At the same time, the often short-term and uncertain security of funding for these positions was seen as a major limitation to the operation of these initiatives. This situation is not unique to connectivity initiatives, it is cited as a common limitation to natural resource management (NRM) activities more generally in Australia (e.g. Robins and Dovers 2007; HC Coombs Policy Forum 2011).

Although the role of leaders or 'champions' was not discussed in detail by many authors, this is potentially in part because the authors themselves were amongst this group. As with so many other endeavours in nature conservation and natural resources management, many initiatives are 'kicked-off' by an individual – a driven person with energy, passion, commitment and a strong personality – a person/s who inspires others to join in and inspires sponsors to part with critical funds and resources. The challenge is to have succession planning in these groups in case the leader 'burns out' (e.g. Sanders, Chapter 4; Walker, Chapter 19) or moves on.

Governance

There is a wide variety of governance models for connectivity initiatives outlined in this book. These ranged from top-down approaches initiated by government (e.g. Great Eastern Ranges, Territory Eco-link, South Australian Nature Links, Wedderburn CMN) to non-government organisation (NGO) and community initiated networks.

As well there is variation in factors such as the formal or informal nature of boards, whether they have been established by incorporation or Memorandum of Understanding, and the size of the secretariat/paid workers. Comparisons are difficult partially because the terminology and nomenclature used can be quite individualist and, while this is probably a case of 'horses for courses' over time, some form of clear continuous governance system, taking into account governance principles (Wyborn; Chapter 26), is probably going to prove to be a crucial element of long-term success.

Governance frameworks for several initiatives have evolved and are still evolving. This evolution is influenced by the 'level' at which the initiative was originally created (e.g. top-down, bottom-up), the growth of the initiative, ability to share responsibility and attract financial resources, to name just a few. For example, some initiatives (e.g. Great Eastern Ranges, Riverland Biosphere Reserve, Wedderburn CMN) have evolved from government-'created' to devolving governance to a range of stakeholders. In contrast, some initiatives that have begun at an informal grassroots level (e.g. Gondwana Link, Gippsland Plains CMN) have evolved to more formal incorporated bodies (see also Fitzsimons and Wescott 2008a). Indeed most initiatives that originally began as more informal networks move in this direction in order to receive funding.

At least one initiative (Habitat 141°) cautioned that overly complex governance arrangements, created too early in the formation of a connectivity initiative can slow progress on the ground, while another (Bunya Biolinks) has made a conscious effort to avoid such complexity.

In summary, large-scale and long-term conservation initiatives need strong, but not necessarily complex, governance arrangements. However, there is a diverse array of systems that reflects a diverse array of reasons and motivations for establishment and the 'personality' of the various groups involved, leading to the conclusion that one size is unlikely to fit all.

Funding

Funding arrangements varied between initiatives but primarily consisted of government, philanthropic, donor and corporate funding to varying extents.

A lack of funding for on-ground actions was identified as a major limitation for many initiatives, while Spooner *et al.* (Chapter 14) also noted that where funding has been received, this had not matched the scale of initiative visions and objectives (of Slopes to Summit) to date.

The Australian Government's *National Wildlife Corridors Plan* is the first national connectivity conservation plan developed by any country and it is likely to guide investment by the Australian Government's funding for restoration of landscapes and biosequestration of carbon nationally (Zammit, Chapter 17). A change of government may alter the basis of this assumption.

Connectivity initiatives face several risks in delivering on their potential due to the need to build investment at scale (particularly if a threshold investment is needed to achieve benefits), the organisational challenge of landscape-scale conservation, scientific and implementation uncertainties and misalignment of goals and opportunities (Whitten *et al.*, Chapter 25). Several initiatives are encouraging new investment tools, such as capitalising on the emerging carbon market (e.g. Bradby, Chapter 3; Carr, Chapter 8; Freudenberger *et al.*, Chapter 16), but as recognised by both Freudenberger *et al.* (Chapter 16) and Edgar (Chapter 22), offsets and credits are still relatively new concepts and these 'monetising-opportunities' at the scale required for landscape restoration are still in their relative infancy in Australia and New Zealand.

A key means of achieving the desired landscape outcomes recognised by most initiatives was 'working together to improve the integration of activities of multiple partner organisations' with limited resources to achieve mutually agreed conservation outcomes. Edwards and Fox (Chapter 11) demonstrated how this could be achieved, with agencies combining funds to invest in the biodiversity works 'wish list' of the owner, with a condition that the land (and thus investment) is protected in perpetuity through a conservation covenant.

The two biosphere reserves featured in this book both highlighted a 'drying up' of government funds as a significant factor in loss of momentum. As time progressed, interest by both the state and Commonwealth governments waned and most of their input became 'in-kind', despite biosphere reserves being a national government responsibility under international agreements (Mackenzie and Fitzsimons, Chapter 7). Sanders (Chapter 4) recommended that planning for sourcing long-term funding is completed early in the project's development.

The final point concerning funding is the security and nature of current government funding arrangements which typically sees funds directed for relatively short-term projects whose continuity is at risk from change in government priorities. Several case studies noted the loss of facilitator or other staff when funds dry up and the difficulty of sustaining organisations under these circumstances. As well there is often little or no provision for operational expenses for NGOs and government agencies participating in connectivity initiatives. Most money is 'soft', short-term and directed at site-specific projects. Without funding for continuing stewardship programs and operational expenses much of the effort on initial on-ground work could be wasted.

Conservation planning and delivery of broader natural resource management objectives

The preparation of a conservation plan that contained goals and identified areas for conservation activities and investment was considered important by most initiatives. In the early days of some initiatives this plan was basic. For example, Gondwana Link (Bradby, Chapter 3) used satellite images and 'Texta' pens to identify areas to target for on-ground conservation investment. Over time plans were improved and usually updated with some using more advanced planning methods. Methods varied depending on the availability of spatial information and analytical skills.

The most common conservation planning framework chosen was Conservation Action Planning (CAP), an adaptive management process developed by The Nature Conservancy and used around the world by that organisation and others to plan conservation projects.

This framework was used by regional-scale partnerships within the large-scale corridors (for example Bradby, Chapter 3; Cowell *et al.*, Chapter 9; Spooner *et al.*, Chapter 14; Freudenberger *et al.*, Chapter 16; Walsh *et al.*, Chapter 18). However, some networks used other conservation planning and analysis techniques instead of CAPs. For example, the Great Eastern Ranges Initiative (Pulsford *et al.*, Chapter 13) used a variety of methods that varied according to planning scale and availability of relevant data. At the continental-scale, satellite remote-sensing data and modelling were undertaken using a range of analytical techniques to identify connectivity corridors, gaps and habitat fragmentation, ecosystem productivity, species richness, endemism and refugia. This approach provided an overarching framework for developing regional-scale CAPs or building on existing conservation plans (Beyer and Baker, Chapter 15).

Connectivity initiatives were considered to be useful delivery mechanisms for other natural resource management (NRM) objectives, at national (e.g. Zammit, Chapter 17) and state levels (Gates and Kondylas, Chapter 6; Mansergh, Chapter 20; Crosthwaite *et al.*, Chapter 21). For example, South Australia's NatureLinks (Chapter 6) aligned well with the principles underpinning NRM planning. The first regional NRM plans, completed in 2008–09, incorporated NatureLinks as a key policy initiative, although the high level aims had not yet been translated into specific priorities to guide on-ground activities in all corridors.

Role of existing and new protected areas

Protected areas and remnant vegetation provide the essential core for any multi-tenure connectivity network (Worboys and Mackey, Chapter 2; Mackey *et al.*, Chapter 23). Connectivity initiatives sought to improve the conservation and management of areas around and between these core protected areas by focusing on protecting or better managing properties with significant ecological value and actively restoring cleared or degraded properties that provide strategic linkages. The mechanisms used varied but included a mix of land purchase, signing of in-perpetuity conservation covenants or shorter-term management agreements linked to stewardship payments, or non-binding voluntary agreements. The mix of these options used depended on the underlying land tenure, types of investment by government and non-government sectors, and social drivers.

Of these approaches, a number of authors reported the catalysing role of land purchase by NGOs in a region (i.e. in Gondwana Link, South Australian NatureLinks, Territory Eco-link, Wedderburn CMN, Gippsland Plains CMN, Riverland Biosphere Reserve, Habitat 141°; see also Fitzsimons and Wescott 2005). However, in some regions, land purchase was not an option for social or legal reasons, and new, innovative approaches were developed to protect freehold (Cowell *et al.*, Chapter 9) or Indigenous lands (Bridges, Chapter 5; Walsh *et al.*, Chapter 18).

Communication

A key element of maintaining and enhancing a connectivity network is to create and maintain a communication system so that disparate members often separated by substantial distance are linked together (McLeish *et al.*, Chapter 12; Crosthwaite *et al.*, Chapter 21). This is also a reinforcement of the critical social elements of networks (Lambert, Chapter 24). Communication with and between landholders in connectivity networks took the form of hardcopy newsletters (e.g. Edwards and Fox, Chapter 11; McLeish *et al.*, Chapter 12), field days and regular meetings (e.g. Morison and Murphy, Chapter 10), and dedicated websites (Bradby, Chapter 3; Pulsford *et al.*, Chapter 13).

Communication *between* initiatives varies. Victoria has a Statewide Advisory Group for conservation management networks that meets quarterly and includes representatives of all CMNs (Crosthwaite *et al.*, Chapter 21). There is also informal communication between Victorian and NSW CMNs.

The Australian Government's Department of Sustainability, Environment, Water, Population and Communities declares it 'acts as the national focal point for biosphere reserves in Australia while the Australian National Commission for UNESCO has overall responsibility for UNESCO activities in Australia' (DSEWPC 2012a). However, active coordination of biosphere reserves is very limited and has declined significantly since the early 2000s following the loss of a dedicated project manager and the discontinuation of the publication *Australian Biosphere Reserve News*.

Coordinators of the larger corridor initiatives identified in the Australian Government's *National Wildlife Corridors Plan* (DSEWPC 2012b) have begun meeting in a more regular and more formalised way, at their own volition, and formed the Australian Connectivity Council, since the release of the draft of that plan.

However, there is little coordination or communication between these different 'types' of multi-tenure conservation networks.

Further research

As research into multi-tenure connectivity initiatives is still in its infancy in Australia and internationally, there remains many areas in need of further research. Further investigation of three areas in particular would complement the work undertaken in this book and elsewhere (e.g. Fitzsimons and Wescott 2007, 2008a, 2008b, 2008c; Worboys *et al.* 2010; Wyborn 2011), increase our understanding of networks and connectivity initiatives, and would ultimately lead to an increase in the effectiveness of multi-tenure conservation networks.

1. Increased research into social dynamics of networks

Further work is needed to understand the social, political and economic dynamics of landscapes and communities. Improved knowledge of the social and demographic characteristics of those landowners participating in connectivity initiatives and those who are not could provide important information and allow approaches to be tailored to attract landowners in the future and to enhance the long-term sustainability of connectivity groups and projects.

2. Longer-term changes in network characteristics

Long-term research and analysis of ecological, social, governance and land use attributes would enhance our understanding of the forces that shape multi-tenure conservation networks. Of particular interest is the identification of reasons for their persistence or failure. The impact that the failure of an established network may have on landowners involved is of particular interest because disenfranchisement may lead to negative outcomes for biodiversity conservation. Longer-term research would also enable a more thorough evaluation of the contribution of networks to biodiversity conservation, the ultimate reason for establishing such initiatives.

3. Comparison with other networks in Australia and internationally

As evidenced from this book, numbers of multi-tenure connectivity initiatives are proliferating in Australia, being established at a range of scales and with increased interest by government. Further comparisons between the types of initiatives and the scales they operate at, both in Australia and with international initiatives (e.g. greenline parks, transboundary protected areas, large-scale wildlife corridors), will also provide greater insight into the characteristics of multi-tenure connectivity networks and their role in protecting biodiversity. Ultimately, this will assist in identifying more effective and efficient models for biodiversity conservation across the landscape. As part of such research, comparisons could be made to experience in the ultimate 'connected' area – the sea. How have marine planners and managers built into their ocean-based projects the essentials of ecological processes of protection and monitoring of actual connectedness between ecosystems?

Conclusion

Australia is in an evolutionary or development phase in terms of experimenting with a range of different approaches to achieving connectivity and landscape-scale conservation, at a whole range of scales from local, ecosystem-based networks to massive continent-crossing linkages.

This is an exciting time for conservation tempered by the need to ensure these initiatives complement – not replace – efforts to halt vegetation loss, establish a comprehensive, adequate and representative reserve system and address the continuing threat of climate change.

Local communities, NGOs and some state governments have led the recent drive for real, on-ground connectivity conservation initiatives. The Australian Government has followed with the *National Wildlife Corridors Plan*. This Plan potentially provides a guiding national framework that in turn may increase funding for connectivity initiatives by the Australian Government. But such a plan will need bi-partisan long-term (decades not years) political support, and support from state and local governments, if it is to meet its lofty ambitions.

While such national-scale plans are vital, it is important to note that most of the on-ground initiatives featured in this book have developed and evolved in the absence of a national policy framework or direct Australian Government support. Considering the current strong interest in establishing connectivity initiatives on the ground, this will continue to pose challenges in balancing demand for support from initiatives in areas which may not be a high priority for conservation at a national level, with encouraging the establishment of initiatives in high priority regions but where there is little local interest.

To achieve more efficient and effective conservation outcomes, there is a relatively urgent need to break down the silos that currently exist between supposed different approaches to connectivity and landscape-scale conservation (e.g. biosphere reserves, wildlife corridors, conservation management networks). A more holistic view of these multi-tenure conservation initiatives is required in policy, planning, research and communication.

While governmental support for a framework of corridors is crucial, other sources of funding (from corporate to philanthropic) will be important to provide the continuity and scale of resources required to operationalise these grand visions. Given the likely increase in investment in these initiatives, providing security for the conservation outcomes achieved from that investment (e.g. through conservation covenants and other binding agreements) will be an increasing focus. This is particularly so considering the evolving, and at times fragile, nature of the initiatives. Finally this increase in investment in connectivity conservation initiatives needs to be underpinned by strong monitoring and research frameworks that ensure that best practice is identified (and rewarded) and by an effective communication network to disperse these findings across all projects in the country.

Biographies

James Fitzsimons is the Director of Conservation for The Nature Conservancy's Australia Program and is an Adjunct Associate Professor at the School of Life and Environmental Sciences, Deakin University. His particular research interests are in the fields of protected area policy, practical conservation planning and mechanisms to integrate conservation outcomes on public and private lands. He has worked in the fields of conservation policy and planning for government environment departments and agencies, and for non-government environment organisations. His PhD research focused on multi-tenure conservation networks and he has been involved with CMN coordination in Victoria and securing large-scale conservation outcomes in Gondwana Link and Territory Eco-link.

Ian Pulsford is a specialist in protected areas and linking landscapes, with over 32 years' experience in conservation policy and practice with the New South Wales Government, including selection, design and management of protected areas. He was a divisional manager with the NSW Department of Environment and Climate Change and then the founding manager of the Initiative from 2007 to 2010. He is a member of the IUCN World Commission on Protected Areas and has served on various government committees, including as an independent expert on a panel advising the Australian Government on the draft *National Wildlife Corridors Plan.*

Geoff Wescott is Associate Professor of Environment at Deakin University's Melbourne Campus. He is a former Chair of the Victorian National Parks Advisory Council, Deputy Chair of Parks Victoria and Chair of the Marine and Coastal Community Network. He is currently Vice President of the Australian Coastal Society and a member of Zoos Victoria's Board of Directors. He is also a member of the Wedderburn Conservation Management Network in Victoria.

References

DSEWPC (2012a). Australia's Biosphere Reserves. Department of Sustainability, Environment, Water, Population and Communities, Canberra. Available: http://www.environment.gov.au/parks/biosphere/ [Accessed 1 September 2012].

DSEWPC (2012b) *National Wildlife Corridors Plan: A Framework for Landscape-scale Conservation.* Department of Sustainability, Environment, Water, Population and Communities, Canberra.

Fitzsimons JA and Wescott G (2005). History and attributes of selected Australian multi-tenure reserve networks. *Australian Geographer* **36**, 75–93.

Fitzsimons JA and Wescott G (2007). Perceptions and attitudes of land managers in multi-tenure reserve networks and the implications for conservation. *Journal of Environmental Management* **84**, 38–48.

Fitzsimons JA and Wescott G (2008a). Evolving governance arrangements in multi-tenure reserve networks. *Environmental Conservation* **35**, 5–7.

Fitzsimons JA and Wescott G (2008b). The role of multi-tenure reserve networks in improving reserve design and connectivity. *Landscape and Urban Planning* **85**, 163–173.

Fitzsimons JA and Wescott G (2008c). Ecosystem conservation in multi-tenure reserve networks: The contribution of land outside of publicly protected areas. *Pacific Conservation Biology* **14**, 250–262.

HC Coombs Policy Forum (2011). *Natural Resource Management Policy and Planning in Australia.* Synthesis of broad issues and opportunities: Document 1. HC Coombs Policy Forum-Fenner School of Environment and Society NRM initiative, The Australian National University, Canberra.

Robins L and Dovers S (2007). Community-based NRM boards of management: are they up to the task? *Australasian Journal of Environmental Management* **14**, 111–122.

Steffen W, Burbidge AA, Hughes L, Kitching R, Lindenmayer D, Musgrave W, Stafford-Smith M, and Werner P (2009). *Australia's Biodiversity and Climate Change.* CSIRO Publishing, Canberra.

Worboys GL, Francis W, and Lockwood M (Eds) (2010). *Connectivity Conservation Management: A Global Guide.* Earthscan, London.

Wyborn C (2011). Landscape scale ecological connectivity: Australian survey and rehearsals. *Pacific Conservation Biology* **17**, 121–131.

Index

www.ingramcontent.com/pod-product-compliance
Lightning Source LLC
LaVergne TN
LVHW070908210926
848482LV00017B/1019

* 9 7 8 0 6 4 3 1 0 7 0 4 5 *